Selection and Use of Engineering Materials

Third edition

J. A. Charles ScD, FEng, FIM
University of Cambridge

F. A. A. Crane PhD, CEng, FIM
Late of Imperial College of Science and Technology

J. A. G. Furness MA, PhD
Quo-Tec Ltd

Butterworth-Heinemann
Linacre House, Jordan Hill, Oxford OX2 8DP
225 Wildwood Avenue, Woburn, MA 01801-2041
A division of Reed Educational and Professional Publishing Ltd

℞ A member of the Reed Elsevier plc group

OXFORD   AUCKLAND   BOSTON
JOHANNESBURG   MELBOURNE   NEW DELHI

First published 1984
Reprinted 1985
Reprinted with corrections 1987
Second edition 1989
Reprinted 1991, 1992, 1994, 1995
Third edition 1997
Reprinted 1999

© Reed Educational and Professional Publishing Ltd 1997

British Library Cataloguing in Publication Data
A catalogue record for this book is available from the British Library

Library of Congress Cataloguing in Publication Data
A catalogue record for this book is available from the Library of Congress

ISBN 0 7506 3277 1

Composition by Genesis Typesetting, Rochester, Kent
Printed and bound in Great Britain

Contents

Selection for Surface Durability

Case studies in materials selection

Preface to the third edition

The continuing success of this book has required reprints of the second edition, and now a third edition. In its preparation great attention has been paid to the invaluable comments made by reviewers and users of the earlier editions. The continuing development of design engineering, the growing importance of plastics, ceramics and composite materials, has required additional text and rewriting in many chapters. Also, since the second edition, there has been a marked growth in the availability of materials databases and in computerized materials selectors. Thus Chapter 14, on the formalization of selection procedures, has been substantially modified to take account of this. Other new features are the explanation of the Weibull modulus in describing the variability of strength to be expected in a material, materials for springs and the influence of hydrogen on the performance of steels and the relevance to sour gas service in the petroleum industry.

As the text has evolved we hope that it will not only be a useful overview of materials usage for students, but suitable also for continuing development in a range of engineering professions. A recommendation was that future editions should provide questions to be undertaken by students. This is easiest to do in a text which is concerned primarily with the aspects of engineering design, giving questions which have a purely mathematical solution. This book is, however, concerned with the understanding of materials usage as well, and many of the questions that could be selected require essay or part-essay answers to reveal that understanding. For this reason, the questions now included, mainly from recent examinations in UK universities, are accompanied by a bibliography of useful texts which should assist response to fully satisfactory answers.

In this regard it is valuable that this book and that by M. F. Ashby, *Materials Selection in Mechanical Design*, come now from the same publisher and could even be regarded as complementary volumes. Ashby's approach is primarily through design considerations, identifying design criteria for different systems and assessing general classes of materials in this respect, whereas the present volume places more emphasis on the details of the materials available and their service.

In preparing this edition, on the recommendation of Mr Rod Wilshaw of the Institute of Materials, J. A. C. has been joined by Dr. J. Furness of Quo-Tec Ltd, and previously the Design Council and the Materials Information Service at the Institute of Materials.

We are most grateful to the various University Departments who gave us permission to reproduce questions from past examination papers. The investigative case studies have again been checked by the manufacturing companies to ensure accuracy in relation to current practice, and we are most grateful to the staff concerned.

February 1997

J. A. Charles
J. A. G. Furness

Preface to the second edition

Sadly, Dr. F. A. A. Crane died at the end of 1984, only a few months after the publication of the first edition. Thankfully by then, however, it was clear that the book on which he had expended so much effort, in a time when effort cost dear, was going to be successful. He was thus able to know the sense of achievement and satisfaction in having our approach to the subject on record and welcomed, which has to be the main reward for the authors of specialized textbooks.

Thus it has been left to me alone to modify and add to the original text where subsequent comment from colleagues and friends and personal reflection on developments has led me. I can only hope Andy Crane would have approved.

The text has been widely modified in relation to the properties and use of non-metallic materials. Joining has been widened to include a more detailed consideration of the weldability of steels, the welding of plastics and adhesion. The sections on high temperature materials and materials for aircraft structures, the latter to include a consideration of aluminium-lithium and magnesium and its alloys, have been revised.

A completely new chapter on materials for automobile structures is now included, mainly as a method of introducing a consideration of a typical field in which there is growing competition between the traditional use of steel and the increasing application of reinforced polymers.

Since writing the first edition there has been a substantial development of data bases for materials and of associated materials selection programmes. This is reflected in the enlargement of Chapter 14, although this is a fast-moving and complex area in which there is, as yet, little integration or cohesion and only general comments are appropriate in this context. Undoubtedly, however, the use of graphical relationships for selection based on computer data bases, is ideally student-friendly and is a valuable aid to understanding and 'grasp'.

I am very grateful for the help of those who have made suggestions for improvements and up-dating. Although others have helped, in particular I wish to thank Dr. J. Campbell, Dr. B. L. English, Dr. J. E. Restall, Dr. C. A. Stubbington and D. A. Taylor for full comments in specialized areas. The investigative case studies have all been checked against present manufacturing practice and I am grateful to R. M. Airey, D. Carr and B. F. Easton for help in this respect.

Cambridge 1988   J. A. Charles

Preface to the first edition

With international competition in every field intensified by industrial recession, the importance of materials selection as part of the design process continues to grow. The need for clear recognition of the service requirements of a component or structure in order to provide the most technically advanced and economic means of meeting those requirements points to the benefits that can follow from better communication between design engineers on the one hand and materials engineers and scientists on the other, most effectively achieved by the inclusion of materials selection as a subject in engineering courses.

When we were students, the teaching of materials selection involved little more than the recitation of specifications, compositions and properties with little comment as to areas of use. It was, regrettably, a rather boring exercise. Much later, faced with the task of lecturing in the same subject at Imperial College and Cambridge University respectively, we naturally tried to provide a more rational, and lively, understanding of materials selection. Although we were unaware of it, we each independently chose to base our teaching method on case studies – discussing how the selection process has worked out in specific examples of engineering manufacture.

Discovering that there were no introductory texts dealing with the subject in the way that we preferred, we independently, and very slowly, started to write our own. It was a mutual friend, Dr D. R. F. West of Imperial College, who suggested that we should join together in a collaborative effort. We are greatly indebted to him for that – left to ourselves we would almost certainly have found the lone task too daunting for completion. Our colleagues and friends have been very helpful in making useful comments on various parts of the text, notably Drs T. J. Baker, J. P. Chilton, C. Edeleanu, H. M. Flower, D. Harger, I. M. Hutchings, W. T. Norris, G. A. Webster and D. Ll. Thomas.

We must thank our wives too, for their encouragement and understanding and for keeping us company at our working meetings, held not infrequently at a hostelry situated conveniently midway between our homes. A mixed authorship can also create problems for the typist, and we are most grateful to Mrs P. Summerfield for her cheerful acceptance of the task and to Mrs Angela Walker who was also very helpful as the deadline loomed. We are also indebted to Mr B. Barber for assistance with the photographs.

Where a book like this is based on lectures given over many years it is not always easy to recall the original sources of materials or attitudes. We have tried throughout to acknowledge the work of others. Where our memories and records have failed we ask forgiveness.

January 1984   F. A. A. Crane
                                                 J. A. Charles

The background to decision

1
Introduction

There are two important principles that should apply to materials selection in engineering manufacture:

(1) materials selection should be an integral part of the design process;
(2) materials selection should be numerate.

It is therefore necessary first of all to examine the nature of the design process and the way in which it is carried out.

Then it is necessary to consider how the selection of materials can be made numerate. We choose to do this by defining and describing all of the individually important properties that materials are required to have and then categorizing the useful materials in terms of these properties.

Initially, this can only be done in quite broad terms but as specific applications come to be considered it emerges that the materials engineer must possess a rather deep understanding of the frequently idiosyncratic ways in which basic properties are exhibited by individual materials, and also of the ways in which those properties are influenced by the manufacturing processes to which the material has been subjected prior to entering service.

The properties of materials

It has been estimated that there are more than 100,000 materials available for the designer to choose from and a correspondingly wide range of properties. Although a material may be chosen mainly because it is able to satisfy a predominant requirement for one property above all others, every useful material must possess a *combination* of properties. The desired cluster of properties will not necessarily be wide-ranging and the exact combination required will depend upon the given application. These may be categorized in an elementary way as shown in Table 1.1.

TABLE 1.1

Category	Typical desirable properties	Main applications
Mechanical	Strength Toughness Stiffness	Machinery Load-bearing structures
Chemical	Oxidation resistance Corrosion resistance UV radiation resistance	Chemical plant Power plant Marine structures Outdoor structures
Physical	Density	Aerospace, outer space Reciprocating and rotating machinery
	Thermal conductivity Electrical conductivity Magnetic properties	Power transmission Instrumentation Electrical machinery Electronics

TABLE 1.2

Plastics	Metals	Ceramics	Composites
Weak	Strong	Strong	Strong
Compliant	Stiff	Brittle	Stiff
Durable	Tough	Durable	Low density
Temperature-sensitive	Electrically conducting	Refractory	Anisotropic
Electrically insulating	High thermal conductivity	Electrically insulating	
		Low thermal conductivity	

Certain types of materials can be broadly generalized as characteristically possessing certain combinations of properties (see Table 1.2).

As always, there are important exceptions to these generalizations. Plastics are indeed frequently extremely durable, but some are subject to stress corrosion. Metals are generally tough; indeed the widespread use of metallic materials for engineering purposes is due largely to the fact that they are mostly able to combine strength and toughness. But there is, nevertheless, a general inverse relationship between strength and toughness, and certain steels are vulnerable to catastrophic brittle fracture.

The brief conspectus of property characteristics given in Table 1.3 offers an overall view of the range that is available.

At an early stage in the design process it should become apparent that several different materials are capable of performing a particular function. It is then necessary to choose between them. This requires that the important properties be measured in an unambiguous, rational manner.

This is easy if a property is well-understood in terms of fundamental science, but not all material properties are of this sort. For example, it is essential to be able to measure the weldability of metals but no single parameter can do this because weldability measures the overall response of a material to a particular process and there are many processes. Other examples of the same type are drawability in the case of forming sheet material and injection mouldability of a thermoplastic. Even so, some attempt has to be made to put a number to any differentiating property, since this is the only way of making the selection process properly rational.

Property parameters are therefore of two types:

(1) *Fundamental parameters.* These measure basic properties of materials such as electrical resistivity or stiffness. They generally have the advantage that they can be used directly in design calculations.
(2) *Ranking parameters.* These generally do not measure single fundamental properties and can only be used to rank materials in order of superiority. They cannot be used directly in design calculations, but could be used in formalized selection procedures.

Failure in service

Since one of the aims of manufacture is to ensure that failure does not occur in service, it is necessary to be clear concerning the possible mechanisms of failure. Broadly, in engineering components, failure occurs either mechanically or by some form of corrosive attack.

There are three main ways in which a component can fail mechanically:

(1) Ductile collapse because the material does not have a yield stress high enough to withstand the stresses imposed. The fracture properties of the material are not important here and the failure is usually the result of faulty design or (especially in the case of high-temperature service) inadequate data.

TABLE 1.3

Strong	Alloys of Fe, Ti and the transition metals				Concrete in compression	
Permissible stress MPa	200–1500				70	

Weak	Plastics	Pb	Alloys of Al		Concrete in tension
Permissible stress MPa	~ 100	10	200		1.5

Stiff	SiC	HM Carbon fibre	Fe	Cu	Diamond	C-fibre composite laminate
Young's Modulus GPa	450	400	200	124	1000	200

Flexible	LDPE	Natural rubber	Butyl rubber	Neoprene
Young's modulus GPa	0.2	0.002	0.001	0.001

Light Specific gravity	Plastics 0.9–2.2		Mg 1.74	Be 1.85	Al 2.7	Ti 4.5	Concrete 2.3

Light
Specific gravity

Material	Plastics		Mg	Be	Al	Ti	Concrete
Specific gravity	0.9–2.2		1.74	1.85	2.7	4.5	2.3

Dense	Fe	Ni	Cu	Pb	Ta	W
Specific gravity	7.8	8.9	8.9	11.3	16.6	19.3

Refractory	Fe	Ti	Cr	Mo	Ta	W	Ceramics
Melting point °C	1537	1660	1850	2625	3000	3380	

Fusible	Lipowitz's alloy		Pb	Sn	Zn	Al	Plastics, glass
Melting point °C	60		327	232	420	660	

Corrosion-resistant	Au	Ta	Ti	Al	PTFE

Conductive	Ag	Cu	Al	Ni	Fe
Electrical resistivity $\mu\Omega\,cm$ (20°C)	1.4	1.7	2.8	7.2	9.8

Non-conductive	LDPE	PTFE	MgO	Mullites	Sialon
Electrical resistivity $\mu\Omega\,cm$ (20°C)	$>10^{15}$	$>10^{18}$	$>10^{14}$	$>10^{13}$	$>10^{12}$

Cheap	Fe	Concrete	Plastics	Pb	Zn	Al
Price/tonne (Fe = 1)	1	0.1	2–80	3	4.5	4
Price/m^3 (Fe = 1)	1	0.03	0.2–16	4	4.0	1.3

Expensive	Cu	Ni	Sn	Ti	Be	Diamond	C-fibre composite laminate
Price/tonne (Fe = 1)	12	40	44	94	700	2,000,000	250

(2) Failure by a fatigue mechanism as a result of a component being subject to repeated loading which initiates and propagates a fatigue crack.

(3) Catastrophic or brittle failure, with a crack propagating in an unstable and rapid manner. Any existing flaw, crack or imperfection can propagate if the total energy of the system is decreased, i.e. if the increase in energy to form the two new surfaces and consumed in any plastic work involved is less than the decrease in stored elastic energy caused by the growth of the crack. The significance of ductile yield in blunting cracks and reducing elastic stress concentration is immediately apparent. Beware, then, materials where there is little difference between the yield stress and the maximum stress.

The evaluation of maximum tensile strength does not indicate anything about the way in which the object is going to fail. It is obviously desirable that failure, should it occur, is by deformation rather than by catastrophic disintegration and this has led to the whole concept of fracture toughness testing: it is vital to know in high-strength materials what size of internal defect can be tolerated before instability develops and brittle fracture occurs, a feature determinable by fracture toughness testing which can then be interpreted with non-destructive testing and inspection.

There are too many different corrosion mechanisms for them to be listed in an introductory chapter, and they will be dealt with later. Generalized superficial corrosion is rarely a problem; greater hazards are presented by specialized mechanisms of corrosion damage such as pitting corrosion in chemical plant, stress corrosion in forgings, fuel ash corrosion in gas turbines, and the introduction of embrittling hydrogen as a result of corrosion. (see Chapter 11.2)

Failure records show that the bulk of mechanical failures are due to fatigue mechanisms. Overall, fatigue and corrosion, and especially the combination of the two, are the most significant causes. Aspects of failure analysis are dealt with in Chapter 4.

Cost

The achievement of satisfactory properties in his chosen materials is only part of the materials engineer's task – it is necessary also that they be achieved at acceptable cost. For this reason cost is sometimes incorporated into property parameters to facilitate comparisons. For example, the expression $C_R \rho / \sigma_{YS}$ relates to parts loaded in tension where C_R is price per unit mass, σ_{YS} is yield strength and ρ is density. It gives the cost of unit length of a bar having sufficient area to support unit load. This is a minimum-cost criterion and examples of corresponding criteria for different loading systems are given in Table 1.4. Some of these materials selection criteria are discussed in later chapters. The example given can also be put equal to C_V / σ_{YS} where C_V is the price per unit volume. Timber and concrete are the only materials sold traditionally in terms of volume, all other materials being sold in units of weight, even though, as the expression shows, P_v is the more meaningful parameter.

Space filling

It is remarkable how frequently cost per unit volume is the sole criterion for materials selection. The usage requirements specify the size the object shall be, and the materials employed are chosen on the grounds of minimum cost at that size: the mechanical properties of the material are then irrelevant. Examples range from pushbuttons to dams. Sometimes, however, the space-filling requirement is met at reduced weight and cost by making the shape hollow – we are then back to the mechanical property parameter, since the thickness of a hollow shell must be determined from considerations of strength and/or rigidity.

Fabrication route

Where there is a competitive situation, particularly with fairly cheap materials – for example on the basis of cost per unit volume – then fabrication costs can be of great significance in

TABLE 1.4. Performance-maximizing property groups

Component	Minimize unit cost for given:		
	Stiffness	Ductile strength	Brittle strength
Rod in tension	$\dfrac{E}{C_R\rho}$	$\dfrac{\sigma_f}{C_R\rho}$	$\dfrac{K_{1c}}{C_R\rho}$
Short column in compression	$\dfrac{E}{C_R\rho}$	$\dfrac{\sigma_f}{C_R\rho}$	$\dfrac{K_{1c}}{C_R\rho}$
Thin-walled pipe or pressure vessel under internal pressure	$\dfrac{E}{C_R\rho}$	$\dfrac{\sigma_f}{C_R\rho}$	$\dfrac{K_{1c}}{C_R\rho}$
Flywheel for maximum kinetic energy storage at a given speed	$\dfrac{E}{C_R\rho}$	$\dfrac{\sigma_f}{C_R\rho}$	$\dfrac{K_{1c}}{C_R\rho}$
Sphere under internal pressure	$\dfrac{E}{C_R(1-v)\rho}$	$\dfrac{\sigma_f}{C_R\rho}$	$\dfrac{K_{1c}}{C_R\rho}$
Rod or tube in bending	$\dfrac{E^{1/2}}{C_R\rho}$	$\dfrac{\sigma_f^{2/3}}{C_R\rho}$	$\dfrac{K_{1c}^{2/3}}{C_R\rho}$
Plate in bending	$\dfrac{E^{1/3}}{C_R\rho}$	$\dfrac{\sigma_f^{1/2}}{C_R\rho}$	$\dfrac{K_{1c}^{1/2}}{C_R\rho}$
Plate in buckling	$\dfrac{E^{1/2}}{C_R\rho}$		
Slender column or tube in buckling	$\dfrac{E^{1/2}}{C_R\rho}$		
Bar or tube in torsion	$\dfrac{G^{1/2}}{C_R\rho}$	$\dfrac{\sigma_f^{2/3}}{C_R\rho}$	$\dfrac{K_{1c}^{2/3}}{C_R\rho}$
Helical spring for specified load and stiffness		$\dfrac{\tau_m}{C_R\rho}$	
Rod or pin in shear		$\dfrac{\tau_m}{C_R\rho}$	
Thin-wall shafts in torsion		$\dfrac{\tau_m}{C_R\rho}$	
Spring for specified load and stiffness		$\dfrac{\tau_m^2}{C_R G\rho}$	
Long heavy rod in tension		$\dfrac{(\sigma_f - lg\rho)}{C_R\rho}$	

Table used with permission from the Fulmer Materials Optimizer

Key:
E	Young's modulus	C_R	Cost per unit mass
σ_f	Yield strength	ρ	Density
K_{1c}	Fracture toughness	l	Length
G	Shear modulus	v	Poisson's ratio
τ_m	Shear yield strength	g	Acceleration due to gravity

determining the final cost in the job. Shape and allowable dimensional tolerances are factors that may play a key role in deciding how, and of what material, a component should be made. The level of tolerances required must be matched up to those that can be obtained readily with the fabrication techniques suited to the material, unless the costs are to escalate. For example, attempts to cast spheroidal graphite cast-iron tuyere nosecaps of awkward design for a blast furnace producing lead or zinc, where the dimensions of the water passages must be uniform to a high degree of accuracy around the nose so as to achieve suitable water flow, will almost certainly result in a high proportion of rejected castings since it is very difficult to position cores with the required accuracy and to be sure that they will not move slightly during casting.

Surface durability

The requirement of surface durability, i.e. resistance to corrosion and surface wear or abrasion is sometimes important enough to determine the final choice, particularly in relation to aggressive chemical attack. More often it is a conditional consideration which indicates the initial range of choice. Further, this range of choice may well include composite structures – i.e. bulk materials coated with a corrosion-resistant or abrasion-resistant surface or chemically treated in such a way that the surface stability is altered.

As an example there is the competition between tool steels and case-hardened or surface heat-treated steel for such components as palls and ratchets, where a cheaper, more easily formed material of lower intrinsic strength is given a hard surface by localized carburizing and a heat treatment. This question is dealt with more fully under 'The Sturmey Archer gear' on p. 316, a component in which surface treatments on steel are widely utilized. Interesting examples also arise in the chemical engineering and food industries, where anti-corrosion linings to plant have frequently to be employed.

Physical properties

There are numerous instances, of course, where materials selection is primarily based on required physical properties. Whilst some instances are quoted in this text, for example in the case of electrical conductors (p. 51) and in components for a high-power gridded tube (p. 320), the thrust of this book is towards structural and mechanical engineering considerations. Within the field of physical properties the development of materials systems for electronic devices, sensor systems, etc. (many of which might be called micro-composites) is a large and rapidly developing area.

Future trends

The pattern of materials usage is constantly changing and the rate of change is increasing. Whereas the succession of Stone, Bronze and Steel Ages can be measured in millennia, the flow of present-day materials development causes changes in decades; there may also be changes in the criteria that determine whether or not a particular material can be put into large-scale use. In the past these criteria have been simply the availability of the basic raw materials and the technological skills of the chemist, metallurgist and engineer in converting them into useful artefacts at acceptable cost, leading to the present situation in which the most important materials in terms of market size are still steel, concrete and timber but supplemented by a constantly increasing range of others. These include metals (copper, aluminium, zinc, magnesium and titanium); plastics (thermoplastics and thermosets); ceramics; and composites (based on plastics, metals and ceramics).

However, two additional criteria may assume increasing importance in the future, arising out of the concept 'Spaceship Earth' (meaning the limited resources of the planet on which we live): these are the total energy cost of a given material and the ease with which it may be recycled. Concrete is a low-energy material but cannot be recycled: in contrast, titanium is a high-energy material which is difficult and expensive to recycle. Only some plastics are recycled to a limited extent at present, but steels and many other metals can be recycled with relative ease. Alexander[1] has calculated the energy content of various materials in relation to the delivered level of a given mechanical property. If, for example, this is tensile strength, the appropriate parameter is $E\rho/\sigma_{TS}$ where E is the energy in kWh required to produce 1 kg of the material, ρ is density in kg/m^3 and σ_{TS} is tensile strength in MPa. He finds that timber is the most energy-conserving material with a value of 24, whilst reinforced concrete is also low at 145. Steels lie within the range 100–500; plastics 475–1002; and aluminium, magnesium and titanium alloys in the range 710–1029. Alexander considers that concrete and timber will retain their predominant position into the 21st century, but foresees intense rivalry between metallic and non-metallic materials. The exhaustion of oil would require polymers to be extracted from coal or biomass, and the increasing scarcity of some metals will limit their usage to certain especially suitable applications. Composite materials will continue to be widely developed. Plus ça change, plus c'est la même chose!

In this text Chapters 1–5 give the background to the materials selection process, and Chapters 6–12 consider specific engineering property and surface durability requirements and how materials relate to these requirements. Selection of a material may frequently be indivisible from choice of fabrication route and the interplay between the two is discussed in Chapter 13. Chapter 14 then considers ways in which the selection procedure may be formalized and quantified.

In the teaching of Materials Selection in the Materials Departments at both Imperial College and the University of Cambridge case studies have been used extensively. Chapters 15–20 present broad studies relating to types of structure or service. Much can also be learnt by dismantling a specific artefact and discussing the selections that have been made, if possible

with the manufacturer. In Chapter 21 three individual case studies that have been used in teaching are included as examples of this approach, chosen to cover a wide range of materials and processing. A set of problems to test the reader is to be found in Chapter 22.

In 1981 the then Department of Industry in the UK agreed to support a proposal by the Royal Academy of Engineering to carry out a study of the use of modern materials with the particular aims of examining the factors which inhibit the wider use of these materials in British Industry and to recommend actions. The approach adopted was to carry out a series of in-depth case studies, which are fully documented in the report.[2] Although the report as a whole was not primarily intended as a teaching document, the case studies in particular contain a great deal of interesting information and challenge for the future and could well be read in support of this text.

Further to this study, the findings of the UK Technology Foresight Materials Sector Panel were published in 1995[3]. The Foresight initiative was set up to bring UK industry, academia and Government together to consider how to take advantage of opportunities to promote wealth creation and enhance our quality of life, focusing on a number of sectors, including materials. The continuous, incremental improvement of materials and processes, rather than the occasional leap forward, was identified as the highest priority. Materials and processes which protect or remedy the environment and which can save lives and alleviate suffering were also perceived as targets for investment.

References

1. W. O. ALEXANDER: *Mat. Sci. Eng.*, 1977; **29**, 195.
2. FELLOWSHIP OF ENGINEERING: *Modern Materials in Manufacturing Industry*, London. 1983.
3. UK OFFICE OF SCIENCE AND TECHNOLOGY: *Technology Foresight: Progress Through Partnership: 10 Materials*. HMSO, 1995.

2
Motivation for selection

Selection of engineering materials, as we have seen in the preceding chapter, is primarily about an awareness and understanding of:

(1) what materials are available;
(2) what processes for shaping these materials are available and how this affects their properties;
(3) the cost of the materials in relation to each other, their processing and their properties.

The materials available to the designer number in their hundreds of thousands. However, they can conveniently be grouped into a number of broad categories, illustrated in Figure 2.1. Processes can also be divided into a number of categories[1], although both materials and processes should be considered in their entirety before selection to avoid the 'but we've always done it this way' approach to design.

There are two basic situations that necessitate materials selection that we shall consider:

(1) development of a new product;
(2) improvement of an existing product.

2.1 New product development

In a fast moving world with the expectations of the customer ever evolving, the successful product must:[2]

(1) meet the needs of the customer;
(2) beat the competition to the market;
(3) offer either better performance, more features, or both;
(4) be perceived to offer value for money in terms of the balance of cost and quality.

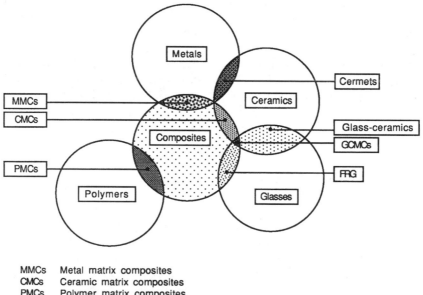

MMCs	Metal matrix composites
CMCs	Ceramic matrix composites
PMCs	Polymer matrix composites
GCMCs	Glass-ceramic matrix composites
FRG	Fibre reinforced glass (experimental)

Figure 2.1 The engineering materials family. (After Ashby[1].)

As well as the customer and the competition driving forward new product development, so pressure may also come from legislation for safer or more environmentally acceptable products. Increasingly, it is realized that consideration of a product life cycle in terms of a 'cradle-to-grave' analysis of the use of the earth's resources is called for.

However, if a product does not satisfy the needs or desires of the customer, it will fail. Establishing what these needs or desires are, means spending time with customers and potential customers early in the product development process. The temptation to think that the needs of the customer are obvious has to be resisted.

Effective management of new technology is critical to success. Steps must be taken to reduce the risk of using new technologies, but at the same time the opportunities arising from these new technologies need to be identified and marketed in the minimum time. The use of a new and innovative material may be interesting in terms of the technology, but if the cost or quality have suffered, the product will fail. Similarly, if in the design the product necessitates the synthesis of a new material, then for engineering structures, it is likely that the design will need reworking to employ existing materials, as the development time for new materials is measured in decades.

It is the job of the materials engineer to evaluate new materials and methods as well as to assess the suitability of existing materials and processing routes for a particular application.

The stages of new product development

Once a market need has been established, the systematic design stages for new product development are summarized in Figure 2.2. The three stages of design identified are[1,3].

(1) *Conceptual design*. Possible designs are produced as block diagrams representing the main components with some idea of layout.

(2) *Embodiment design*. This stage involves refining the conceptual designs so that a version exists suitable for marketing and manufacturing teams to visualize the product, usually from computer-generated images. Overall dimensions and shape are emerging at this stage, as are the generic classes of material and processing techniques to be used.

(3) *Detailed design*, at which stage the preferred layout design is fully dimensioned. Materials and process selection is also now refined to approach a specification.

Product design is an iterative process, as shown in Figure 2.2. Once an item goes into production and sales are generated, the reaction from customers is analysed and the design and manufacturing stages can be reviewed. Improvements can then be made by re-appraising the detailed design (see Section 2.2).

Product development costs

The costs actually spent in developing a new product at any given stage in the product development cycle are lower than the costs committed by that same stage. This is illustrated in Figure 2.3. The expenditure in the early stages of product development relate to the man-hours spent in planning, designing and developing prototypes. This is minimal compared to the overall project costs, but during this process, the future expenditure for the product's entire life are determined. By the end of the conceptual design stage, the materials selection has largely been made, along with the fabrication route, including, for example, expensive tooling. Hence, it is this conceptual design and planning stage that is crucial to a product's, and sometimes a company's success, and the benefits of a 'right first time' approach are manifest. This issue is also highlighted in Table 2.1.

The nearer a product is to production, the greater the cost of making any change. The impact of the early stages in the product development cycle are clear.

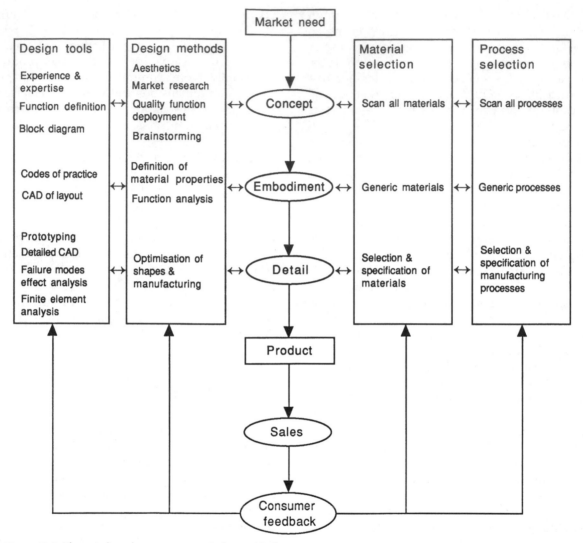

Figure 2.2 The product design process. (After Ashby[1].)

Block diagrams

Block diagrams are important at the conceptual design stage. After product function definition, the product can be broken down in terms of its main components and arranged as a block diagram, with purely functional information and no 'aesthetic' information. However, information concerning physical connections between the main components will also be contained in the block diagram.

Codes of practice

Many industry sectors have specific product legislation or codes of practice to follow and which therefore influence product design. For example, medical devices must meet certain safety requirements and electronic components must meet electromagnetic interference requirements.

Computer-aided design (CAD)

CAD involves the use of computer software packages, most of which are capable of allowing the creation of a design onto the computer screen. A number of manipulations can then be performed, allowing the design to be viewed from different angles and confirming that different components are not occupying the same space.

This becomes particularly important when a large team of design engineers is involved. The Boeing 777 jetliner, for example, was the first Boeing jet to be designed using solely data input into a CAD system[4]. Previous jets had been built by a mixture of CAD systems and expensive mock-ups, but it was often only on the assembly line that all the design flaws were finally ironed out. This is hardly surprising given the large number of designers of the job, with, at its peak, 238 design teams working on the 777, with over 2000 computer workstations networked via eight mainframe computers. However, the use of CAD not only enabled the designers to make sure that there was no interference between parts, but also that sufficient space was left for maintenance, both in terms of human access and part removal.

CAD is often linked with computer-aided manufacture (CAM). The digital data generated by CAD can be used to control the manufacture of components on computer numerically-controlled (CNC) machines.

Failure modes effect analysis (FMEA)
FMEA is concerned with assessing the impact on a product of the failure of a particular component. Once the vulnerability of the various designs has been determined, process and quality control can be organized to reduce the risk of failures.

Finite element analysis (FEA)
This is another computer-based technique. In FEA, a component is broken down into small three-dimensional elements and mathematical models are applied to analyse mechanical or physical properties. The technique can be useful in identifying stress concentrations but is more frequently used in the analysis of materials processing techniques such as injection moulding, forging and rolling.

Function analysis
Function analysis is concerned with the study and definition of the primary and secondary functions of each component, which can then be expressed simply as a verb and a noun. For example, the primary function of a flywheel is to *store energy*, whereas secondary functions would include *resist fracture*. This allows for numerate materials selection.

Prototyping
Recent developments have enabled the use of CAD data in the prototyping stage of product development. These 'rapid prototyping' techniques can reduce the product development times significantly[5]. The computer data is essentially converted into a series of slices; these slices can then be built up, layer by layer, into a model of the component. This can be achieved in a number of ways, including laser scanning to cure a layer of resin or sinter a layer of powder, using UV light to cure a layer of resin, extruding a layer of polymer or cutting out sheets of special paper by laser (processes referred to respectively as: stereolithography, selective laser sintering, solider, fused deposition modelling and laminated object manufacturing). The models produced by these rapid prototyping techniques can be used to check form and fit and to give the product development team something solid to show potential customers and suppliers.

For certain components, it is possible to use the models produced by these above techniques to check functional performance, for example, the air intake manifold for an engine, although they can be too fragile for some applications. Current efforts are focused on developing models produced by rapid prototyping for use as mould patterns. This can reduce the lead times significantly when compared with the more traditional techniques for manufacturing patterns and tooling.

Quality function deployment (QFD)
The term QFD is a derived term from Japanese and is concerned with ensuring that the customer gets what he wants rather than what the design team thinks he wants. The methodology analyses customer needs and also how these needs are currently satisfied by benchmarking both the products of the company and its competitors. QFD fits well with the concept of simultaneous engineering (see 'Teamwork' below).

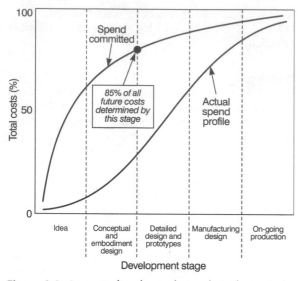

Figure 2.3 Committed and actual spends in the typical product development cycle[2].

TABLE 2.1 The cost of design change[2]

Development stage	Relative cost of design change
Concept	1
Detailed design	10
Tooling	100
Testing	1000
Post-release	10,000

Teamwork

In the development of a new product, it is typical for a multidisciplinary team to be established. It takes time and effort to build an effective team, however once in place, it is possible to perform as many as possible of the product development tasks at the same time rather than sequentially, a technique now known as simultaneous or concurrent engineering. This term only describes what has often been the practice before, particularly in medium to small engineering companies where communication is usually rapid and interactive.

The underlying philosophy behind simultaneous engineering is to develop a process where activities can be overlapping and also starting earlier. This is illustrated in Figure 2.4. The activities of marketing, design, materials engineering, manufacturing and purchasing must all communicate effectively. Information must be used in parallel rather than through 'autonomous islands'. Simultaneous engineering has been shown to reduce the timescales for the manufacture of a new product, as well as the development costs in the longer term.

Of course, the kind of team involved in the process depends on the complexity of the product. For example, a totally new product utilizing new technologies would almost certainly require a team drawn from all company departments, and possibly also involving help from outside including consultants, sub-contractors and possible customers. It must be the objective of the design team to reduce the lead time and the need for multiple prototyping as far as possible, perhaps through the use of computer modelling techniques.

2.2 Improvement of an existing product

Redesign can become crucial if your product is losing out to the competition. Of course, it may be the case that, following a design review, even during the development of a new product, it is clear that the design is not viable and work should be refocused in other areas. However, redesign can be the key to the improvement of sales and profit margins, through parts reduction, use of parts common to other products in the range, reducing the assembly time and just plain better design, improving the aesthetics and user-friendliness, increasing the added value.

The financial risk of introducing new products can be considerable. This risk can be reduced by introducing product updates at regular intervals, so-called 'incremental innovation'. The financial risk is lessened by making small step changes of lower cost at the introduction of each new

Time →

Activity	Concept	Design development	Prototype	Production start-up
Marketing Product planning	▓▓▓			
Engineering		▓▓▓▓▓▓		
Validation			▓▓▓▓▓	
Manufacture				▓▓▓▓▓

'Conventional' engineering

Time →

Activity		Concept	Design development	Prototype	Production start-up
Marketing Product planning		▓▓▓▓▓			
Engineering	Feasibility	▓▓▓			
	Production design	▓▓▓▓▓			
Validation	New technology	▓▓▓▓			
	Main programme		▓▓▓▓▓▓		
Manufacture	Feasibility	▓▓▓▓▓			
	Tooling			▓▓▓▓▓▓	

Simultaneous engineering

Figure 2.4 Conventional and simultaneous engineering.

product, rather than dramatic developments which are much more risky. For example, individual new components of a product can be tried out at separate introductions. Frequently, electronic consumer products are 90% the same as their predecessors, despite being labelled 'new'. This methodology also results in these new products being released at regular intervals, thereby maintaining visibility in the marketplace.

2.3 Problem situations and constraints on choice

Even if the design philosophy already presented has been followed a critical situation may arise, for example by the failure of supplies of an already optimized or used material, or by the failure of components or plant in such a way that a previous choice was clearly not successful. Immediate replacement then has to be arranged to maintain customer confidence.

If, for example, a manufacturer of light engineering products buying in bar stock for the production of gear trains encounters a spate of cracking during hardening after machining, the fault can lie either with the material chosen in terms of its quench sensitivity, the nature of the particular batch delivered, or with the control of the hardening process in relation to the material generally. Since outgoing deliveries have to be effected on schedule an immediate decision has to be made, perhaps even without any chance of obtaining the wide range of information necessary for a full assessment of the situation. Another form of this general type of situation

15

arises through materials availability. The original material specified is no longer obtainable or deliveries prove unreliable, and a substitution has to be made urgently. The failure of a specific item on a plant may shut down the complete production and an immediate replacement has to be made – with these situations the economic basis for selection is concerned with getting the manufacturing line or equipment working again in the minimum time, and usually means the employment of what is reasonable and available rather than necessarily an optimum choice.

Another difficulty may arise in terms of the scale on which materials will be delivered. In a complex engineering artefact there may, for example, be small critical components such as palls, lifters, ratchets, etc., where wear resistance is of prime importance and for which a particular tool steel is recommended. Such a steel will often not be held by stockists and manufacturers will only quote for deliveries of, say 12 tonnes, inflicting an unacceptable stock-holding requirement for the user. Only fairly standard steels are generally available from stockists or factors, and even where a technical advantage will almost certainly be lost the optimum solution may sometimes have to be discarded for the best available.

It cannot be stressed strongly enough that changes in design can often change the range of suitable materials, and must be integrated with the possible fabrication methods for those materials. If, for example, gears that were normally heat-treated and flame hardened are increased significantly in size it may not be possible to hold distortion within the tolerances for grinding to the final dimensions. In such a case surface nitriding, but using a different steel, may be the solution since the process can be carried out after machining with allowance for growth but negligible distortion (see p. 316).

There are even constraints as regards fashions in materials, particularly for the domestic consumer market, which produce resistance to the use of perhaps technically superior or even cheaper alternatives. These may have their roots in an earlier association of a specific material with poor response to a particular fabrication route which has now been overcome, or where the fabrication route has been changed. Aesthetic values may be ascribed to a particular usage, or a reduced personal maintenance content may be involved which is not strictly accountable, but none the less valued by the consumer market. As an example, few would now choose carbon steel table knives or garden tools because of the care required in cleaning, drying and greasing, and yet for their purpose they would be cheaper and more readily maintained to a sharper cutting edge than the normal stainless steel alternatives.

In other extreme cases we have situations where the use of technically superior materials would be discouraged for what has come to be regarded as a short-term replacement item by the public, even if the long-term economics were favourable. An interesting example here is the exhaust system of cars, where the long-term economy of stainless steel cannot be disputed, but where in many countries the increase in initial capital cost or early replacement cost is not generally found acceptable, since the first ownership is usually short.

References

1. M. F. ASHBY: *Materials Selection in Mechanical Design*, Butterworth-Heinemann, 1992.
2. THE DESIGN COUNCIL: *Successful Product Development – Management Case Studies*. HMSO, London, 1994.
3. M. F. ASHBY: in *The Engineers Guide to Materials Selection – Modern Methods and Best Practice*, AEA Technology, 12 March 1996.
4. BOEING: Commercial Airplane Group Fact Sheet *Boeing 777 Computing Design Facts*.
5. N. A. WATERMAN and M. F. ASHBY: eds. *Materials Selection*. Chapman and Hall, 1996.

3
Cost basis for selection

The process of selecting a list of promising candidate materials for a given application will be carried out initially in terms of the required properties, but final decisions will always involve considerations of cost which in many cases will be the dominant criterion. Placing a product on the market inevitably involves risk, and in a capitalist economy calculations prior to marketing must aim at the certainty of profit within a foreseeable period of time. The allowable margin of error associated with these calculations, and thus the vigour with which they are carried out, depends upon the state of the market and the activities of competing manufacturers. Increase in costs from superior materials or components has to be offset by substantial improvement in performance, as previously indicated, if it is not to appear finally as an increased increment of cost for the project as a whole. A change of material also brings in-house costs such as those associated with changes of instruction and stocking, particularly in the latter where the variety of materials being used is increased by the change.

Whilst in any given set of circumstances the competition between materials or components may be finally decided on costs where otherwise similar performance is obtainable, the precise level of performance and cost must depend on the type of application involved.

In the interaction between performance and cost it is possible to see a continuous spectrum stretching from, at one end, applications which demand the maximum achievement of performance (i.e. performance oriented products) to, at the other end, applications in which considerations of cost must be predominant, (i.e. cost-oriented products).

Typical examples of fully performance-oriented products would be advanced armaments (e.g. atomic submarines) and space vehicles. In these cases the over-riding need for complete reliability in service means that, once the decision to manufacture has been made, considerations of cost will frequently be subordinate. However, expenditure which does not improve the level of performance and reliability will only lead to reduced sales or increased resistance to project funding even where the level of cost is not the most important consideration. Such funding may well be politically controlled and external sales may not be involved, although for many advanced armaments there is still a competitive market.

A less clear-cut example is a train for a commuter network. Although the level of performance required is not as high as in the previous two examples, it is still at a substantial level, or should be, to provide a reliable service on crowded networks. Yet the builder of trains is faced with the fact that there is hardly a railway system throughout the world that is not running at a loss. Nevertheless, wherever the money is to come from, once the decision to build is taken performance must be provided to the required degree and this fixes the level of cost.

Examples of cost-oriented products are a mass-market motor car and a washing machine. The mass-production industries must market their products at a price the public will pay so that once an acceptable performance has been achieved, i.e. once it has been established that a design is able to function to meet the perceived market need, it then has to be decided what level of performance can be offered for the required price. The essential point here is that the manufacturer does not *have* to provide the maximum level of performance of which he is technologically capable. He has merely to ensure that his 'value-for-money' parameter is no worse, and preferably better, than that of his competitors; he therefore seeks to provide the level of performance which is economically right, i.e. the optimum rather than the best performance. This must, of course, be acceptable to the consumer. As well as varying from product to product, the acceptable level of optimum performance may vary from time to time as the general climate of public opinion changes.

But how do you measure a 'value-for-money' parameter? The current trend is to move away from the volume manufacturing of uniform products towards products meeting the needs of the individual. The 'mass market' is becoming a mass of 'niche markets'.

Whereas in the 70s and 80s the price of a product may have been of paramount importance to the customer when it came to the decision to purchase, now, it seems, the consumer is becoming more educated in terms of the real value of quality, good design and, in particular, the benefits of the sensible and responsible use of our finite resources. We are, slowly, moving away from being a throw-away society.

3.1 Cost-effectiveness and value analysis

In the present context it is convenient to give special meanings to the terms value and cost:[1]

(1) value is the extent to which the appropriate performance criteria are satisfied;
(2) cost is what has to be paid to achieve a particular level of value.

The properties of a given design and material may be regarded according to the extent to which they are cost-effective; that is to say, the extent to which they may be dispensed with in the interests of reducing costs.

The designer will be prepared to incur costs for the provision of a certain property in proportion to the penalties that will result when it is absent. Thus, the civil engineering contractor will not regard toughness as a cost-effective property when designing a bridge, since if his bridge breaks then his professional reputation is destroyed with it.

On the other hand, the automobile manufacturer has traditionally treated corrosion resistance in the average motor car as a highly cost-effective property because, provided progressive rusting of the bodywork does not reach a critical stage before the motor car has reached second-time or third-time buyers, he suffers no penalty from the eventual, inevitable, failure.

One of the contributions that the materials engineer can make as a member of a design project team is his ability to distinguish between material-sensitive and design-sensitive properties. A tough material is one that is resistant to the initiation and propagation of cracks, whereas a tough design is one that is free from notches and stress-raisers. It may be quite expensive to obtain an especially tough material for a critical application but relatively cheap to free a design from stress-raisers. It is technical incompetence to solve a problem more expensively than is necessary.

Cost-effective decisions should only be made in the light of full knowledge relating to:

(1) the special requirements of anticipated service;
(2) the properties of all available materials and their relationship to those requirements.

An important aspect of the service requirement may be formal regulations laid down by an appropriate Safety Board.

Inevitably, cost-effective decisions act to inhibit technological advance. Every commercial product is required to give a satisfactory return on capital expenditure in the shortest possible time, so that the cost of any improvement in technology must be more than recouped from corresponding savings resulting from improved performance. As an example, current designs of coal-fired power plant give efficiencies of 45–46%. This has been made possible by the development of improved ferritic steels (T91 for superheaters and P91 for thick section components)[2] with maximum steam conditions of 300 bar and 580°C giving cost-effective operation. For improved efficiencies, say 50% at 325 bar and 650–700°C, a material must be developed with greater creep and corrosion resistance for the water panels, capable of being welded without pre-heat or post-weld heat treatment as a result of the construction method. More highly alloyed ferritic steels are being developed for this purpose (e.g. HCM12A and HCM2S). The more advanced plant will also require austenitic steel superheaters with improved creep strength and corrosion resistance. However, if the increased efficiency and any improved environmental performance over the lifetime of the power plant were unlikely to make up for the increased material costs, then such improvement in materials may not be worthwhile.

In the case of coal-fired power plant, it is generally accepted that research into materials for power generation is still important; indeed, a preliminary study of materials suitable for plant operating at 375 bar and 700–720°C is under way[3]. It is anticipated that nickel-based and austenitic materials will have to be developed for many components.

3.2 Analysis of cost

The total cost of a manufactured article in service is made up of several parts, as shown in Figure 3.1. Whether or not a manufacturer operates in a competitive market, but particularly if he does,

reduction in the cost of products to the consumer should be the aim, and in this it is as important to reduce the costs of ownership as it is to reduce the purchase price. Unfortunately, most attention is usually directed towards reduction of purchase price since this is the simplest and most direct way of increasing sales of cost-oriented products. Although reducing the costs of ownership is equally valuable to the consumer, there is often less emphasis in this direction since it will usually increase the basic purchase price. The justification is, of course, long-term in that when spread over a reasonable life the decrease in running costs more than compensates for the increase in purchase price.

Thus in the automobile field, the wider use of galvanized steel for motor car bodies would help eliminate the rust problem and greatly extend the life of the whole car, which at present in the bulk sales market tends to be limited by the body rather than the mechanical components. Similar remarks apply to the use of stainless steel for silencers. In both of these cases the necessary technology is available, but there is often little incentive for the manufacturer to use the more expensive materials because by the time failure has occurred he is no longer involved, and the case that the initial consumer would be willing to pay more for a longer life product is not always clear-cut.

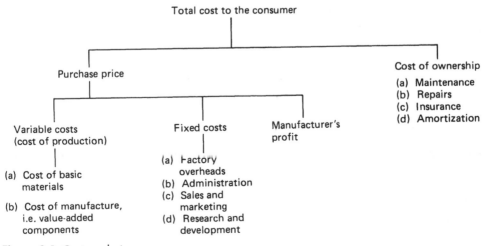

Figure 3.1 Cost analysis

At the quality end of the automobile market, however, it is possible to take a longer-term view, and Jaguar Motors, for example, fit stainless steel silencers on standard production models. The use of galvanized steel for structural purposes in cars, by such manufacturers as Rolls-Royce to give greatly longer body life, has been established for a long time and this approach is now being followed also by some of the better bulk manufacturers such as Audi and BMW. This may well reflect a growth in a more performance-oriented purchasing sector, but with safety and reliability of increasing importance. While galvanized steel costs approximately £0.50 per kg (US$0.70), aluminium costs approximately £2.70 per kg (US$4.00) and glass-reinforced sheet moulding compound (SMC), generally considered the most cost-effective plastic for body panel applications, costs approximately £1.20 per kg (US$1.75). However, costs have to be calculated on the basis of properties for a particular design criterion and not merely on costs based on weight or volume, and for reasons explained in Chapter 18, these materials are gaining acceptance among automobile design engineers.

The variable costs (i.e. production costs) arise, of course, in the primary raw material costs and the conversion margins in the fabricated product to cover the cost of the intermediate operations to the finished form. The primary cost can be markedly affected by supplies, marketing methods, international politics (including tariffs), metal stocks (strikes, dumping, etc.). Fabricating industries for the most part are limited in outlook to their own countries and do not possess effective price-regulating organizations or mechanisms, which, in any case, may be banned by the State anti-trust laws. Frequently the fabrication costs are low in relation to the value of the material (particularly for non-ferrous metals and plastics) and the scope for manipulation and influence is small. The main cost worries in components made from the more expensive metals are caused by the variations in base metal price, and by abrupt changes in trade activity.

Basic material costs

Many factors can influence the cost of a basic raw material.

Compound stability

In metals, the more stable the compound in which the element is found, the greater will be the amount of energy and thus cost in the process of reducing that compound for the recovery of the metal value. Interestingly the history of metal usage relates to the stability of its compounds, i.e. the ease with which it may be extracted. Gold, silver and copper occur in the elemental state, and copper and lead are relatively easily reduced from accessible minerals.

Relative abundance

Relative abundance, and the degree of complexity in mineralogical association, are obviously important factors since the less concentrated a material source is, the more effort must be devoted to its extraction. Thus iron, where the reduction from oxides is only marginally more energy-consuming than copper and which has also the richest and most easily recovered ores, is the cheapest metal.

A typical iron ore contains 60–65% Fe (lower grades down to 25% have been employed but are now considered uneconomic).

A typical copper ore contains 1–1.5% Cu
A typical uranium ore contains 0.2% U
A typical gold ore contains 0.0001–0.001% Au.

In the 1980's a widespread and deep-seated excess capacity developed in the mineral industries with utilization of capacity exceeding 75% in less than one quarter of the minerals, and many of these were fairly insignificant. In at least the medium term there is ample capacity to meet present and prospective demand for nearly all minerals, even with due allowance for typical disruptive influences on supply. Whilst the overall world production of metals has continued to grow, almost certainly the fall in demand for metals in industrialized countries outside of commercial reccessions is the result of challenge from competing materials. Whilst cost savings and productivity improvements can to some extent offset weak prices, a satisfactory situation for the mineral industries can only come about when supply and demand are restored to an approximate balance, and endemic excess capacity is eliminated. In a recent survey[4] the productivity as measured by the value added per head was considered high in the metals sector of UK industry, but the forecast for growth for these industries in the UK over the period 1990–2004 was considered low, unlike the manufacture and processing of plastics, where over the same period the forecast for growth is high.

For plastics, the raw materials cost is dependent on the prepolymers, derived largely from oil. This is only part of the picture, however, as there are many stages between the oil platform and a product for sale in the shopping precinct, including: oil recovery, oil refining, base chemical production, polymer manufacture, compounding, processing, assembly and, finally, sale. The price at each stage is affected by both subsequent and preceding stages in a complex manner.

Supply and demand

The elementary theory of economics considers that the price of a commodity is fixed by a unique equilibrium between supply and demand. This price is given by the point at which the demand curve intersects the supply curve (curves D and S in Figure 3.2). Prices vary as a result of horizontal shifts in one or other of the demand and supply curves. When demand rises, prices tend to rise because a buoyant market lessens the keenness of competition between different suppliers and enables them to maintain wider profit margins. Although the consumer is then paying more for the product this is not necessarily disadvantageous overall if it leads to improved capital investment and efficiency, which thereby adds to the future stability of the company concerned, with a maintained contribution to national wealth and employment.

When there is surplus productive capacity, prices should fall as competing producers pare their profit margins to avoid shutting down large-scale plant. This simple market mechanism does not, of course, always operate, and there are considerable incentives to maintain prices at artificially high levels by arrangement.

When, for any particular product, there is only one major producer in the field then it is easy for price control mechanisms to be distorted away from the public interest and most countries have anti-monopoly laws to prevent this. On a national basis this may work well, but internationally it is more difficult. There is little to be done about the fact that if a single country is the sole large-scale producer of a certain commodity for which there is a large and continuing demand throughout the rest of the world then that country has the ability to maintain the price of the commodity at a level which is quite inappropriate to its true value. Even when two countries are involved, they can arrange to control its marketing to the benefit of them both. When a number of major producers join together to control prices, this is known as a cartel. The prices of some metals still appear to be controlled in this way.

Level of consumption is important because when production is low, unit costs are high. Reducing unit costs requires high-volume production methods which are only obtainable with large-scale plant and equipment. But however much it is desired to reduce prices, the rightward limit of any supply curve is set by the productive capacity of available plant. A large jump to a position such as S1 in Figure 3.2 requires the construction of a larger, or technologically updated, plant. Such a project requires the investment of substantial risk capital and calls for considerable confidence in the level and consistency of future demand. This can be done. For example, when a new material becomes available it is usually produced at first in small quantities and the price is correspondingly high. There is then a production barrier which must be surmounted before the price can be significantly reduced

Figure 3.2 Curves of supply and demand: S = supply; D = demand.

because when the level of production is low the price is too high for the consuming industries to place large orders, but the producer cannot drop his price until he is sure that large orders will be forthcoming. However, once this barrier has been surmounted the price should fall sharply and remain steady so long as there is no further major change in the equilibrium between supply and demand. We have only to look at the history of aluminium and titanium to see materials move from being rare and expensive exotics to being relatively moderately priced items of everyday industrial use under the influence of demand in a few decades. There is currently much interest in the metallocene catalysts being used to improve the properties of polyolefins (e.g. polypropylene and polyethylene), but the balance between scaling-up of production of new materials using these new catalysts and developing new markets is a sensitive one, with several large manufacturers currently either operating production units or pilot plant. There must be a clear commercial benefit, both for producer and end-user.

Cost fluctuations

When a material is in general short supply its price may sometimes fluctuate violently as a result of non-technical factors. In 1969 the pro-

ducer price for nickel was £986 per tonne. There was then a strike at Falconbridge which brought production to a halt. Immediately the price of nickel on the open market rose to £5500 per tonne. As a direct result of this, the British Steel Corporation raised its prices for austenitic stainless steels by 14%. The consequence of such an increase was to cause traditional applications of austenitic stainless steels to be examined to see if there was any possibility of using low nickel ferritic stainless steels instead. The combined basin and draining sections incorporated into kitchen sink units had normally been made wholly of austenitic stainless steel. The ferritic variety of stainless steel is capable of functioning in the draining section of the unit but had not been widely used because of its being less amenable to the forming method and slightly inferior performance as regards corrosion resistance. Modern steel-making methods enabling the control of interstitial solutes at lower levels improved the formability of the material and widened its application. Such ferritic steels could be purchased at prices around 25% lower than austenitic steels and there was therefore considerable incentive to avoid the problems associated with the fluctuating price of nickel by the substitution.

The incentive to use substitutes has been even stronger in the case of copper and its alloys, where the price situation for copper has, for many years, been extremely fluid and unstable. In the mid-1950s the price of copper on the London Metal Exchange fell from £436 (1945) to £160 (1958) as a result of overproduction against more general depressed economic growth. Since then the 'normal' slope of the approximate price curve, reflecting inflation, has been frequently swamped by massive oscillations due to political factors, industrial strikes, local wars and world recessions (see Figure 3.3). Similar effects may be found with other commodities, prices tending to collapse during recessions and rise if it happens that production difficulties coincide with increased demand at the end of a recession (Figure 3.4). In the case of tin, trading was suspended over a substantial period, as a result of large stocks

Price of Copper 1968 -1995

Figure 3.3 Fluctuations in prices of copper. (Data from London Metal Exchange.)

Lead Prices 1976 - 1995

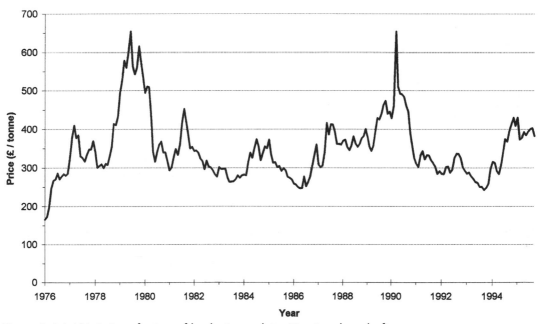

Figure 3.4 (a) Variation of prices of lead, zinc and tin. (Continued overleaf)

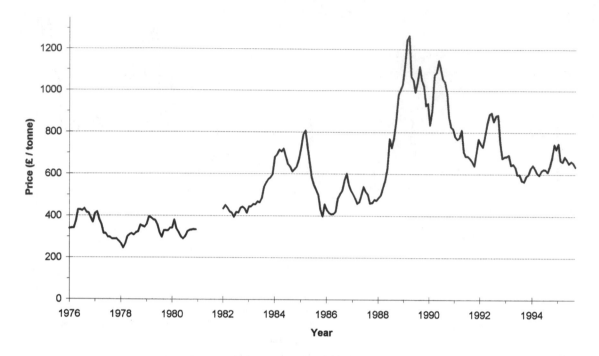

Zinc Prices 1976 - 1995

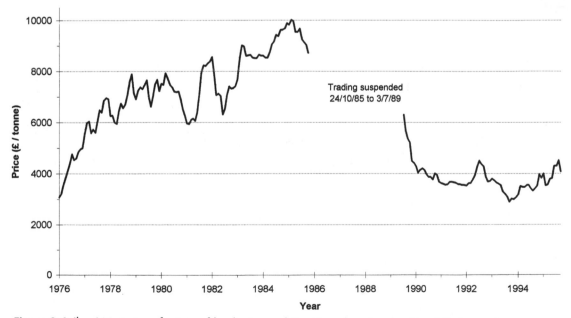

Tin Prices - 1976 - 1995

Trading suspended
24/10/85 to 3/7/89

Figure 3.4 (b, c) Variation of prices of lead, zinc and tin. (Data from London Metal Exchange.)

and a breakdown of producers' agreements. The average price of commodity plastics in the UK for the period 1990–95 is shown in Figure 3.5. This illustrates how plastics are similarly affected by fluctuating prices. Increasingly, end users are involved in direct discussion with polymer producers, agreeing on grades and forecasts of demand. This may allow a longer term view of prices, but the scene may be just too complicated for this to be effective. The classical market response to plunging prices is for the producers to lower their production rate or otherwise restrict supplies. However, this does not always happen; sometimes because the economy of a whole country is dependent on the revenue from a single commodity, or perhaps because severe cutbacks in state-owned companies would be expected to produce unacceptable political and social consequences.

There are several options open to the manufacturer who must buy a material which is subject to severe price instabilities. Three possibilities are: (1) advanced stock control, (2) material substitution and (3) diversification of operations.

The classic advice given to investors on the stock market (which they hardly ever take) is to buy when prices are low and sell when they are high. The analogous advice to a manufacturer would be to stock up when prices are low and de-stock when they are high. Whether or not he does this depends upon his perception of the time scale over which the price fluctuations occur, because money in the bank earns interest whereas metal in the warehouse earns nothing. This is mainly a matter of confidence and, in fact, companies seem to de-stock during a recession, probably because cash flow becomes a problem when sales are low.

Material substitution is sometimes possible. For example, aluminium is an obvious substitute for copper in many electrical and heat conduction applications but there are problems. The inherent low strength of aluminium can be overcome by the use of steel-cored cables but difficulties associated with the joining of aluminium by soldering have been particularly significant in maintaining the use of copper in many cases. Again, copper has maintained its position as the principal material for the

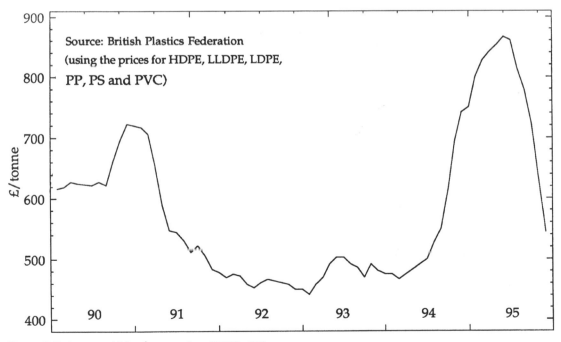

Figure 3.5 Average UK polymer prices (1990–95)

manufacture of small-bore tubing in central-heating systems, largely because of resistance to the use of hard-drawn stainless steel where bending is more difficult and where joining has to be by compression fittings. Copper has also maintained its position for low-temperature heat exchanger applications as in water heaters, again because of the ease of assembly by soldering as compared with aluminium.

Diversification of some proportion of a company's operations into some other less sensitive area is another way of lessening the problem.

Within the conditions of a volatile market, large users of a metal may prefer to negotiate a future supply price with the producers and risk a change in market forces. But for many purchasers there will always be a need to buy directly or indirectly through the commodity exchange, where the dealing reflects the supply and demand position and fixes prices.

Commodity exchanges – the London Metal Exchange (LME)

As pointed out by Gibson–Jarvie,[5] under the impetus of the industrial revolution, Britain moved from a net exporter to net importer of metals on a large scale. The result was that prices began to fluctuate with shipments of ore or metal arriving at very irregular intervals and the value of the cargoes varied greatly as supplies temporarily exceeded or lagged behind demand. It is a characteristic of the different industries that producers would like to see a steady, smooth demand or a predictably smooth increase (or decrease), whereas stockists and consumers are operating at a different rhythm.

Fairly soon, fast packets and eventually the telegraph, made it possible for a merchant in London to know of the departure of a particular ship some time before she could be expected to dock in this country. By making use of this intelligence a merchant could to some extent iron out the wider of these fluctuations in price by dealing in a cargo while it was still at sea, or selling it forward. The result was a smoother price characteristic although there were still

major difficulties in that metal was arriving in all sorts of shapes and sizes and at different purities. This could make the non-physical buying and selling of cargoes difficult, and it was clearly necessary to insist on standard forms and purities (assays). Dealings therefore became standardized on Straits Tin and Chile Bar copper; lots were at fixed tonnages and the forward trading period was settled at 3 months, this being the average time for a voyage from Chile or the Malay Straits.

From forward dealing it was an obvious step to 'hedging'. Hedging is used as an insurance against adverse price movements. For every physical transaction when there is an interval of time between the commencement and completion long enough for prices to move appreciably, a hedging contract will be entered into such that a possible loss on the one will be offset by a profit on the other.

As an example, a cable manufacturer may contract to supply cables using 100 tons of copper wirebars. As he starts the order and draws copper from his stock, he will buy forward on the LME 100 tons, this price being used for his quote for cable supply. When the cable contract is completed, he replenishes his physical stocks by buying 100 tons at the then current cash price on the LME. Finally, he sells his forward-bought copper also at the LME cash price for that day, and so closes his hedge. Note that the cable manufacturer has not only protected himself from an adverse movement in the copper price, but he was also able to establish a firm price with his own customer, as to its copper content for the order, the moment it was accepted.

Such forward dealing also attracts speculators. It is always said that the presence of such professional risk-takers serves to make the market more flexible. There is, of course, considerable risk, since delivery is explicit in all contracts, and a dealer must be prepared to deliver against a forward sale, either by delivering warrants of purchase on the market on the forward date at the market price or physically delivering from his warehouse.

Where the supplies are plentiful the forward price tends to be at a premium over cash, the difference being known as a contango. Should supplies be scarce, or should there be a heavy demand for nearby metal, the cash price may rise above that for 3 months forward and the market is said to have gone into backwardation. The extent of a contango is, in practice, limited to the cost of financing and carrying metal for the 3-month period. An interesting aspect is that a consumer can take advantage of a contango as an opportunity to build up stocks, at the same time selling forward. The difference in selling forward, being the extent of the ruling contango, will cover his costs of finance and storage.

Effects on cost of composition and metallurgical complexity – effect of purity

A metallic alloy is made up of a basis metal of a certain purity to which is added the required range of alloying elements, either as pure metals or as 'hardeners' (concentrated mixtures of the element and the basis metal produced independently which enables more ready solution and distribution in the melt under normal foundry conditions). The degree of purity required in the basis metal will vary with the type of alloy being produced. In the aluminium alloy field material intended to be used for general purpose, moderately stressed castings is able to tolerate higher quantities of impurities than, say, a high strength casting alloy for use in aircraft. The higher the purity of the basis metal the more expensive the alloy will be as shown by the approximate costs given in Table 3.1.

TABLE 3.1 Typical alloy prices (£/tonne 1996)

Aluminium ingots	
99.5% purity	1060
99.8% purity	1110
99.99% purity	1910
Magnesium ingot	
99.8% purity	2500
LM4 (SAE 326) aluminium diecasting alloy (3 Cu–0.15 Mg–5 Si–0.8 Fe–0.4 Mn)	1150
LM10 (SAE 324) aluminium-magnesium casting alloy (0.1 Cu–10 Mg–0.25 Si–0.3 Fe)	1950

The commonest impurities in aluminium alloys are iron and silicon. In the LM10 aluminium-magnesium casting alloy the silicon impurity reacts with the magnesium in the alloy to form the intermetallic constituent Mg_2Si, which has a serious embrittling effect if present in excessive amounts, and 0.25% Si would be considered a normal silicon content. The basis aluminium used for manufacturing the alloy must therefore be of at least 99.7% purity as compared with the 99.5% or even 99.2% purity which is acceptable for many other alloys.

The wider specification of LM4 with regard to certain elements will not only permit the use of lower-grade aluminium for virgin ingot, at reduced cost, but will also more easily enable the composition to be achieved by the melting of scrap to produce secondary ingot, again with reduced costs.

The composition of LM2 (SAE 303) (1.5 Cu–0.3 Mg–10 Si–1 Fe–0.5 Mn–0.5 Ni–2 Zn–0.5 Pb–0.2 Sn–0.2 Ti) is even wider, giving a great deal of tolerance towards a range of impurities, and is thus widely cast from secondary ingot material supplied at a still lower price (£1090/tonne, 1996).

Costs of alloying

If an alloying element costs more than the basis metal to which it is being added then it is self-evident that the alloy must cost more than the metal, and vice-versa. Thus a cryogenic steel containing 9% Ni costs more than mild steel, and brass costs less than copper (Table 3.2).

TABLE 3.2. Comparison of typical basis metal and alloy costs (£/tonne, 1996)

Mild steel	240
9% Nickel steel	1000
Nickel	5535
Copper	1780
Brass bar (65/35)	1480
Zinc	750

Although the figures in Table 3.2 are in the anticipated direction there is not a strict quantitative relationship. Many other factors, such as the scale of the alloy usage and the practical difficulties in alloying to a tight specification in complex systems, can have a marked effect on costs. If, for example, an alloy contains small quantities of a readily oxidizable element, expensive melting procedures to avoid losses on melting may be required.

Consider the relative costs of the aluminium alloys 5083 and 7075 (Table 3.3).

TABLE 3.3. Typical aluminium alloy costs (£/tonne, 1996)

5083 Al–4.5Mg	2900
2024 Al–4.5Cu–1.5Mg	3700
7075 Al–5Zn–3Cu–1.5Mg	3800

It is not possible here to account for the variations in cost in terms of the individual alloying elements. Clearly, other factors are operating; in this case one of the most important being metallurgical complexity. 5083 is a binary solid-solution-hardened alloy, whereas 7075 is a complex high-strength precipitation-hardened alloy often used for critical application. The complexities of behaviour of the last-mentioned alloy are such that the rejection rate during manufacture could on occasions exceed 60% unless high levels of metallurgical control are exercised. This is almost equivalent to saying that a given order has to be made three times before deliverable quality is attained, and it is therefore not surprising that the alloy is expensive.

Filling and blending of plastics

Fillers have been used in plastics ever since wood flour, asbestos, mica or cotton fabric were added to Bakelite's phenolic resin to enhance toughness. These fillers frequently had another advantage, that of lowering the price of the material. Of course, not all fillers will result in cheaper plastic components. Glass- and carbon-fibre filled plastics are used in demanding applications where their improved strength and stiffness is required. Not only will the fibre-filled materials be up to ten times more expensive than the virgin polymer, but processing will also be more demanding.

A number of plastic blends have also been successful commercially. The resulting materials display properties not attainable with the unblended starting polymers, and occasionally the blend properties can be better than those of the base resins. Blends have frequently opened up new markets, filling gaps in the properties of engineering thermoplastics. Examples of polymer blends include: ABS and polycarbonate (connectors and housings), nylon and ABS (sports goods, gears, housings) and nylon and polyphenylene oxide (automotive mouldings). Typical properties are listed in Table 3.4.

TABLE 3.4 Typical properties of plastics and their blends

Plastic	Density (Mg/m³)	Tensile strength		Elongation to failure %	Cost £/kg	Trade names
		MPa	ksi			
ABS	1.05	50	7	8	1.6	Cycolac, Novodur
Polycarbonate (PC)	1.20	65	9	110	3.0	Lexan, Makrolon
Polyamide 6/6 (PA 6/6)	1.14	60	9	60	2.8	Ultramid, Durethan
Polyphenylene oxide (PPO)	1.05	65	9	60	2.6	Noryl
ABS/PC	1.10	50	7	80	2.5	Bayblend, Proloy
PA/ABS	1.06	45	7	270	2.5	Triax
PPO/PA	1.10	55	8	100	3.2	Noryl GTX

Effect of quantity

The cost of basic material is also a function of size of order. The larger the size of an order for material the smaller will be the unit cost. Even in the case of common, well-established materials the surcharge to be paid on small quantities can be alarming. It is to be emphasized that the additional charges are not necessarily levied to offset the cost of special manufacture, since it is usually the case that completion of a small order still has to await the passage through the factory of normal quantity. The higher charges result from the fact that the irreducible administrative procedures and delivery charges represent a higher proportion of the total cost of the order. Clearly the highest costs will be paid when buying from a small local stockist.

Value-added costs

The usual industrial procedure is for a manufacturer to buy in material in a form which is suitable for his purpose, process it and then sell it in its new form. The manufacturer is not selling material but rather the value that he has added to the material in its passage through his factory. Whatever the precise nature of the processes that are operated in the factory, the quantities that are added to the material, and which will determine its final price on exit, must include the variable costs of skilled and unskilled labour, energy, technical development and supervision, royalty payments, etc. as well as fixed costs of the factory and an acceptable profit. The more a material is altered the greater the value-added component of its final cost should be.

The extent of fabrication costs is frequently not appreciated. For example, in a low-cost material such as mild steel, the working cost to produce annealed thin sheet, or complex girder section, may approach that of crude steel supplied from the steelworks to the rolling mill. The more complex the section in rolling, with higher roll maintenance and general operating costs, the higher the price of mechanical reduction. Hot stampings and drop forgings, generally involving higher labour costs and die replacements, are more expensive per unit weight than rolled products, particularly for non-repetitive parts. As with all fabrication techniques which involve the expense of shaped dies, the longer the run up to the full life of the die, the lower will be the component of die cost in the product (see Chapter 13).

An inspection of typical product prices will sometimes indicate a higher cost of castings as compared with wrought products per unit weight, dependent on the complexity of shape and quantity. In the case of steel this is partly a function of the normal foundry costs of mould preparation, sand reclamation, etc. and partly the higher intrinsic costs of steelmaking on a smaller scale in foundries, where the operating costs are much higher than in large furnaces for primary steelmaking feeding material to rolling mills.

It may be that the properties or the shape required in the product favour a particular fabricating route, but more often the required level of performance may be achievable by more than one method of fabrication, with direct competition in cost. At one time it was taken as axiomatic that wrought products were always more reliable and gave greater toughness than castings, but there has been such an improvement in the quality of high-grade castings that this is not now necessarily the case. In some instances the use of a particular fabrication route is built into the product specification. As an example, the British Standard for domestic gas appliances requires that gas handling components in, for example, water heaters, are produced as brass hot-stampings, although aluminium alloy castings would be satisfactory other than for the possibility of lack of pressure tightness if there were undetected macro- or microshrinkage.

The dimensional tolerance required is also an important factor in the choice of both material and fabrication route, since it controls part of the cost accruing during manufacture. The level of tolerance required must be matched up to those that may be readily obtained with the fabrication techniques best suited to the material, otherwise costs will escalate.

Where a material is to be heat-treated, any tendency to distortion can lead to a high reject rate or, in the case of alloy steels, to the need for a final expensive grinding operation of the already heat-treated component. In machining, the higher the degree of accuracy required, the more expensive the operation, since finishing cuts become protracted. Frequently, the material chosen will dictate the quality of finish obtainable and the speed with which it can be achieved.

Such precise control and high accuracy implies rigid inspection and quality control. This is an expensive procedure requiring extra staff, space, equipment and held-production (i.e. material in the factory representing tied cash), and contributes very largely to the cost of manufacturing to stringent and rigid specification.

Stock control aspects

Holding stocks of materials represents tied-up money, and clearly the narrower the range of materials required within a factory the easier stock-holding becomes. It enables larger orders of, say, steel bar stock, to be negotiated at lower unit cost and simplifies storage and identification. At the same time, of course, it is seldom attractive to use an expensive low-alloy steel for applications equally well served by a carbon steel; the point is that each main area of application should be studied in order that ideally one material should be employed for any one type of usage, minimizing the range of specifications employed overall. This also helps the workforce accumulate greater expertise in fabrication with fewer switches in material.

It may be that there are specific dangers in the use of materials which have the same fabrication function and appearance, but which have greatly different properties in the service conditions of a component. As an example, in the manufacture of radio transmission valves external soldering of the fins to an anode was achieved by a silver-solder containing cadmium. The use of such a solder at hotter points internally, within the glass envelope, would have led to cadmium vapour formation, possible melting, and breakdown of valve operation. In order to be sure that such a mistake could not be made the physical form of the solders taken into the factory was very different – the silver–cadmium in slug form, and those without cadmium as wire. This distinction greatly eased identification of stock until final use.

References

1. H. J. SHARP: *Engineering Materials: Selection and Value Analysis*. Iliffe, 1966.
2. *Materials Technology Foresight for the UK Power Generation Industry*, a Report of a Working Party of the Institute of Materials, April 1995.
3. CEC, Cost 501
4. UK OFFICE OF SCIENCE AND TECHNOLOGY *Technology Foresight: Progress Through Partnership: 10 Materials*. HMSO, 1995.
5. J. R. T. GIBSON-JARVIE: *IMM Bulletin*, Section A 1971; **80**, 160.

4
Establishment of service requirements and failure analysis

4.1 Selection and design in relation to anticipated service

The majority of decisions on materials selection should be taken by the design team. It would be satisfactory to be able to say that such decisions are always based on a quantitative analysis of the form and extent of all the various demands anticipated in service for a particular design, and it might seem that, provided the designer has a clear idea of the properties required in his materials and the modes of failure to be avoided in service, this should be a simple process. Unfortunately, simple situations arise only rarely in engineering practice. It has already been made clear that any application requires its own special combination of properties, and usually the demands are conflicting. We may, therefore, be seeking a combination of properties which it is impossible to achieve fully in any one material and a compromise has to be reached. Pick[1] has said that 'Material (and process) selection always involves the act of compromise – the selection of a combination of properties to meet the conflicting technical, commercial and economic considerations.' It is, for example, difficult to choose a material which would combine high yield strength and high fracture toughness, or to combine the highest fatigue strength with high-temperature creep resistance. Frequently, all that can be done is to take account of the relative importance of various service requirements and pitch the compromise accordingly. Thus it is that in formalized quantitative selection procedures, weighting factors are applied to individual properties in reaching the best compromise. The apportionment of such weightings may be difficult, as discussed in Chapter 14.

It is not surprising, therefore, that frequently the engineer has tended to play safe. Often he has stuck with a material which he (and others) have used in the past in contexts similar to the new design, which, in itself, is often only a development of an existing form. Progress, albeit slow, has occurred as he took note of feedback in relation to service performance, particularly as regards any form of total failure that occurred, and incorporated design modifications to take account of this. In parallel with this, bringing new materials into use has often depended upon the development of existing, well-tried formulations with similar but improved combinations of properties so that they can be introduced with confidence. In recent years the situation has changed markedly. Design engineers have had to move into areas where there was no past experience to draw upon and with a very incomplete knowledge of the service requirements. Frequently, such designs have to be developed with inadequate data, both as regards the details of service conditions and sometimes in relation to possible materials of construction. Nuclear power engineering and space technology are two outstanding examples. It is true that experience is being accumulated, but the record in these areas is impressive. Enormous effort has had to go into the analysis of structures in relation to these new service conditions, and new and improved materials have had to be developed in some cases to meet these conditions.

In all fields, moreover, and particularly with cost-oriented products, fierce competition has brought marked change to traditional materials usage and fabrication methods, to reduce costs. As an example, the introduction of injection-moulded engineering plastics to replace often complex composite components of steel, aluminium, brass, resin-impregnated laminate, etc. can markedly reduce the number of components in an overall design and the manufacturing cost. Such part-integration is possible because it is relatively easy to mould plastic parts with complex sections. The front end of the Peugeot 405 used to be manufactured from about 30 separate parts; it is now one plastic moulding. Similarly, the part-count for the tail plane of the Airbus A310 aeroplane was reduced from over 2,000 in the metal version down to under 200 when plastic composites were utilized.

4.2 The causes of failure in service

A selection process must be greatly influenced by the analysis of experience in similar applications, specifically an analysis of the causes and mechanisms of failures. Failures can be classified as arising from a number of main origins.

Errors in design

This obviously includes errors in terms of the material selected, or of the condition in which a given material should be supplied and the emphasis in earlier sections of this book has been that the materials choice has always to be closely integrated with the geometric and functional design. If a particular component is grossly overdesigned (by which is meant the use of an excessively high factor of safety) this is not only economically disadvantageous but may result in overloading other parts of a composite structure. Underdesign will lead to premature failure and the attendant consequences. Choice of the most appropriate factor of safety depends upon a correct assessment of service conditions both in terms of the type and severity of duty together with the influence of the environment, and this

must be aided by a proper analysis of any previous failures that may have occurred.

Inherent defects in a material properly selected

This is an important area. It is vital to know every feature of a material which in service could become a critical defect; the ability to inspect and evaluate such defects within the whole economic framework of the material use is also essential. In this category come, for example, casting defects in foundry products, and non-metallic inclusions in wrought steel.

Defects introduced during fabrication

During the manufacture of a component using the material and fabrication method selected, defects in fastening and joining (e.g. welding), poorly controlled heat treatment giving quench cracks and internal stresses, poor machining, incorrect assembly and misalignment producing unexpected stress levels, may result in subsequent failure in service. Ideally, the original design will anticipate and incorporate the effects of the fabrication route, on which the design may even depend, but the degree of modification of the intrinsic mechanical and chemical properties may sometimes go beyond that originally envisaged, wholly or locally. This is why field testing of a realistic prototype is so desirable.

Deterioration in service

The resistance to environmental conditions of chemical attack or corrosion and wear, or the stability of the microstructure on which mechanical properties depend (as in elevated temperature operation), will have been part of the initial design context, but unusual conditions are sometimes encountered which give rise to a change in performance and premature failure. Overload in relation to the mechanical stresses anticipated would be similarly classified.

A major factor in this area will be the quality of maintenance during use, for example, lubrication or the renewal of corrosion protection where this has been specified or the adherence to instructions concerning component replacement. Anderson,[2] quoting Holshouser and Mayner,[3] instances the analysis of 230 laboratory reports on failed aircraft components where, in spite of the high standard set by airline companies, 102 could be attributed to improper maintenance (mostly taking the form of undesirable changes in geometry such as nicks and gouges), 52 of these occurring as a result of a fatigue mechanism.

4.3 The mechanisms of failure

In the identification of a cause of failure, so that information can be fed back to design or manufacturing control stages, it is, of course, first necessary to recognize the failure mechanism and any relationship with the structure, compositional characteristics or design of the material component which may be revealed.

The possible mechanisms of failure are brittle and ductile fracture, fatigue (high or low cycle), creep, buckling or other forms of instability, gross yielding, corrosion, stress corrosion, corrosion fatigue, wear processes (e.g. erosion, fretting, galling).

In the recognition of these mechanisms a range of investigative techniques may be involved, particularly full metallographic examination of the region of deformation and the fracture surface, but also encompassing checks on the composition and mechanical properties of associated material and in cases of reaction with the environment, analysis of corrosion products.

For fractography, as well as the normal binocular microscope, the scanning electron microscope, particularly with energy dispersive elemental analysis facility, is an extremely powerful tool. Every effort should be made to protect fracture surfaces from deterioration or adulteration prior to examination so that not only physical features are retained, but compositional aspects can be accurately determined.

Intercrystalline failures

Intercrystalline failures of a brittle form, with a total lack of macroscopic deformation, usually indicate some form of grain boundary heterogeneity, precipitate or segregate, which is controlling the failure mechanism. Hydrogen embrittlement is one frequently observed cause of this form of failure, and into the same category come grain boundary carbides in tempered steel, and the influence of segregates such as phosphorus in steel. Stress corrosion also results in brittle intercrystalline failure, but usually with multiple cracking associated with, but not directly part of, the main failure path. The use of scanning electron microscopy (SEM) or electron probe techniques for identification of corrodent is invaluable in some instances.

Ductile intercrystalline failure may be observed as the result of the plastic linkage of microvoids. The latter will have developed around second-phase particles at grain boundaries, particularly where the interface with the matrix is weak. Whilst prior particle boundaries in powder-forged products in steel need not necessarily relate to grain boundaries they frequently do, and fracture in these materials is normally by microvoid linkage along these boundaries, the voids developing in association with oxide non-metallic inclusions, initially oxide formed on the powder surface. The lower the oxide content the tougher the steel will be[4]. In overheated steels grain boundary concentration of sulphide may result in a similar form of failure.

The effect of stresses at temperatures in excess of $0.5T_m$ may be creep, where creep rupture again occurs by microvoid linkage but with the voids generated by cavitation mechanisms and not necessarily associated with second-phase particles.

Transcrystalline failure

This may again be brittle or ductile. If brittle the fracture will be flat, but may appear granular or crystalline, frequently with a 'chevron' pattern pointing back to the site of initiation. Such forms of cleavage failure are typical of very brittle materials, but stress corrosion along crystallographic directions, and thus across crystals, is similarly indicated as cleavage. In this latter case the detection of corrodent is, of course, confirmatory.

Many fractures may be transcrystalline and flat but still exhibit ductile behaviour on a very fine scale. The fine-scale ductility is revealed in markings, sometimes in association with dimples. The most significant mechanism in this class is fatigue, where the familiar striations appear, related to the cyclic growth of the crack. In corrosion fatigue, where the rate of crack growth is accelerated, there are usually multiple cracks developed, the mechanism fully defined by the presence of corrosion product. Another class of microductile failure could be said to be the wear processes, galling (where some surface-to-surface joining has occurred), abrasion, and fretting (where corrosive processes are also introduced).

In ductile transcrystalline failure, where a large amount of plastic deformation (e.g. necking) may have occurred before fracture with slow crack growth, the fracture surface will generally present a fibrous appearance and on close examination will show ductile shear lips associated with microvoid coalescence. Where shear lips are absent on such a fracture this will often indicate the point at which the fracture started.

The shape of the dimples on such ductile fracture surfaces will indicate the stress system responsible. Equiaxed dimples between shear walls indicate a normal tensile system, elongated dimples pointing in opposite directions in the top and bottom halves indicate a shearing system with the elongation of the dimples in the direction of shear. If dimples are elongated, but of the same shape and direction in both top and bottom halves, this suggests a tearing process. A schematic representation of equiaxed, shear and tear dimples is shown in Figure 4.1, after Broek[5] and Pelloux.[6]

Considerable care has to be exercised in the examination of dimples and directional features, which can appear to change with the angle of tilt

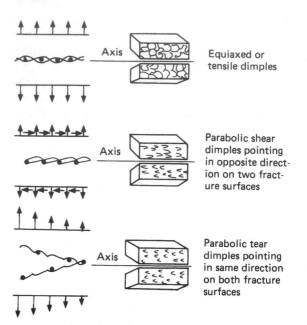

Figure 4.1 Plastic fracture: equiaxed, shear and tear dimples and associated second-phase particles (after Broek[5] and Pelloux[6]).

in relation to the optical axis. This applies in particular, of course, to direct scanning electron microscopy study and the use of replicas in the transmission electron microscope.

With the identification of the mechanism of a failure, through inspection of the failed component, and with the acquisition of all the necessary background data on design function and method of fabrication location, conditions of use, contaminants, etc. it will usually be possible to identify the cause of failure.

Having identified a cause, the next stage is the identification of an appropriate solution. It may be that the failure is due to a component being out of specification as regards the material (e.g. cleanliness) or fabrication method, in which case tighter quality control might need to be specified; not a cheap solution. The material specification may need to be changed or the fabrication route altered to match the demands of service more closely; the original data may have been inadequate. In many cases the failure may be traced to a stress concentration which can be eliminated by a modification of design. This applies frequently in the case of fatigue failure. Corrosion or corrosion-assisted failures may call for the introduction of a surface coating not previously incorporated. It is interesting to note that environment-sensitive fracture (ESF) in general, and that associated with hydrogen embrittlement in particular, constitutes over 25% of failures experienced in the oil and gas industry[7] and, it would be anticipated, in the petroleum industry overall.

In some circumstances a failure may reveal a potential hazard in relation to other components still in service, which cannot be replaced or treated so as to ensure the initially expected life. Under these circumstances there is no alternative but to restrict the service loads to a lower level, or else to shorten the life to replacement. Such a procedure is also often the most economical.

Non-destructive testing

Non-destructive testing can be important in detecting early signs of failure, and, of course, as the means of quality control for flaws arising in fabrication, which may have been identified as the cause of failure in earlier production. Cracks are detected by magnetic particle methods, penetrant tests, ultrasonic scanning and by eddy current. Radiography by X-rays is used for internal flaws such as shrinkage and blowholes as well as cracks. Gamma-ray radiography is useful for heavy section components, outdoor work and in confined spaces, since it requires neither electric power nor cooling water supply.

Failure of plastic parts

Factors that influence the durability of plastics include: creep, stress relaxation, fatigue, wear, thermal degradation, uv degradation, chemical attack and environmental stress cracking.

Internal stresses are particularly important in stress cracking, and many thermoplastics will exhibit crazing and possibly fracture in the presence of solvents. These internal stresses can be relaxed by ageing the moulded part at elevated temperatures, although this may result in some distortion. The alcohol test can be used to detect internal stresses – the part is immersed into a solvent mix containing alcohol and then examined for stress-cracking. Creep and stress relaxation will be manifest by a loss of dimensional stability as the polymer chains tend to coil up to their lowest entropy (see Chapter 6). Thermal and uv degradation are often manifest by a yellowing and embrittlement of the plastic. Mechanical properties such as impact strength and fatigue strength are badly affected by such degradation.

The behaviour of plastic parts is very dependent on the manufacturing conditions (e.g. moulding temperature, moisture content of feedstock, alignment of polymer chains, frozen-in stresses, conditioning regime) as well as the basic polymer chemistry and structure.

4.4 Corrosion

Data for specific corrosion resistance may emanate from three distinct sources.

(1) Existing data from reference sources: data tables, corrosion charts and stability maps. These may be either put together by disinterested authors as reviews, or they may form part of the technical sales publications of a company promoting a particular product. The data put forward in the latter case are still reliable although clearly since corrosion is finally very much related to individual local circumstances, only general guidance is achieved.
(2) From existing plants where direct comparison can be made as before; the full background of local conditions must be very carefully considered to be sure that the comparison is valid. For example: is the water of the same type?; have the vessels, pipes, etc. the same flow characteristics?; and is the material of construction in the same condition?; etc.

(3) Corrosion test data can be obtained from a wide range of tests, either in simulated service in the laboratory, or in service. Laboratory tests, of which there is a wide range (full or partial immersion, accelerated tests using rotating disc specimens or impinging jets) may be followed simply by weight loss and by change in appearance, possibly with the measurement of pits, or they may be electrochemical in nature. In the latter case the corrosion potential and the polarization occurring can be determined. To an increasing extent electrochemical measurements are employed to give greater understanding of the reactions occurring. The scanning reference electrode technique (SRET) enables corrosion activity to be monitored in real time, and variations in activity can be spatially resolved, to evaluate, for example, the tendency to pit generation, from which a crack may develop.[8,9] Laboratory corrosion tests are valuable in indicating the range of materials and heat-treated condition which may be usable, and in checking whether a material or a protective system does, in fact, behave as the published data suggest, or that it is going to meet some specification laid down for the component.

In simulated service (or field tests) a number of replicas of different types of specimens are subjected to the service conditions. These may take the form of coupons of the material or materials inserted into the environment, or of sample components fitted in the service required. For example, to test the efficiency of a new marine paint, racks of individual coupon samples may be exposed to a marine environment or, alternatively, sample panels on buildings or boats in different materials may be painted, with the performance in both cases being regularly monitored. Service or field tests are useful in taking the materials selection or the corrosion prevention system on to a final stage where there is no comparable previous experience, and in the estimation of probable life in the actual service conditions.

Bearing in mind the sensitivity of corrosion processes to changes of acidity, aeration, temperature, etc. efforts are now being made to develop predictive computer programs for specific materials, so that the behaviour of a material with a given combination of conditions can be assessed. Such a computer program may be designed to display the behaviour in terms of the anode and cathode polarization curves (see Chapter 11) if this is preferred, from which the corrosion current, if any, can be predicted. Such programs are built up on a theoretical framework with data supplied from corrosion tests and from plant data. There has been considerable activity in modelling and data analysis techniques.[10]

References

1. H. J. PICK: *Met. Mater.*, 1968; **2**(9), 271.
2. W. E. ANDERSON: *Int. Met. Rev.*, Dec. 1972; **17**, 240.
3. W. L. HOLSHOUSER and R. D. MAYNER: *Fatigue Failure of Metal Components as a Factor in Civil Aircraft Accidents*. US National Transportation Safety Board, Washington, 1971, p. 9.
4. F. L. BASTIAN and J. A. CHARLES: *Proc. International Conf. on Fracture*, Cannes, 1981; **1**, 209–216.
5. D. BROEK: *A Study on Ductile Fracture. NLT-TR 71021 U*. National Aerospace Laboratory, Nederlands, 1971, p.81.
6. M. N. PELLOUX: *ASM Met. Eng. Q.*, 1965; **5**(4), 29
7. M. B. KERMANI in *Hydrogen Transport and Cracking in Metals*, ed. A. Turnbull. Inst. Materials Pub. No. 605, 1995, 1.
8. R. AKID: *Materials World*, 1995; 3, 11, 522–5.
9. H. N. MCMURRAY, S. R. MAGILL and B. D. JEFFS: *Iron and Steelmaking*, 1996, 23, 2, 183.
10. J. E. STRUTT and J. R. NICHOLIS: *Plant Corrosion – Prediction of Materials Performance*. Ellis Horwood, Chichester, 1987

5
Specifications and quality control

5.1 The role of standard specifications

There is considerable complexity of communication in transmitting the details of materials and component selection and usage to fabricators and suppliers. It is just not possible to carry out manufacture without some form of widely understood shorthand, and this shorthand is the standard.

Specifications of materials form just one of four broad categories of standards, as described by Weston.[1]

Glossaries

One of the first requirements in developing a programme of standards in any area of technology is to establish its terminology and the units for the quantities associated with it. This is such a basic requirement that instinctively everyone tries to establish a terminology for himself, and this is particularly true in developing fields. There always exists the danger of one term for more than one concept or, more subtly dangerous, varying shades of meaning for one particular term. A standard terminology, related to precise and generally accepted meanings, is a safeguard against confusion and the lack of precision so essential to proper development and exchange of technical knowledge.

Methods of test

Clearly standardized methods for determining composition, mechanical and physical properties are fundamental for specification purposes and also to ensure coordinated technological development of new materials.

Specifications

These may relate to materials, components or end products, some of which may be complex units manufactured from standard materials and components. They fall into two main groups: *dimensional*, which define form and shape, and *quality*. The main function of the former is to achieve interchangeability – bolts, nuts, ball-bearings, etc. Such standards also serve to simplify or reduce the range of sizes needed to cover general usage – the standards for steel sections are an outstanding example of this – amongst other reasons it was the enormous multiplicity of section shapes and sizes that resulted in the first British Standards to be issued in the early years of this century.

Quality specifications can follow either of two patterns: one specifying the detail of the process of manufacture, e.g. the composition of a steel to precise limits, how it should be made and fabricated, and the other specifying the level of performance required and leaving it open to each manufacturer to decide how to achieve it. Both have their limitations and advantages and historically were illustrated by the different approach of American and British specifications for steel. In the USA precise chemical compositions were given, and the mechanical properties to be expected were only provided as a guide to what could be expected and not as a specification requirement. The British specifications gave broad limits for composition as a guide, but laid down the properties which must be achieved. Increased flexibility and rationalization has developed however, as exemplified by the British Standard for Wrought steels, BS 970, where supply may either be to close limits of chemical composition (A grades), where no mechanical properties or hardenability are specified, or to a combination of mechanical properties (M grades) or hardenability requirements (H grades) and chemical composition. Considering BS 970[2] in detail:

000–199   C and C-Mn steels.
200–240   Free cutting steels
500–999   Low alloy steels
Letters: A Specify by analysis
          M Specify by mechanical properties
          H Specify by hardenability
Numerals = Mn% × 100
2nd and 3rd numerals = S% × 100
Last two numerals = C% × 100

The tensile strength level expected in hardened and tempered material is designated by a further letter.

	MPa	[= ksi]
D	550–700	80–100
Q	625–775	91–112
R	700–850	101–123
S	775–925	112–134
T	850–1000	123–145
U	925–1075	134–156
V	1000–1150	145–167
W	1075–1225	156–178
X	1150–1300	167–188
Y	1225–1375	178–199
Z	1550 min	225 min

Thus a specification BS 970: 080 M40 R indicates that the steel containing 0.36–0.44% carbon, 0.60–1.00% Mn is to be supplied on the basis of mechanical properties to a strength level of 700–850 MPa [101–123 ksi] in the hardened and tempered condition.

The most widely used system internationally for designating carbon and alloy steels is that of the American Iron and Steel Institute (AISI). The numerical designations summarized in Table 5.1 are used by both the AISI and the Society of Automotive Engineers (SAE). The basic analysis requirements for the 10XX series plain carbon

steels with 1.0% Mn max. are given in Table 5.2. From this it can be seen that the AISI equivalent of BS970:080 M40 is 1040. The two tables were derived from those in the ASM (American Society of Metals) Handbook[3], which is to be recommended as a very valuable source of information. Most helpfully it also contains tables of equivalent specifications in different countries where such equivalence exists. The Materials Selector[4] is also a very valuable source.

The situation with plastics is very different, with the formulations proprietary to the manufacturers. Property values in data sheets must be used with caution, paying particular attention to the test standards used. There are, however, developing international standards for single and multi-point plastics data, ISO 10350 and ISO 11403 respectively.

Codes of practice

The fourth class of standard is a *code of practice*. In British terminology this embodies the accepted principles of good practice and sets out the methods of erection or installation, or the suitable conditions of use to give performances as intended. They take care of a great deal of the complexities which arise in the application of a particular material to a specific job.

Reverting to material specifications, the primary function is to specify with sufficient accuracy, but not with restrictive 'overkill', the essential requirements of a material or component so that anything produced to that specification is completely satisfactory for the intended application. It should be unambiguous, concise, readable and give due emphasis to the more important requirements with the minimum of cross-reference. At the same time there should be realistic flexibility in alternative methods of manufacture and tolerances, balancing the economics of production and the requirements of the user.

In so far as the use of standards makes the rapid translation of instructions possible, and assists in restricting the number of possible

TABLE 5.1. AISI-SAE system of designations

Numerals and digits	Type of steel and nominal alloy content	Numerals and digits	Type of steel and nominal alloy content
Carbon Steels		**Nickel-Molybdenum Steels**	
10XX(a)	Plain carbon (Mn 1.00% max)	46XX	Ni 0.85 and 1.82; Mo 0.20 and 0.25
11XX	Resulphurized	48XX	Ni 3.50; Mo 0.25
12XX	Resulphurized and rephosphorized	**Chromium Steels**	
15XX	Plain carbon (max Mn range–1.00 to 1.65%)	50XX	Cr 0.27, 0.40, 0.50 and 0.65
Manganese Steels		51XX	Cr 0.80, 0.87, 0.92, 0.95, 1.00 and 1.05
13XX	Mn 1.75	**Chromium Steels**	
Nickel Steels		50XXX	Cr 0.50
23XX	Ni 3.50	51XXX	Cr 1.02 C1.00 min
25XX	Ni 5.00	52XXX	Cr 1.45
Nickel-Chromium Steels		**Chromium-Vanadium Steels**	
31XX	Ni 1.25; Cr 0.65 and 0.80	61XX	Cr 0.60, 0.80 and 0.95; V 0.10 and 0.15 min
32XX	Ni 1.75; Cr 1.07	**Tungsten-Chromium Steels**	
33XX	Ni 3.50; Cr 1.50 and 1.57	72XX	W 1.75; Cr 0.75
34XX	Ni 3.00; Cr 0.77	**Silicon-Manganese Steels**	
Molybdenum Steels		92XX	Si 1.40 and 2.00; Mn 0.65, 0.82 and 0.85; Cr 0.00 and 0.65
40XX	Mo 0.20 and 0.25	**High-Strength Low-Alloy Steels**	
44XX	Mo 0.40 and 0.52	9XX	Various SAE grades
Chromium-Molybdenum Steels		**Boron Steels**	
41XX	Cr 0.50, 0.80 and 0.95; Mo 0.12, 0.20, 0.25 and 0.30	XXBXX.B denotes boron steel	
Nickel-Chromium-Molybdenum Steels		**Leaded Steels**	
43XX	Ni 1.82; Cr 0.50 and 0.80; Mo 0.25	XXLXX.L denotes leaded steel	
43BVXX	Ni 1.82; Cr 0.50; Mo 0.12 and 0.25; V 0.03 min		
47XX	Ni 1.05; Cr 0.45; Mo 0.20 and 0.35		
81XX	Ni 0.30; Cr 0.40; Mo 0.12		
86XX	Ni 0.55; Cr 0.50; Mo 0.20	(a) XX in the last two digits of these designations indicates that the carbon content (in hundredths of a per cent) is to be inserted.	
87XX	Ni 0.55; Cr 0.50; Mo 0.25		
88XX	Ni 0.55; Cr 0.50; Mo 0.35		
93XX	Ni 3.25; Cr 1.20; Mo 0.12		
94XX	Ni 0.45; Cr 0.40; Mo 0.12		
97XX	Ni 0.55; Cr 0.20; Mo 0.20		
98XX	Ni 1.00; Cr 0.80; Mo 0.25		

variants, it is essential to the improvement of productivity and good delivery times.

Standards serve to disseminate proved data, distilled from experience or agreed after the careful study of research results, and they pro-vide a common basis of design and of produc-tion. This standard conservative approach is sound because the average designer would not have the time, nor possibly the background training, to appreciate and critically examine all

TABLE 5.2. Composition ranges and limits of AISI-SAE standard carbon steels containing less than 1.00% manganese–semifinished products for forging, hot rolled and cold finished bars, wire rod and seamless tubing

AISI-SAE designation	Heat composition ranges and limits, %		AISI-SAE designation	Heat composition ranges and limits, %	
	C	Mn		C	Mn
1005	0.06 max	0.35 max	1049	0.46–0.53	0.60–0.90
1006	0.08 max	0.25–0.40	1050	0.48–0.55	0.60–0.90
1008	0.10 max	0.30–0.50	1053	0.48–0.55	0.70–1.00
1010	0.08–0.13	0.30–0.60	1055	0.50–0.60	0.60–0.90
1011	0.08–0.13	0.60–0.90	1059	0.55–0.65	0.50–0.80
1012	0.10–0.15	0.30–0.60	1060	0.55–0.65	0.60–0.90
1013	0.11–0.16	0.50–0.80	1064	0.60–0.70	0.50–0.80
1015	0.13–0.18	0.30–0.60	1065	0.60–0.70	0.60–0.90
1016	0.13–0.18	0.60–0.90	1069	0.65–0.75	0.40–0.70
1017	0.15–0.20	0.30–0.60	1070	0.65–0.75	0.60–0.90
1018	0.15–0.20	0.60–0.90	1074	0.70–0.80	0.50–0.80
1019	0.15–0.20	0.70–1.00	1075	0.70–0.80	0.40–0.70
1020	0.18–0.23	0.30–0.60	1078	0.72–0.85	0.30–0.60
1021	0.18–0.23	0.60–0.90	1080	0.75–0.88	0.60–0.90
1022	0.18–0.23	0.70–1.00	1084	0.80–0.93	0.60–0.90
1023	0.20–0.25	0.30–0.60	1085	0.80–0.93	0.70–1.00
1025	0.22–0.28	0.30–0.60	1086	0.80–0.93	0.30–0.50
1026	0.22–0.28	0.60–0.90	1090	0.85–0.98	0.60–0.90
1029	0.25–0.31	0.60–0.90	1095	0.90–1.03	0.30–0.50
1030	0.28–0.34	0.60–0.90			
1035	0.32–0.38	0.60–0.90			
1037	0.32–0.38	0.70–1.00			
1038	0.35–0.42	0.60–0.90			
1039	0.37–0.44	0.70–1.00			
1040	0.37–0.44	0.60–0.90			
1042	0.40–0.47	0.60–0.90			
1043	0.40–0.47	0.70–1.00			
1044	0.43–0.50	0.30–0.60			
1045	0.43–0.50	0.60–0.90			
1046	0.43–0.50	0.70–1.00			

Typical limits on phosphorus and sulphur contents are 0.040% maximum phosphorus and 0.50% maximum sulphur. Steels listed in this table can be produced with additions of lead or boron. Leaded steels typically contain 0.15 to 0.35% lead and are identified by inserting the letter 'L' in the designation – 11L17; boron steels can be expected to contain 0.0005 to 0.003% boron and are identified by inserting the letter 'B' in the designation – 15B41.

the relevant facts from more specialized coverage of, say, alloy properties and usage.

At the same time it must be said that all is not perfect in the world of standards. Each standard is produced by a committee of involved technologists hammering out the fruits of their combined experience in a narrow field. Any one committee must contain representatives from users and producers with inevitably conflicting interests, and it is not surprising if the result is sometimes a rather unsatisfactory compromise. Often a national standard in a new field is intended as a replacement for a proprietary specification which has become well known. If the latter refuses to die and, although associated, is still different from the former, then confusion is caused. It is generally agreed that there are too many specifications – they are certainly not always directly internationally exchangeable. A good deal of work has been done by the International Standards Organization but ISO standards have not yet won general acceptance. A country which exports considerably will generally have to manufacture in conformity with the standards of the client countries. An extra burden is placed on design and production staff by the need for familiarity with a multiplicity of different standards. The proliferation of standard materials in production in the more advanced industrial societies is sometimes blamed as a major cause of uncompetitiveness.

A converse criticism is that standards, by their nature, tend to reduce all to a common denominator, thereby restricting enterprise and the development of new materials. In addition, the production of specifications is erratic and slow as a result of the nature of the committee method. These criticisms have a certain validity, but are probably unavoidable in modern industry.

A complicating factor is also that, in small engineering companies in particular, the specification of materials tends to be historic, even to the use of long-since superseded proprietary alloys and out-of-date national standards. The old B.S En. specifications, long-since replaced by BS 970 for steels, still feature widely both in conversation and correspondence.

5.2 Inspection and quality control

Every manufacturer has a responsibility to ensure that his product is of merchantable quality. The precise meaning to be attached to this term must be agreed between manufacturer and client at the time the order is placed, and ideally should be written into a specification, which is a description of that which is to be provided in terms of properties, freedom from defects, tolerances, etc. To ensure that a material, product or treatment conforms to the specification, testing is necessary, and this is known as *inspection*. The producer must inspect his product for two reasons. First, he must test his completed product to ensure that it will not be rejected by the customer at the point of

sale for being below the required standard, but, second, he must also do so to safeguard his interests in the future. A customer may accept an unsatisfactory product through ignorance or error, but if in service it fails, causing material damage or loss of life, the manufacturer continues to bear responsibility even though the client had accepted his product. But he need provide no more than he is asked since, as a general rule, he cannot be expected to know what the user is intending to do with his product. The law of product liability is complex and varies from one country to another. A useful summary has been given by Vane.[5]

In the UK, product liability is tied in with the Consumer Protection Act (1987). This places the responsibility onto the designer to ensure that the user has not been unreasonably endangered, for example, by premature failure or by the release of toxins. Failure to provide warning of a potential hazard is also a serious omission. At the detailed design stage, it is incumbent upon the designer to use methods such as failure modes effect analysis and finite element analysis, as well as to manufacture and test prototypes and final designs.

It is commercially important not to manufacture to standards that are excessively high since this not only causes a direct increase in production costs but also makes the inspection more costly. Inspection is inherently an expensive process because, in addition to the costs of responsible and technically qualified inspectors and their equipment, there are costs in the enabling aspects – space requirements, cleaning, moving, etc. Most important of all, elaborate inspection procedures delay the passage of components through the factory, thereby increasing the factory holding, tying up capital and extending delivery times.

However, the manufacturer generally knows the foibles of his products better than the customer and in the case of quality products he may, for the good of the industry as a whole, offer advice at the negotiation stage if he feels that the customer is pitching his requirements at an unrealistically low level. A manufacturer has to maintain his reputation, often against strong competition, and in the case of exports the costs of overseas attendance in negotiations relating to disputes caused by customer rejections or service failures may well warrant a high level of inspection prior to shipment. The need to obtain agreement about every conceivable aspect of the product before manufacture commences means that contracts and specifications are frequently very complicated documents. Herein lies the value of the standard specification in which much of the work has already been done.

As well as testing the end product, the manufacturer generally finds it wise to carry out inspections at various intermediate stages in the manufacturing process. Costs are saved thereby because unsatisfactory products can be rejected at an early stage before too much work has been done upon them. This is known as *quality control* because the measurements taken provide information not only about the product itself but also about the satisfactory operation (or otherwise) of the manufacturing processes. It is in the field of high-performance products such as aircraft components or nuclear plant that quality control assumes the greatest importance and correspondingly absorbs the greatest costs. The ultrasonic inspection of every one of a batch of already expensive Incoloy tubes for a chemical plant can increase the cost by 30%; in more critical examples, inspection costs can equal manufacturing costs.

It is open to the customer, also, to inspect the article of sale, if he wishes to do so. For run-of-the-mill standard products he may do this after delivery has been effected, but in more special cases he may elect to send his own inspector to the manufacturer's works to examine inspection records, witness some proportion of the tests and generally assure himself that quality control procedures are satisfactory. In such cases it is essential to ensure that the intended division of responsibilities and costs is properly established so that everybody knows what has to be done and who has to do it. This should be written into the contract.

References

1. G. WESTON: *Function of Steel Specifications*. Iron and Steel Institute Special Report 88 (1964) pp. 1–14.
2. BS970: *Wrought Steels for Mechanical and Allied Engineering Purposes*, Part 1. 1983, British Standards Institution.
3. ASM Handbook, Tenth Edition, American Society of Metals, 1990.
4. N. A. WATERMAN and M. F. ASHBY: eds. *Materials Selector Vol 1*. Chapman and Hall, 1996
5. H. C. VANE: *Metallurg. Mater. Technol.* July 1983, 331.

Selection for mechanical properties

6
Static strength

The term 'strength' is often used rather loosely. There are three distinct usages:

(1) *static strength* – the ability to resist a short-term steady load at normal room temperature;.
(2) *fatigue strength* – the ability to resist a fluctuating or otherwise time-variable load;
(3) *creep strength* – the ability to resist a load at temperatures high enough for the load to produce a progressive change in dimensions over an extended period of time.

In this chapter it is the first of these properties that will be discussed.

A proper understanding of the strength of a material generally requires the determination of its stress-strain curve either in tension, compression or shear: from this several parameters of strength can be taken, according to the relevant mode of failure.

Because of the relative ease with which the tensile test can be carried out, most strength data for metals are obtained in tension; relative to these, compression data are sparse. However, concrete and ceramics are commonly tested in compression. Plastics, on the other hand are frequently tested in flexure.

For most ductile wrought metallic materials, mechanical properties in compression are sufficiently close to tensile properties as to make no difference for the purposes of materials selection (although there are a few exceptions). In other classes of materials they may be different (Table 6.1). Frequently, an important factor is the presence or otherwise of internal flaws from which cracks can propagate readily in tension but less easily in compression. Metals in the cast condition may be stronger in compression than in tension. In cast irons and concrete there is the additional factor of brittleness – they are much

TABLE 6.1. Tensile and compressive strengths of materials

	Tensile strength		Compressive strength	
	MPa	ksi	MPa	ksi
Low-strength grey cast iron	155	22	620	88
High-strength grey cast iron	400	58	1200	174
Portland cement	4	0.6	40	6
Concrete	3	0.5	40	6
Wood	100	15	27	4
Polyether ether ketone (PEEK) + 30% short carbon fibres	233	34	240	35
Epoxy + 50% unidirectional E-glass fibre prepreg laminate – parallel to fibres	1100	160	900	130
(perpendicular to fibres)	(40)	(6)	(150)	(22)
Epoxy + 60% unidirectional high strength carbon fibre prepreg laminate – parallel to fibres	2000	290	1300	190
(perpendicular to fibres)	(80)	(12)	(250)	(36)

(% by volume)

45

stronger in compression than in tension. Wood, however, which fails in compression by separation and buckling of its fibres (a quite different failure mechanism from that in metals) is much weaker in compression than it is in tension. A similar mechanism of failure in compression can be observed in wrought materials containing elongated non-metallic inclusions, and in some directional fibre composites. Aramid fibres in particular have poor compression properties.

Variability in strength – the Weibull modulus

Materials vary greatly in the consistency of strength which is attainable. For ceramics and cement, for example, the scatter in value of tensile strength in considerable, whereas for steel a fairly closely defined value can be given and the scatter is small. For ceramics, therefore, there is no single 'tensile strength', the value obtained depending on the size (length), distribution and orientation of internal flaws in relation to the loading direction. In such a system the strength is also dependent on the volume of the test piece, as reflecting the statistical chance of encountering a large flaw. The statistics of strength are then reflected in the Weibull modulus, which indicates the proportion of similar samples of the same volume (V_o) which will survive loading to a given level of tensile stress (σ_o):

$$P_s(V_o) = \exp\left\{-\left(\frac{\sigma}{\sigma_o}\right)^m\right\}$$

where m, the Weibull modulus, and σ_o are constants. This equation gives a curve of the form shown in Figure 6.1. When $\sigma = \sigma_o$ $P_s(V_o) = 1/e = .37$. Thus the constant σ_o is simply the stress level at which 37% of test specimens survive and m indicates the scatter in strength performance. Typically, bonded or sintered particulate assemblies such as brick, cement, pottery and ceramics have low values of m, in the range 5–10, whereas steel has a high value of ~100 and the 37% survival value is very close to the normally defined value of tensile strength.

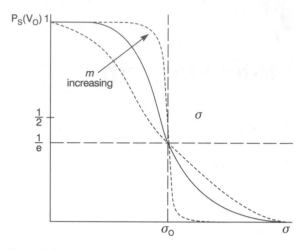

Figure 6.1

It is important to recognize, however, that the Weibull modulus applied to a component will depend on the manufacturing method as well as the material itself. If a steel is welded, then the quality of the weld metal may be more variable than the bulk, and the value of m for that situation will dominate the design. Similarly, if critical aspects of the design can vary somewhat from component to component, through difficulties in control of the manufacturing process, then the value of m for the component will reflect this.

6.1 The strength of metals

Most metallic materials used for engineering purposes exhibit a significant degree of ductility: thus, in determining the stress-strain curve it is possible to apply strains that are well into the plastic regime. It is still true to say that all engineering metals are crystalline so that plastic deformation occurs by crystallographic slip. Since the properties of metallic crystals are in general anisotropic it follows that the strength of an engineering metal may depend upon the direction in which it is measured. This is certainly true of single crystals but in most engineering metals, which are polycrystalline, the degree of anisotropy of strength which is encountered depends very greatly on the manufacturing history of the piece concerned. Thus, a casting which has solidified with equiaxed grains will have random grain orientation and there will not be the directionality of properties that is found with fully columnar grain structures. Advantage is taken of directionality in some specialized castings such as turbine blades.

Wrought materials vary in the degree of anisotropy that is developed. It is necessary to distinguish between crystallographic and microstructural anisotropies. The latter results from banding and elongated non-metallic inclusions and whilst it is important in relation to toughness and limiting ductility it does not greatly influence strength. Anisotropy in wrought materials is strongest when the processing history of the part concerned consists of repeated large geometrically similar deformations, as in sheet, plate, rod, bar and sections.

Directionality in sheet materials may be sought or avoided, depending on the application: control may be effected by regulation of impurity contents and annealing schedules. However, the directionality of strength is much less marked in metals than it can be in some other materials, such as composites. Nevertheless, caution should be exercised when employing test data: measurements should be taken in a sense that is appropriate to the design under consideration.

Grain boundaries, and thus grain size, are an important feature in determining the progress of plastic deformation, by impeding the dislocation movement by which deformation occurs, and the progression of cleavage during subsequent fracture. The smaller the grain (crystal) size the stronger and tougher the material at temperatures which are low in relation to the melting point. At high temperatures a fine grain structure may lower strength and be a disadvantage (see Chapter 10).

Grain refinement by aluminium addition is, in fact, the most effective method of producing structural steels with high toughness at moderately lower than ambient temperatures. In normalizing heat treatments for improved properties in steel, to give a fine-grained ferritic structure, advantage is taken of the allotropic transformation, where a small austenite grain size, produced by heating to a little above the transformation temperature, gives an even finer ferritic grain size on cooling, since several ferrite grains will nucleate at the boundary of a previous austenite grain.

Grain size may also be controlled by the degree of strain given prior to recrystallization. Repetition of low strain and recrystallization was, in fact, the earliest method by which very large and single crystals were originally produced, enabling fundamental studies of single crystal properties to be undertaken.

Assessment of strength in metallic materials

The true criterion of strength in a metallic crystal is the critical shear stress resolved into the plane of crystallographic slip, but this parameter is of no value to the engineer dealing with bulk specimens. Traditionally, therefore, various points taken from the stress-strain curve determined on a bulk polycrystalline specimen have been used as measurements of strength.

Stress-strain curves typical of several metallic materials are shown in Figure 6.2. In materials which yield discontinuously there are clear measures of strength at the lower yield stress (Figure 6.2a) and yield point (Figure 6.2b). The elastic limit and limit of proportionality are difficult to measure accurately and are little used today.

Figure 6.2c is typical of materials which undergo continuous yielding. The smooth transition from the fully elastic to the elastic-plastic regime means that there is no clear singularity available to provide a definition of general yielding. The usual procedure is to measure the stress for a certain plastic strain and call this the proof stress. Values may be reported for plastic strains of 0.5, 0.2, 0.05 or even 0.01%, but for general engineering purposes 0.1 or 0.2% proof stresses are preferred.

Figure 6.2 Stress–strain curves for metals: (a) impure iron, (b) medium-carbon steel, (c) hard brass, and (d) grey cast iron. Dotted line indicates true stress, taking into account the ductile reduction of cross section.

Tensile strength, as determined on ductile materials by dividing the maximum load by the original cross-sectional area of the testpiece (ultimate tensile strength in earlier terminology), should not be used as a design parameter because it refers to a strain which should not remotely be approached in service. However, it occurs widely in data compilations because of the ease with which it may be determined: often it is the only figure available for a given material.

Figure 6.2d refers to a flake graphite cast iron. This is a brittle material and the particular stress-strain behaviour is governed by the graphite phase. There is no linear part to the curve and no observable yield stress (the total elongation at fracture may be no more than 0.5%). The usual strength criterion is the tensile strength, obtained by dividing the fracture load by the original cross-sectional area of the testpiece. Metal matrix composites can exhibit similar behaviour.

Level of strength

The level of strength that is attainable in metallic specimens may be viewed either (a) in absolute terms or (b) as an indication of the degree of development of the alloy concerned.

A yield strength of 1000 MPa (145 ksi) is a high strength in absolute terms, i.e. in any commercial engineering context it would be agreed that this is a high strength. But a strength of 500 MPa (72 ksi) would be a low strength for a heat-treated steel, yet a high strength for an aluminium alloy. This distinction enables strength to be regarded as a measure of metallurgical complexity or alloy development. Thus, because a strength of 500 MPa (72 ksi) in a heat-treated steel is readily obtained, such a material will therefore be reasonably cheap and extremely reliable, i.e. it will be subject to few metallurgical hazards in service. On the other hand, an aluminium alloy with a strength of 500 MPa (72 ksi) will be highly developed and therefore metallurgically complex, often with production difficulties, and thus expensive. Advanced alloys of a given metal are frequently more sensitive to service characteristics than their simpler brethren, and would not be chosen for use if strength were the only important criterion for materials selection. This leads to the idea that strength can be considered in terms of the theoretical lattice strength of the basis metal concerned. This is about $E/10$, where E is Young's modulus, but of course the strength actually obtained in commercial materials is much lower than this and values of yield stress of $<E/300$ would be considered as low strength and $>E/150$ as high strength. These divisions are, of course, quite arbitrary but offer a useful basis for comparison.

Relative strength

Whilst the concept of relative strength is valuable in helping to choose between candidate materials, initial consideration must be in absolute terms since this will be the basis of design calculations. It is convenient, then, to define, equally arbitrarily, four levels of absolute strength:

(1) Low strength $\sigma_{YS} < 250\,\text{MPa}$ (< 36 ksi)

(2) Medium strength $250 < \sigma_{YS} < 750\,\text{MPa}$ (36–110 ksi)

(3) High strength $750 < \sigma_{YS} < 1500\,\text{MPa}$ (110–220 ksi)

(4) Ultra-high strength $\sigma_{YS} > 1500\,\text{MPa}$ (>220 ksi)

Low strength (yield stress = 0–250 MPa; 0–36 ksi)

Most annealed pure metals are of low strength but the actual value obtained is highly dependent upon purity. Thus the tensile yield strength of 99.0% purity aluminium is 39 MPa (5.7 ksi) whereas that of 99.99% aluminium is only 12 MPa (1.7 ksi). None of the annealed commercial-purity face-centred-cubic metals can attain yield strengths of 100 MPa (14.5 ksi) and many exhibit only a small fraction of this. The strength of close-packed-hexagonal magnesium is greatly dependent upon manufacturing method. At 100 MPa (14.5 ksi) annealed sheet is some five times stronger than the sand-cast product, presumably due to directionality effects. Commercial-purity titanium, which also has a close-packed-hexagonal structure at room temperature, is strong enough to enter the medium-strength range (there is some justification for regarding this material as an alloy of titanium and oxygen).

The body-centred-cubic metals are stronger than face-centred-cubic metals, but again widely varying strengths may be obtained and although annealed yield strengths of 150–200 MPa (22–29 ksi) are readily attained much higher values may be obtained in products which have been heavily worked prior to annealing.

Most metals, except perhaps lead and tin, can be alloyed to give strengths that lie in the upper two-thirds of the low-strength range, and they

are mostly very simple materials, the most important in terms of volume production being mild steel in all its diversity of forms.

There are two reasons for using metallic materials at such a low level of developed strength. One, as previously discussed, is connected with cheapness and processing simplicity, the other with the ever-present need for a small amount of strength to back up some other property which is more important in the application considered.

Cheapness and processing simplicity lead to the almost automatic choice of mild steel for articles which can easily be produced by cutting or punching hot-rolled or cold-rolled and annealed sheet or forging hot-rolled rod, bar or drawn wire. Such articles include small levers, toggles, low-duty bolts, rivets, paper clips, fencing wire, motor car body panels, etc.

In fact, yield strengths of 200 MPa (29 ksi) and upwards, well into the medium-strength range, are readily obtainable by control of microstructure, principally grain refinement, without the need for extensive alloying.

Almost the only disadvantage of mild steel for many of these applications is the ease with which it rusts so that there is often a need for some form of protective treatment, at the least painting, but more often, and especially for outdoor applications, galvanizing or electroplating with tin, cadmium, zinc or even chromium. The obvious example here is tinplate for foodstuffs. The extra processing involves extra cost, of course and because of this drawback there is considerable competition from aluminium, aluminium alloys and injection-moulded thermoplastics.

The rusting problem, particularly in marine situations, enables aluminium alloys to offer competition. The non-heat-treatable aluminium alloys, the binary Al-Mg series, are basically low-strength materials although some of them can be strain-hardened sufficiently to enter the medium-strength range. Thus, Al-1.5Mg gives similar properties to Al-1.25Mn with annealed yield strengths around 50 MPa (7 ksi) but strain-hardens up to yield strengths of 200 MPa after 75% cold reduction. Higher magnesium levels raise these values towards the medium-strength boundary: 5% magnesium raises the corresponding yield strengths to 150 and 400 MPa (22 and 50 ksi) respectively. This is a valuable range of alloys because they combine metallurgical simplicity with excellent corrosion resistance and weldability. Applications include general sheet metal work, anodized containers, brackets, and door furniture in the form of die-castings, but more specialized uses are superstructures for ships and vessels for cryogenic containment.

The heat-treatable Al-Mg-Si series of alloys occupies a position at the top of the low-strength range although a few members are capable of edging into the medium-strength category. Possessing excellent corrosion resistance, and also being weldable (with difficulty) they are contenders for a few of the major structural applications such as small bridges, but find more general use for less important appliances such as domestic ladders and greenhouse frames.

The wrought non-heat-treatable coppers, brasses, and bronzes are low-strength materials in the soft condition but can all be strain-hardened well into the medium-strength range. It must be remembered that strain-hardening greatly reduces formability. Non-heat-treatable copper base castings, not being strain-hardened, sit firmly in the low-strength category.

The need for strength to supplement some other more important property frequently involves conflict because it often happens that any attempt to increase strength diminishes the major property. There are many examples, as follows:

Formability

This is a major requirement in articles which must be pressed or deep-drawn from sheet or bent from strip. In some applications strength must be at a practical minimum otherwise it militates against easy forming. Material for these purposes must often be purchased in versions which have been developed especially to maximize formability. This may involve controlled anisotropy (as in extra-deep-drawing quality steel or low-earing aluminium) or low levels of second-phase particles (as in steel strip for bending). There is frequently also a need for controlled surface finishes.

Machinability

Machining becomes more difficult as strength increases. Many small and intricate lightly-loaded parts such as cycle components, screws, nuts and gears are machined from bar-stock of steel or other materials in high-speed automatic lathes. This material needs to have its machinability enhanced by additions such as sulphur (to form manganese sulphide in steel) or lead (in steel, aluminium alloy or brass). Although these additions tend to embrittle the material to which they are added, this disadvantage is tolerated because of enhanced productivity.

As most machining is carried out on rolled stock, it is essential to preserve, as far as possible, the as-cast globular shape of the inclusions produced by 'free-cutting' additions through to the final product before machining. This involves careful control of the rolling conditions: for example, steel containing Type I MnS inclusions for free-cutting is rolled at 1200°C, at which temperature the relative plasticity of the sulphide is low as compared to the matrix. The inclusions therefore elongate less during production of the stock and this minimizes the adverse effect on mechanical properties and maximizes the improvement in machinability. During machining, deformation of inclusions will then take place in the compressive shear zone of the chip under temperature conditions where the relative plasticity of the inclusion is high. This reduces the ductility locally and gives pick-up on the tool surface, reducing friction.

Wear resistance

Such items as camshafts, tappets, and cycle gear components are not heavily loaded but nevertheless require high surface hardness which, if present throughout the bulk of the part, would correspond to quite excessive levels of strength and brittleness. The solution is often to take a low-strength steel and alter its chemical composition at and just below the surface, so that after suitable heat treatment the surface is hard whilst the core retains its low strength and high toughness. The chemical treatment methods include pack carburizing, cyaniding, gas-carburizing and nitriding and, depending upon the method employed, the depth of the hardened skin varies from about 0.5 mm (nitriding) to 1.5 mm (pack carburizing) (see p. 316). It is also possible to apply coatings of hard alloys to the surface of a component by methods such as fusion welding.

Electrical conductivity

The materials which exhibit the highest electrical conductivity are silver, copper and aluminium. In each of these materials conductivity is maximized when the material is of highest purity and in the fully annealed condition, a combination which, unfortunately, corresponds to minimum strength. In many electrical applications this is of no consequence but for conductors strung between poles or towers in electric power distribution and transmission lines (catenary wire),

the ability of the wire to support its own weight is important: the weaker the conductor, the closer together must the poles or towers be. Unfortunately, the standard methods of strengthening metals all result in a decrease in electrical conductivity and a compromise is therefore necessary: the acceptable strengthening method must be the one which produces the smallest decrease in electrical conductivity for unit increase in strength. Figure 6.3 shows the effect of some elements in solid solution on the electrical conductivity of copper. This establishes the suitability of silver and cadmium for strengthening conductivity copper: both are used, and copper with 0.8% cadmium is one of the standard materials for conductors used in overhead distribution lines, with an annealed electrical resistivity of 1.8 $\mu\Omega$ cm as compared with 1.7 $\mu\Omega$ cm for annealed copper.

Aluminium-base materials are also used for electrical purposes and in this case where strength is needed there is the option of a precipitation-hardening system: aluminium with about 0.5% magnesium and 0.5% silicon can be heat-treated to a tensile yield strength of 140 MPa (20 ksi) whilst retaining an electrical resistivity of 3.2 $\mu\Omega$cm as compared with 2.8 $\mu\Omega$cm for commercial purity aluminium.

Figure 6.3 Effects of added elements upon the electrical conductivity of copper. (Courtesy: Copper Development Association.[1])

Of course, for major power transmission this method of providing strength is not adequate. In the United Kingdom transmission lines (400 kV, 275 kV and 132 kV) are equipped for the most part with steel-cored aluminium conductors. High-tensile steel provides the strength: hard-drawn 99.45% purity aluminium wire provides the electrical conductivity. Many different types of lay-up are available; for example, a small conductor would consist of a single steel strand surrounded by six aluminium strands, whereas a large conductor, as widely used, has a seven-strand steel core surrounded by fifty-four strands of aluminium wire in three layers. Not only are such conductors stronger and lighter than equivalent copper conductors, they are also cheaper.

All-aluminium-alloy conductors and aluminium alloy-reinforced conductors are now being introduced in the United Kingdom and have been used in overseas countries for many years. The best conductor to select depends upon the detail of the balance of electrical considerations and the requirement to meet mechanical loading conditions which are heavily dependent upon local meteorology. Down-droppers from overhead lines to sub-stations (switchyards) are almost universally made of softer, higher-conductivity, grades of aluminium.

On the other hand, aluminium alloys are sometimes used for rigid bus-bars in 400 kV switching stations. To support self-weight in spans of up to 14 or 15 m extruded tubes are used in the fully precipitation-hardened aluminium alloy containing 0.9% magnesium and 0.6% silicon.

Medium strength
(yield stress = 250–750 MPa; 36–110 ksi)

In this category lie, for the most part, alloys that are moderately well developed for strength. Of the major engineering metals, only titanium can enter this range as a commercial-purity metal in the annealed condition, although work-hardened copper is well within with a yield stress of 340 MPa (49 ksi). Strain-hardened

commercial-purity aluminium fails to qualify, and so do the lower-alloyed members of the solid solution alloyed aluminium alloys.

The heat-treatable 2xxx- and 7xxx-series aluminium alloys are genuine medium-strength alloys, although even the most highly-developed members reach little more than half-way through the range with the highest-strength members having difficulty in exceeding yield strengths of 500 MPa (72 ksi). Of the two main groups the 2xxx-series based on the Al-Cu-Mg system, is the older, having evolved from the original precipitation-hardening alloy discovered by Wilm in 1911. The 7xxx-series, which are alloys of Al-Zn-Mg-Cu, produce the highest static strengths of all aluminium alloys but have needed extensive metallurgical development to become truly competitive with the 2xxx-series. The major application of both groups is in aerospace and they are discussed in more detail in Chapter 15.

Copper-base alloys are able to cover the range with strain-hardened copper-tellurium alloy at the bottom developing a yield strength of 265 MPa (38 ksi) and the fully precipitation-hardened copper-nickel-silicon alloy at the top with a corresponding figure of 480 MPa (70 ksi). Alloys such as copper-5% tin in the strain-hardened condition are used for springs.

The medium-strength range includes the enormous diversity of the low alloy (HSLA) steels typically used in building construction, offshore oil rigs, pressure vessels, oil and natural gas transmission lines, with yield stresses varying from 200 MPa (29 ksi) to 1000 MPa (145 ksi) depending upon alloy content and treatment. Also referred to as microalloyed steels, these materials contain less than 0.1% each of elements designed to grain refine or precipitation harden, namely niobium (columbium), vanadium and titanium. A tensile strength of about 500 MPa (73 ksi) is typical for microalloyed steels. The lower-strength members of this group are strengthened mainly by grain refinement achieved by normalizing or controlled rolling assisted by micro-alloying with niobium (columbium). The higher-strength members may be lean-alloyed and sometimes develop strength by quenching and tempering.

Interstitial-free (IF) steels are in this medium strength class but also exhibit good formability. Vacuum degassing prior to casting reduces the interstitials carbon and nitrogen to very low levels (<0.002% each). As with HSLA steels, titanium and niobium are added to react with the remaining carbon and nitrogen, forming TiN and Nb(CN) respectively. TiN reduces grain growth and Nb(CN) inhibits recrystallization. The carbon content of an IF steel is typically 20 ppm. These steels were developed for the automobile market with the aim of combining high strength with formability.

The usual methods of strengthening steel can result in 'spring back' during deep drawing operations. The formability of steel sheet is frequently quantified by the 'r value', with higher r values for steel sheet that can undergo more severe deformation without fracture or anisotropic deformation. The typical r value for plain carbon steel sheet, cold rolled and annealed is 1.6; microalloyed steels display r values of about 1.2; Interstial-Free (IF) steels can be produced giving a cold-rolled tensile strength of 320 MPa (46 ksi) with an r value of 2.0.

From the middle of the range upwards come the engineering/automotive bar steels, from the 0.40 carbon-manganese steel up to the low-alloy steels such as the manganese-molybdenum variety. These steels are mostly intended to be used in the quenched and tempered condition and since the production of a sufficient proportion of martensite is an essential prerequisite to the development of optimum properties, the selection of a suitable steel from the many available compositions must be based on a consideration of hardenability and ruling section. A disadvantage of this method of strengthening steels is the high cost of the alloying elements needed to confer sufficient hardenability on steels intended for sections greater than 10–12 mm. There are also technical disadvantages, not least the distortion and residual stresses caused by the quench-hardening treatment. Accordingly, for some steels needed in thick sections, e.g. pressure vessel steels, the strength is developed by direct transformation from austenite to ferrite-plus-carbides, the latter in this case being molyb-

denum and vanadium carbides produced by lean additions of those elements. The problem of hardenability is thereby avoided and yield strengths up to 750 MPa (110 ksi) may be readily obtained.

Where corrosion resistance or resistance to oxidation are required it is sometimes necessary to use a stainless steel. The simple chromium-nickel non-heat-treatable austenitic stainless steels in the annealed condition exhibit yield strengths little more than 200 MPa (29 ksi) but they can be strengthened by simple strain-hardening to 650 MPa (94 ksi). If the composition is such that plastic deformation causes the austenite to transform to martensite, as with the 17 Cr–7 Ni alloy (AISI 301), strengths into the ultra-high range can be obtained. Strain-hardened austenitic stainless steels are available only as sheet, strip and wire. Formability is very poor and joining is difficult.

Probably the most straightforward and useful method of obtaining intermediate strengths in austenitic stainless steels is by the use of nitrogen in interstitial solid solution. The addition of about 0.2% nitrogen gives proof stresses up to 420 MPa (61 ksi), depending upon base composition, and the steels have proved useful for pressure vessels.

The classical quench and temper method of hardening can be applied to the straight martensitic stainless steels with chromium contents around 11–14%. The yield strength is largely dependent on the carbon content and in a 13% chromium steel varies from about 400 MPa (58 ksi) with 0.1% carbon to 650 MPa (94 ksi) with 0.25% carbon. For best toughness and corrosion resistance the carbon content should be low. This, and the need to temper at high temperatures to avoid a brittleness trough, limits the available strength of these simple compositions.

High strength (yield strength = 750–1500 MPa; 110–220 ksi)

High-strength metal alloys are highly specialized and in all cases the volume production is very low. This is especially true of the more highly developed examples such as the controlled transformation stainless steels which are not only highly alloyed but also require complex treatments to develop their properties. However, although simpler materials than these can occupy useful positions within this range it is beyond most of the non-ferrous materials. Only one copper-base material qualifies, the precipitation-hardened copper-2% beryllium alloy, with a yield strength of about 900 MPa (130 ksi), but all of the titanium alloys, with the exception of titanium-oxygen (CP titanium) and titanium-2.5% copper alloys, can be included.

Otherwise, this strength range is populated mainly by steels of one sort or another. For yield strengths not very much greater than 1000 MPa (145 ksi) it would be normal to select one or another of the fairly run-of-the-mill medium-carbon low-alloy steels which are always used in the quenched and tempered condition. The upper levels of strength in this range would typically be required for heavy-duty springs, truck transmission components, heavy-duty crankshafts, gearing, etc. and a classic British example of this type of steel is 817M40 (formerly En24) in BS970 which nominally contains 0.4C–1.5Ni–1.2Cr–0.25Mo. Higher-strength versions of the same basic type of steel are 826M40 (En26) with 0.4C–2.5Ni–0.7Cr–0.5Mo and 835M30 (En30B) containing 0.3C–4Ni–1.2Cr–0.3Mo. The corresponding American steel is AISI 4340 which has similar composition and properties.

When strengths in the upper part of the high-strength range are required the medium-carbon low-alloy steels present problems. This is because in order to obtain maximum strength the tempering temperature must be low – perhaps as low as 200°C if sensitivity to temper embrittlement or 'blue brittleness' means that tempering temperatures around 350°C cannot be employed. Thus toughness tends to be inadequate. There is also the problem that hardenability is limited: the low-alloy steels exhibit deteriorating properties as section size rises above about 12 mm. When such steels are being used at the limits of their capabilities there is a need for considerable expertise in manufacture and processing.

Although many of the high-strength steels must contain around 0.4% carbon to obtain the required strength, this is detrimental to toughness and especially to weldability, so that the carbon content should always be the minimum needed to develop the required strength. Thus 835M30 and AISI 4130 contain only 0.30%. However, for applications in which weldability is essential, and in addition the material is called for in thick sections, then a quite different type of steel is needed, since for good weldability carbon contents must be significantly lower than 0.30% and deep hardenability requires heavier alloying. HY130 is an example of such a steel: it was developed especially for deep submersibles with the carbon adjusted downwards towards 0.10% and the hardenability provided by 5% nickel together with chromium, molybdenum and vanadium. These steels provide excellent toughness at yield strengths around 1000 MPa (145 ksi).

The use of secondary hardening by precipitation of chromium and molybdenum carbides is capable of producing yield strengths up to around 1500 MPa (218 ksi) and with high alloying the actual properties obtained depend upon the carbon content. In the 9Ni–4Co steels, the high M_s (martensite start) temperature obtainable by the combination of cobalt and a low carbon content of 0.2% produces a self-tempering martensite so that as-welded structures exhibit an excellent combination of strength and toughness. However, to obtain the higher strengths the carbon must be raised to 0.30% and the resultant loss of toughness makes the steel less suited to welding.

The limitations of the straight C–Cr martensitic stainless steels were mentioned in the previous section. Their properties can be improved if minor additions of molybdenum, vanadium and niobium (columbium) are used to confer tempering resistance: with carbon down to 0.13% and added nickel to improve toughness, yield strengths up to 900 MPa (130 ksi) can be obtained.[2] As with all quenched and tempered steel sheets, forming must be done in the annealed condition, the full heat treatment being carried out after fabrication – considerable

TABLE 6.2. Compositions of precipitation-hardened martensitic stainless steels

	C	Si	Mn	Cr	Ni	Mo	Cu	Nb	Al
FV520B	0.04	0.4	0.5	13.5	5.5	1.6	1.8	0.2	–
PH13-8	0.04	0.03	0.03	12.7	8.2	2.2	–	–	1.1

scaling and distortion may occur. Forming problems are considerably relieved if the carbon content is still lower: this can be achieved if the strength is developed during tempering by precipitation hardening of the martensite with additions of nickel, copper, niobium (columbium) and others: commercial examples include 17–4PH,* PH13–8,* Custom 455,† and FV520 (B),‡ (Table 6.2). The very low carbon content and careful balance of alloying elements mean that processing and machining can be carried out in the soft solution-treated condition, the final yield strength of up to 1500 MPa (218 ksi) being developed by a single ageing treatment at between 450 and 620°C (840 and 1150°F). Steels of this type are used for forgings, fasteners, pump and valve parts. In the UK long spans over the M1 motorway are supported on FV520 (B) rollers and the central tower of York Minster is strengthened by very large bolts of this material.

There are several precipitation-hardening stainless steels in which the matrix is fully austenitic before and after heat treatment.[3] These can produce yield strengths up to 900 MPa (130 ksi) by precipitation of carbides in the presence of phosphorus; gamma prime $(Ni,Fe)_3$ (Ti,Mo) has also been proposed as a strengthening agent.

There are very few titanium alloys which do not enter the high-strength range – most of them are capable of developing strengths around or above 800 MPa (116 ksi), and selection is dictated mainly by other considerations. For example, the need for high creep resistance restricts selection to alpha or near-alpha alloys; the beta phase is characterized by a low elastic modulus; some alloys cannot be made commercially available in the form of sheet; many alloys are not weldable.

However, no titanium alloy is capable of reaching the top of the high-strength range: the strongest titanium alloy cannot exceed a value of 1250 MPa (181 ksi) for 0.2% proof stress and titanium alloys can only compete with the high-strength steels on the basis of density-compensated parameters – the density of titanium is 4.5 tonne/m^3 as compared with 7.8 for steels.

Ultra-high strength (yield strength >1500 MPa; >220 ksi)

The pre-eminent ultra-high-strength quenched and tempered steel is 300M (ASTM A579 Grade 32) a 2Ni–Cr–Mo–V steel with the addition of 1.6% Si to displace the blue-brittleness trough in the tempering curve.[4] The 0.2% proof stress of this steel is 1560 MPa (226 ksi) and for many years it was the almost automatic choice for the most demanding applications such as aircraft landing-gear components. To ensure adequate reliability in service, especially in respect of fracture toughness and fatigue, the steel has to be double-melted and double-tempered.

The 5 Cr–Mo–V steels originally used for hot-working dies can deliver similar levels of strength (e.g. H11) and have been occasionally used for structural purposes: their high alloy content makes them too expensive for widespread use.

*Armco Steel Corporation; †Carpenter Technology Corporation; ‡Firth-Vickers Special Steels Ltd. (trademarks).

TABLE 6.3. Properties of maraging steels

	0.2% PS		K_{IC}	
	MPa	ksi	MPa.m$^{1/2}$	ksi.in$^{1/2}$
18Ni–8.5Co–3Mo–Al–Ti	1400	203	110–176	100–160
18Ni–8Co–5Mo–Al–Ti	1700	247	100–165	91–150
18.5Ni–9Co–5Mo–Al–Ti	1900	276	90–100	82–91

The principal disadvantage of 300M and similar steels is that in the fully heat-treated condition machining is extremely difficult and expensive, and whilst rough machining can be carried out in the softened condition there is always some finish-machining to do after final heat treatment. The maraging steels, which are capable of developing much higher strengths than 300M, have the advantage that the relatively low-temperature (480°C, 900°F) ageing treatment by which their strength is developed can be carried out after final machining. Thus, although their high nickel content means that their basic material cost is much greater than that of 300M, the lower machining costs may make them competitive.[5] Another advantage which maraging steels have over conventional quenched and tempered steels is their full weldability: it is possible to join together prefabricated fully-hardened pieces without any further heat treatment other than local heating to age the weld zone. They are generally regarded as extremely tough materials although some doubt has been thrown on their ability to maintain high toughness in thick sections. Typical properties are given in Table 6.3.

The controlled transformation stainless steels are austenitic after solution treatment and are fabricated in this condition. Transformation is effected by destabilization of the austenite either by refrigeration or, more usually, by low-temperature heat treatment. The precipitation of $M_{23}C_6$ reduces the carbon content of the matrix and raises the M_s temperature so that on cooling from the heat-treatment temperature the austenite is no longer stable and transforms to martensite. Possible yield strengths are up to 1250 MPa (181 ksi) or, when cold-rolled prior to tempering, 1800 MPa (261 ksi). Although these steels are used quite widely in aerospace (see Chapter 15), they seem to find little application elsewhere.

Very high strengths of up to, or in some cases above 2000 MPa (290 ksi), can be conferred on suitable steels by certain thermomechanical processes such as ausforming and marstraining. Steels of the 4340 type are suitable subjects but commercial exploitation of these materials is limited by practical problems.

Finally, it is necessary to mention music-wire, the hard-drawn high carbon steel which in the form of thin wire is still the strongest commercially available material. The wire is patented by passing through tubes in a furnace at about 970°C (1780°F) with care to avoid decarburization, giving a uniform large grain size austenite. Rapid cooling in air or molten lead follows to a low transformation temperature, so that the final structure is of very fine pearlite with no separation of pro-eutectoid ferrite. This structure enables the wire to withstand very large reductions, as compared to annealed material with separated ferrite cells, or tempered martensite where the carbide does not develop into the same fibrous structure. The wire is usually worked to tensile strengths in the range 1600–1850 MPa (232–268 ksi). Patented wire is frequently used for springs which can be cold formed and for wire rope for haulage purposes.

The information from Section 6.1 is summarized in Table 6.4

TABLE 6.4. Strength of metals and metal alloys

Low yield strength (0–250 MPa)	Annealed pure metals Mild steels Non-heat-treatable Al–Mg alloys
Medium yield strength (250–750 MPa)	Heat-treatable 2xxx/7xxx Al alloys High strength structural steels Engineering steels Commercially pure (CP) titanium Stainless steels
High yield strength (750–1500 MPa)	Titanium alloys Cu–2% Be precipitation hardened Medium-carbon low alloy steels High strength low alloy steels Precipitation hardened stainless steels
Ultra-high yield strength (>1500 MPa)	Maraging steels Patented wire Tool steels

6.2 The strength of thermoplastics

All unreinforced thermoplastics must be regarded as low-strength materials (Table 6.5). However, because of the generally low density exhibited by plastics, the strength–weight ratio is much more favourable. Thus nylon with a yield strength of 80 MPa (12 ksi) and a density of 1.14 Mg/m^3 has a strength–weight ratio equal to say,

a medium-carbon low-alloy steel heat-treated to a yield stress of 550 MPa (80 ksi).

The short term tensile strength versus temperature for a number of thermoplastics is shown in Figure 6.4. However, there are two characteristic features of thermoplastics which make the evaluation of strength more difficult than is the case with metallic materials. One is the time-dependent nature of their mechanical properties and the other is their temperature sensitivity. The first of these characteristics means that proper design procedures for any given thermoplastic, even for service at normal room temperature, must employ the appropriate data for that material as obtained under creep conditions. The second characteristic means that quite small changes in temperature can produce significant alterations in properties.

The time-dependent nature of the properties of thermoplastics is clearly seen in Figure 6.5. They behave very differently under applied loads compared with metals. Many design formulae have been derived for structural components dependent on the elastic modulus as a fundamental measure of the response to stress, the assumption being that the material behaves in a linear, elastic manner until yielding. For most metals this is true; for plastics, the stress-strain curve is rarely linear, there is no true proportional limit and the behaviour is affected by strain rate and temperature. Because they combine the

TABLE 6.5. Short-term tensile strengths of unfilled thermoplastics at 23°C

	MPa	ksi		MPa	ksi
PPO	66–85	10–12	uPVC	50–60	7–9
Polyethersulphone	85	12	GP polystyrene	40–50	6–7
Acrylic	60–80	9–11.6	ABS	25–50	3.6–7
SAN	75	11	Polypropylene	25–35	3.6–5
Polyacetal	60–70	9–10	HD polyethylene	25–30	3.6–4
Polysulphone	70	10	High impact polystyrene	40	6
Nylon 66 (dry)	60	9	PTFE	20	3
Nylon 6 (dry)	40	6	LD polyethylene	8–10	1–1.5
Polycarbonate	60	9			
Thermoplastic polyester	60	9			

Data from Powell.[6]

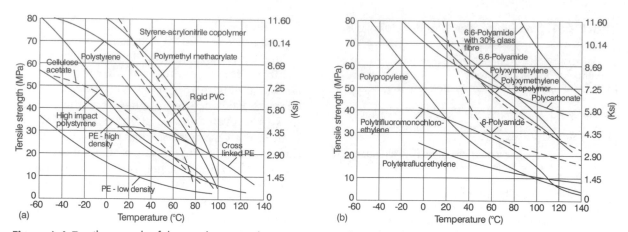

Figure 6.4 Tensile strength of thermoplastics in short term tests in the range −60°C to 140°C.[5] From: *Saetchling International Plastics Handbook*, Carl Hanser Verlag, 1987.

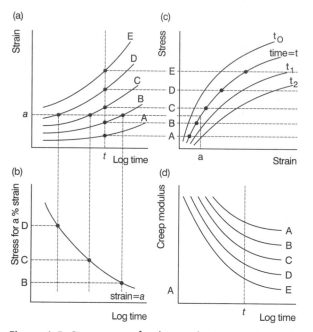

Figure 6.5 Creep curves for thermoplastic materials: (a) creep curves at various stresses (A<E); (b) isometric stress–time curve; (c) isochronous stress–strain curves and (d) creep modulus–time curves at various stresses

mation. Viscoelastic materials creep under steady stress and stress relax under steady strain. (N.B. creep and stress relaxation are not the same.)

To design a component that will not creep by more than a certain amount over its projected lifetime, the relevant creep curves should be referred to.

However, since the procurement of long-term creep data is both time-consuming and expensive there is a natural tendency to employ short-term data for preliminary comparisons although, even here, the expression 'short-term' can take a variety of meanings. If a short-term tensile test is carried out at a constant rate of elongation (as in BS2782), then different stress-strain curves will be obtained at different rates of elongation. More meaningful data can be obtained by carrying out proper creep tests and then extracting from the results isochronous stress-strain curves for the appropriate time (Figure 6.5).[6] A 100-second isochronous curve will not differ greatly from a fairly slow constant-rate curve.[7]

The thermal behaviour of thermoplastics depends on their molecular structure. For materials like polyethylene, where the backbone of the structure is made up of carbon-carbon single bonds, the molecular chains tend to be very flexible and they can fold up on themselves or with other polyethylene chains to form areas of reasonably ordered crystalline structure. It is unusual to find 100% crystalline materials because of the size of most polymer molecules,

behaviour of an elastic solid and a highly viscous material, they are termed 'viscoelastic'.

As shown schematically in Figure 6.6, the deformation of plastics can be split into three elements: instantaneous (reversible) elastic deformation, time-dependent viscoelastic deformation and time-dependent (irreversible) plastic defor-

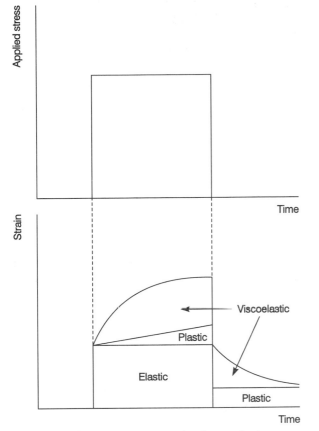

Figure 6.6 The strain response of a thermoplastic as a function of applied stress and time

Figure 6.7 Molecular arrangement of thermoplastics

as there is simply not enough mobility in the polymer. Semi-crystalline materials are more typical, with areas of crystalline and amorphous molecular arrangements, the latter where the molecular chains are coiling randomly in space (see Figure 6.6). Time and temperature during processing have major influence on the degree of crystallinity in a semi-crystalline polymer. For example, such a polymer cools from its melting temperature very slowly, the molecules are mobile for longer and thus the degree of crystallinity is greater than for a rapid cooling rate. For polymers where the molecular chains contain more rigid bonding or where there are large, bulky side groups at regular intervals, the chains will want to coil up, but there will be no regions of crystallinity. These are amorphous polymers (see Figure 6.7).

The processing of semi-crystalline and amorphous polymers is very different. Semi-crystalline polymers must be molten for sufficient flow in processes like injection moulding. In the molten state, the structure is amorphous and it is when the structure is cooled through the melting point that crystallization takes place. The processing window for amorphous polymers is much wider as they soften when heated and can attain sufficient fluidity for processing. This softening range is generally greater than the melting range for semi-crystalline polymers.

This behaviour is illustrated in Figure 6.8 where the tensile modulus of an amorphous and a semi-crystalline polymer is plotted against temperature. For amorphous polymers, the modulus is highest at low temperatures, in the 'glassy' region. Here little intermolecular movement is possible. As the temperature increases, secondary intermolecular forces are overcome, the polymer chains have sufficient thermal energy to rotate, and the modulus falls off rapidly through a transition region, characterized by the 'glass transition' temperature, T_g. The modulus then levels off with a further increase in temperature as the polymer chains remain entangled although local sections are mobile. In this region, the material can exhibit large elongations under very low load, hence it is often referred to as the 'rubbery' region. With a further increase in temperature, large scale molecular movement can take place and the polymer becomes a viscous melt, and the melt temperature, T_m, is often indistinct. Semi-

59

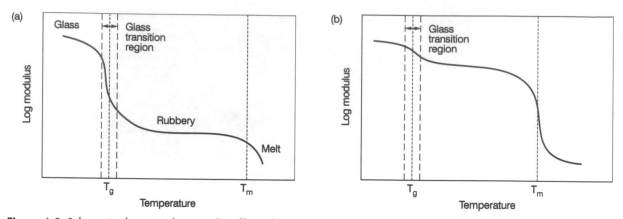

Figure 6.8 Schematic diagram showing the effect of temperature on the modulus of (a) amorphous and (b) semi-crystalline thermoplastics

crystalline polymers exhibit a similar shape to the curve, although the presence of a T_g depends on the presence of amorphous regions in the microstructure. As the crystalline regions break down, the melt temperature, T_m, is more distinct compared with amorphous polymers.

These regions are significant in determining the useful temperature range of thermoplastics. Amorphous plastics, like ABS, must be used below the T_g (80–100°C for ABS). Semi-crystalline polymers are useful above their T_g; for these materials it is the melting temperature, T_m, which limits their structural stability, although this does depend on the level of crystallinity.

The general properties of amorphous and semi-crystalline thermoplastics are compared in Table 6.6.

Since different plastics exhibit different modes of failure it is easy to make invalid comparisons. There is a rough analogy with metallic materials since some plastics seem to fail in a ductile manner whilst others are brittle, but of course the molecular processes and macroscopic behaviour are very different from those that occur in metals. Thus in plastics the point of failure may be a yield point (as in polycarbonate), necking rupture (as in unplasticized PVC) or brittle fracture (as in polystyrene)[8] and these do not

TABLE 6.6. Comparison of amorphous and semi-crystalline thermoplastics

Property	Amorphous	Semi-crystalline
Mould shrinkage	Low 0.4–0.8%	High 1.5–3.0%
Chemical resistance	Attacked by most solvents Susceptible to stress cracking	Resistant to many solvents Resistant to stress cracking
Optical	Transparent	Translucent unless thin section or highly oriented crystal structure
Processing	Softens gradually as temperature increased	Processed at temperatures above the melting point
Wear	Usually modified for use in wear applications	Low friction and good wear properties

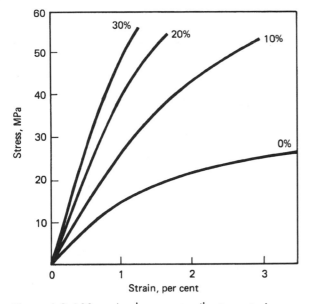

Figure 6.9 100 sec isochronous-tensile stress-strain curves for polypropylene having different percentages by weight of glass fibres. (Courtesy: The Design Council.[7])

exhaust the possibilities. For the purposes of preliminary materials selection, therefore, it is essential to ensure that like is compared to like.

When a thermoplastic is reinforced by fibres the strength can be greatly increased (Figure 6.9 and Table 6.7), although at the expense of some ductility and, sometimes, impact strength. The most commonly used fibre reinforcement is glass, and in articles produced by injection moulding the fibres are of necessity short, little more than a few millimetres in length. This naturally introduces variable anisotropy of properties and published figures for strength and stiffness frequently fail to take this into account, so that the properties exhibited by real articles may be lower than anticipated. The glass content may vary from 15 to 40% by volume. Fibre reinforcement frequently exacerbates problems of shrinkage and distortion and where good maintenance of shape and dimensional accuracy is essential the fibres may be replaced, partially or wholly, with glass beads. The mechanical properties are then, in general, less good, since reinforcement with glass beads is less effective than that produced by fibres (Figure 6.10).

The importance of strength in thermoplastics design depends very much upon the application,

TABLE 6.7 Short-term tensile properties of short glass–fibre reinforced thermoplastics

| | Tensile strength | | Elongation at failure |
	MPa	ksi	%
Polyamide 6/6	60	9	60
+10% glass fibre	95	14	8
+33% glass fibre	120	17	4
+60% glass fibre	220	32	2
+10% carbon fibre	130	19	5
Polycarbonate	65	9	110
+20% glass fibre	90	13	4
+30% glass fibre	110	16	4
+30% carbon fibre	165	24	3
Polyethylene terephthalate	55	8	300
+15% glass fibre	80	12	6
+30% glass fibre	125	18	3
+45% glass fibre	200	29	2

(% by weight)
Data from: The Plastics Compendium[9].

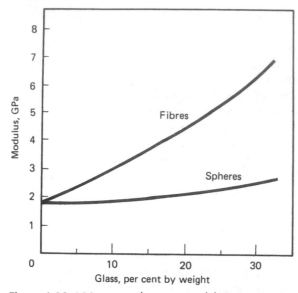

Figure 6.10 100 sec tensile-creep modulus at 0.2% strain of polypropylene with glass fibres or spheres versus percentage by weight of glass. (Courtesy: The Design Council.[7])

The general intention is to improve the strength and stiffness of a material by incorporating within it fibres that are stronger and stiffer than the matrix. Matrices may be of thermoplastics, thermosets, ceramics, metals or concrete: fibres may be of glass, carbon, metal, ceramic or polymeric materials. Fibres, and also composites, can be produced covering a wide range of properties and costs – the span of applications is correspondingly wide.

Table 6.8 provides some representative figures for the strengths and stiffness of some commercially available fibres and reinforced polyester or epoxy resins.

The best-known fibre composites are the glass reinforced plastics (GRP) which usually comprise various types of arrays of glass fibres in a polyester resin matrix. The glass employed for the fibres is most commonly the cheap low-alkali borosilicate E-glass and although the stronger and stiffer S-glass is available it is rarely used with polyester. Although glass is as strong as steel (much stronger in density-compensated terms), its stiffness is much less, being similar to that of aluminium. This means that when GRP is used as a structural material its design is generally dominated by considerations of stiffness rather than strength. The conventional method of manufacture – the laying and impregnation by hand of various types of cloth or mat – means that where local variation in strength of the product is desired, this can be accomplished by varying the overall thickness of the composite rather than changing the strength of the materials used. This is most readily accomplished by local alteration of the number of laminations employed. Indeed, most of the methods of thermosetting composite manufacture, (including hand lay up, prepreg moulding, compression moulding, reinforced reaction injection moulding and filament winding) allow for the reinforcement levels to be varied locally, thereby tailoring the strength and stiffness of the component to the required levels. Nevertheless, the ability to vary the proportion of fibre to matrix means that in principle the strength and stiffness of the composite can take values ranging from those of the pure

and whilst it is clear that the primary functions of articles such as pipes, tanks, and gears demand a good ability to bear loads, it is equally clear that many engineering components such as housings, bearing frames, and stackable containers must above all maintain shape. Thus in thermoplastics, the elastic modulus is relatively more important than it is in metals and designing for rigidity is often more important than designing to bear loads (see Chapter 8).

6.3 The strength of fibre-reinforced composites

The properties of fibre-reinforced composites are determined by five factors: (a) the relative properties of the matrix and the fibres, (b) the relative proportions of fibre and matrix in the composite, (c) the length of the individual fibre particles, (d) the nature of the fibre–matrix interface, and (e) the geometrical arrangement of the fibres within the composite.

TABLE 6.8. Mechanical properties of various fibres and composites

Fibre/composite		Tensile strength		Tensile modulus		Density
		MPa	ksi	GPa	10^3 ksi	Mg/m^3
E–glass (specially prepared)		3000	435	70	10	2.54
E–glass (standard)		1500–2000	220–290	70	10	2.50
S-glass (specially prepared)		4300	624	80	12	2.49
S-glass (standard)		2600	377	86	12.5	2.52
Carbon fibre (high modulus)		4000–5000	580–725	390–450	57–65	1.91
Carbon fibre (high strength)		4000–7000	580–1015	230–300	33–44	1.74
Carbon fibre (intermediate modulus)		3000–5000	435–725	290–300	42–44	1.81
Carbon fibre (mesophase pitch)		2200	320	830	120	2.20
Boron fibre		3500	510	420	60	2.65
Aramid (Kevlar™ 29) fibre		2700	390	60	9	1.44
Aramid (Kevlar™ 49) fibre		2700	390	130	19	1.45
Polyethylene fibre		3000	436	115	17	0.95
Polyester matrix		30–40		3		1.2
Epoxy matrix		40–90		3.5		1.2–1.8
PEEK matrix		100		3.5		1.3
GRP chopped strand mat 20% glass		100	15	8	1.2	1.5
GRP woven fabric 30% glass		250	36	16	2.3	1.7
Epoxy prepreg woven fabric 50% E-glass		600	87	20	2.9	1.8–2.0
Epoxy prepreg	∥	1100	160	45	6.5	1.8–2.0
unidirectional 50% E-glass	⊥	35	5	8	1.2	
Epoxy prepreg woven fabric 50% Aramid		500	73	30	4.4	1.2–1.4
Epoxy prepreg	∥	1100	160	60	8.7	1.2–1.4
unidirectional 50% Aramid	⊥	35	5	8	1.2	
Epoxy prepreg	∥	2000	290	140	20	1.5–1.6
unidirectional 60% High Strength carbon	⊥	80	12	10	1.5	
Epoxy prepreg unidirectional 60%	∥	2400	348	160	23	1.5–1.6
Intermediate modulus carbon	⊥	80	12	10	1.5	
High carbon steel		2800	406	210	30	7.8
Aluminium alloy 7075		570	83	72	10.5	2.8
Ti-6 Al-4V		950	138	115	17	4.5

% by volume

Data from various sources, including:

E.A. BRANDES, ed: *Metals Reference Book.* Butterworth, London, 1983.
A. KELLY, ed: *Concise Encyclopedia of Composite Materials.* Pergamon, Oxford, 1994.
M.F. ASHBY and D.R.H. JONES: *Engineering Materials 1,* Butterworth-Heinemann, Oxford, 1996.
Ciba Composites, *Prepreg Technology,* 1995.
R.M. MAYER: *Design with Reinforced Plastics.* The Design Council, London, 1993.
D. BLOOR, R.J. BROOK, M.C. FLEMINGS and S. MAHAJAN: *The Encyclopedia of Advanced Materials,* Pergamon, Oxford, 1994.

matrix to those of the pure fibre. In the case of an E-glass/polyester resin composite this means tensile strengths from 20 to 2000 MPa (2.9–29 ksi) and stiffness from 2 to 76 GPa (290–11 025 ksi). In practice, the need for workability during the laying-up process imposes constraints on the permissible volume fraction of fibres and upper limits are also influenced by the geometry of the reinforcement. Three possible modes of fibre arrangement can be visualized: (a) random, (b) woven, and (c) uniaxial, although each of these can be sub-divided many times.

Random distribution often implies short, chopped, fibres and this arrangement can be encountered in glass-filled thermoplastics such as glass-filled nylon. Short fibres are essential if injection moulding is the method of manufacture. However, in practice there will be regions of aligned fibres and regions of random fibre distribution in such injection moulded parts due to the flow of material during processing. In random reinforcement of thermosets the fraction of fibres rarely exceeds 40% by weight and is frequently much less. The common mode of reinforcement in polyester-resin-based composites is chopped strand mat and a typical tensile strength at 30% by weight of fibres would be 110 MPa (16 ksi).[10] Chopped strand mat can be bought in roll form, where the chopped glass fibres are held together by a matrix-compatible binder. For larger components (for example, a boat hull) the process is semi-automated with a spray gun which incorporates a chopper, so that catalyzed polyester resin and chopped fibre are sprayed together onto the mould. This process is followed by consolidation with a hand-roller, removing any unwanted air. The need for proper impregnation ensures fairly low volume fractions of fibre.

Woven reinforcement of continuous fibres permits higher volume fractions of fibres – up to about 65% by weight. The directional variation in properties then depends upon the arrangement of the fibres – an orthogonal arrangement is most common although others are possible. At 50% weight fraction of fibres a typical tensile strength would be 240 MPa

(35 ksi). Woven fabrics can be laid by hand into the mould and impregnated with catalyzed resin, typically polyester or epoxy. For applications where control of the fibre volume fraction is more critical, it may be cost-effective to use materials where the desired amount of resin has been pre-impregnated into the fabric – so-called prepregs. For prepregs, it usually makes sense to cure under atmospheric pressure using a vacuum bag technique and to cure at elevated temperatures in an oven. Prepregs are used where quality control is crucial, such as in the aerospace industry.

Unidirectional reinforcement allows the greatest volume fraction of fibres: 90% filling is possible although it is usually less. This type of reinforcement offers the greatest potential for enhancement of properties (Figure 6.11) but it also imposes on the composite maximum anisotropy of properties. Of course, truly unidirectional reinforcement is almost impossible to lay onto a mould dry, and unidirectionals are

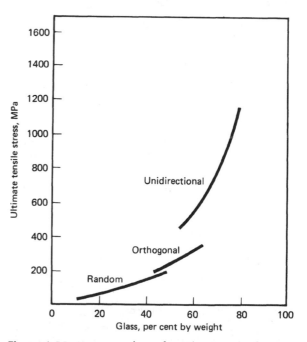

Figure 6.11 Average values of tensile strength of polyester–glass fibre composites versus percentage by weight of glass for the three major modes of arrangement of fibres. (Courtesy: The Design Council.[7])

invariably supplied as prepreg tapes. With these materials it is possible to design their lay up so that there is the minimum amount of redundant reinforcement to maximize properties. In a woven fabric, the fibres will be interlacing and therefore they will not lie in a straight line across the directions of principal stress; there will always be some crimp in the fabric. Attempts have been made to use a thin sacrificial thread to bind rows of fibre together without crimp, but their use has been limited to date. Hence their predominant use in the aerospace industry where weight is a premium.

In the field of advanced composites, boron fibres were early contenders. Their initial promise has not been maintained, principally because of high cost (they are up to five times the cost of carbon fibres) but also because they are manufactured by vapour deposition of boron on to a tungsten filament. The tungsten contributes little to the properties but adds to the weight of the fibres and makes them difficult to cut and drill. However, because they are thicker than most other fibres, boron fibre composites exhibit a higher compressive strength than do carbon fibre composites.[11] A few commercial applications have employed composites of boron and epoxy resin.

Carbon fibres are most commonly produced from precursor fibres of polyacrylonitrile (PAN) or from various forms of mesophase pitch (MPP). Using PAN precursor fibres, carbonization in the 1300–1700°C range will produce a high strength fibre, whereas further heat treatment in the range 2000–3000°C will result in high modulus fibres. With MPP-based precursors, highly crystalline graphitic fibres result from the high temperature carbonization process. These crystalline graphite layers are capable of exhibiting stiffness values close to the theoretical maximum (~1000 GPa). However, most MPP-based carbon fibres have relatively low tensile and compressive strengths. Vapour-grown (VG) carbon fibre is a relatively new type of fibre. With good growth rates resulting in fibres with high tensile strengths it is anticipated that they will represent a low-cost route to carbon fibres.

Other fibres of interest for reinforcing polymer composites include uncarbonized organic fibres, notably aramid and polyethylene fibres. Aramid fibres are spun from a solution containing a liquid crystal aromatic polyamide (poly (p-phenylene terephthalamide)). All the commercially available aramids exhibit tensile strengths of about 3 GPa, with modulus varying from about 60–120 GPa. Low modulus fibres (e.g. Kevlar 29) are used in ropes, cables and ballistic protection. Higher modulus fibres (e.g. Kevlar 49) are used in high performance epoxy composites. Aramids have poor compressive properties and are frequently used in hybrid constructions with glass or carbon to compensate for this. The creep properties of aramid fibres, however, are very good. Because of the low density of polyethylene (0.95 Mg m^{-3}), the specific tensile properties of polyethylene fibres are among the highest achieved by any organic material. The fibres have poor compressive properties, however, and they are limited to an upper use temperature of about 100°C, with poor creep characteristics. Examples include Spectra (Allied Signal) and Dyneema (DSM).

Ceramic fibres have also found application in composite materials, but mostly with ceramic and metal matrix composites. Examples include α-Al$_2$O$_3$ fibres of diameter 10–20 μm. Tensile strengths of 1.5–1.7 GPa have been found, with stiffnesses greater than 400 GPa. Silicon carbide fibres are produced by chemical vapour deposition onto carbon or tungsten cores and have found application in titanium matrix composites.

Thermoplastic composites

Composites manufactured from thermosetting resins such as epoxy and polyester exhibit good mechanical properties and lend themselves to low volume manufacture of large mouldings. However, the materials used have limited shelf life and the cured components cannot be recycled. There are considerable advantages in moving to a thermoplastic matrix reinforced with long or continuous fibre (i.e. not injection mouldable thermoplastics). They can be formed into 'prepregs' with indefinite shelf life, they can be moulded in semi-automatic stamping and thermoforming operations and they can be recycled. They are split broadly into two categories: glass mat thermoplastics (GMTs) and advanced thermoplastic composites (ATCs). GMTs were developed for the automotive industry and initially comprised glass-fibre-reinforced polypropylene. The front end of the Volkswagen Golf is currently manufactured from GMT. ATCs are more specialized materials with high fibre volume fractions and aligned continuous reinforcement. Matrices are typically PEEK or polyphenylene sulphide. These latter materials are finding niche applications in the aerospace, automotive and sports goods markets.

Hybrid composites

It is possible to use mixtures of different fibres in the same composite – such composites are termed hybrids.[12] The Ford GT40 racing car incorporated a web of carbon-fibre weighing 1.4 kg (3 lb) into its glass fibre body, thereby decreasing the weight by 27 kg (60 lb) to 42 kg (92 lb) yet simultaneously increasing its stiffness. The torsional stiffness of the rotor blades in the Aerospatiale SA360 helicopter is increased by a structural skin of carbon fibre over a glass and carbon fibre framework and honeycomb core.

Whilst the stiffness of glass fibre composites can be increased by the addition of carbon fibre, the impact resistance of carbon fibre composites can be increased by the incorporation of glass fibres. It has been suggested that the increase in modulus caused by the addition of carbon fibre to glass fibre composites is greater than that which would be predicted by the law of mixtures – the so-called 'hybrid effect'. The compressive strength of hybrids is similar to that of single fibre composites. It can also be efficient to use a fabric woven with the required fibre mix to form the hybrid composite, so that rows of carbon and aramid fibre, for example, are woven together.

6.4 Cement and concrete

Cement

A cement is a material normally used for uniting other materials: it is initially plastic but hardens progressively over a period of time. In terms of volume production the most important type of cement is Portland cement, so called because it produces a concrete which is supposed to resemble the natural limestone found at Portland in Dorset. It is made by firing a suitable mixture of clay and chalk and then comminuting the resulting calcine to a fine powder. In the North American classification there are five types of Portland cement (ASTM C15):

Type I: Ordinary Portland cement for general construction
Type II: Moderate resistance to sulphate action or moderate heat of hydration
Type III: Rapid hardening cement
Type IV: Low heat cement
Type V: High resistance to sulphate action

North American Type I cement consists mainly of calcium silicate together with a smaller quantity of calcium aluminate. Other cements are available for special purposes, such as ciment fondu (aluminous cement).

New plastic-containing cements

Hydraulic cements are generally accepted as being weak in tension and bending, and particularly susceptible to impact loads, but there is nothing inevitable in the relatively poor mechanical properties of such inorganic materials. Mother of pearl, for example, consists almost entirely of aragonite ($CaCO_3$) and has a flexural strength of about 150 MPa (22 ksi), i.e. about ten times that of cement, and a fracture energy significantly greater than 1 kJ m^{-2}, 50 times that of cement.[13] The mechanical properties of the material are attributed to the remarkable microstructure in which flat aragonite plates are glued together with a tenuous layer of protein polymer.[14]

During the setting of hydraulic cement a large variety of interconnected hydrate structures are produced of tubular, sheet and honeycomb form. There are many discontinuities present in such a complex structure and there is an average pore diameter of ~10 nm. Such discontinuities can only reduce mechanical properties, but it is normally the larger flaws arising from the entrapment of air, or the incomplete packing of the cement particles that gives rise to the presently accepted level of mechanical properties.

New work aimed at developing a hydraulic cement which would be as strong as plastics or aluminium, without using fibre reinforcement, has produced results which indicate that strength depends on the size of the largest flaws present in what is, otherwise, an extremely high-strength system where reinforcement should be unnecessary. The dependence on the largest flaw size is well known in other brittle materials as the Griffith relationship $\sigma = \sqrt{E\,\gamma/\pi c}$, where c is the length of the hole in the material and γ is the fracture surface energy. Strength falls as flaw size rises. The significance of the total porosity or number of flaws is secondary in deciding the statistical chance of a flaw of critical size being in a region of maximum strain. As with other ceramics, there will be a variation in strength and a single value cannot be defined. The Weibull modulus will indicate the variability, as previously discussed (p. 46) This approach differs from the traditional view of the strength of cement and concrete where it would have been associated with total pore volume. Investigation[15] showed that the presence of these limiting flaws was a consequence of three factors – poor rheology of the slurry due to flocculation, inefficient packing of cement grains by the mixing machine, and air entrainment in the mix. By selecting two size fractions of cement powder, ~5 μm and 75–125 μm to give denser packing and adding to a proportion of aqueous polyacrylamide gel, after mixing, pressing and drying (RT) a cement was produced from which larger flaws had been removed – called macrodefect-free (MDF) cement. Reducing the total pore volume in MDF raises the flexural strength so that at 0.5 mm defect size it can reach well over 100 MPa

(14 ksi) for <1% porosity and strengths in excess of 200 MPa (29 ksi) are thought possible.

Having eliminated the gross reasons for low strength it is now possible to consider whether the chemistry of the hydration process can be altered so as to improve the density and bond strengths within the finer microstructure. It is claimed that a new generation of MDF cements are emerging that have useful levels of fracture energy and bond strength, although still not tough by comparison with metals and wood. These cements can be regarded as either cold-formed ceramics (although much tougher) or as inorganic 'plastics' (although with a much higher modulus at 40 GPa than even reinforced plastics). Even high-tension springs have been produced in the material.[13,16] MDF cement sheets could be widely used as a cladding material, competing with steel. A 2 mm thick sheet not only competes in the mechanical sense with a 0.7 mm steel sheet (seven times stiffer in bending, and supporting four times the flexural loading before breaking) but it is also readily embossed with patterns during moulding, it is self-coloured and corrosion-resistant. It also does not burn, has a low thermal conductivity and provides fire protection. The rheology of the MDF mix is so much improved that it will mould extremely well and compete in fast-moulding operations with thermosetting and thermoplastic resins for such items as bottle caps. However, over time the polymeric component will absorb water, soften and swell, resulting in property degradation.

This is clearly an important and rapidly developing area. So far effort appears to be restricted to the use of the idea in cement as such, and not in cement as a constituent of concrete. Here the influence of the aggregates could reduce the advantage and the substantial extra cost might not be justified.

Concrete

A true concrete is mixed from four components: cement, sand, stones and water. The sand is sometimes called fine aggregate and the stony

Figure 6.12 Relationship of compressive strength to free water–cement ratio. (Taken from Orchard.[19])

also need to resist chemical attack or possess a low thermal conductivity.

The strength of concrete depends upon the fact that when the cement and aggregate are mixed with water the constituents of the cement are hydrated and progressively crystallize as previously described to form a gel which surrounds the particles of aggregate and binds them together to produce a conglomerate.

It is generally considered that the strength of a moderate-strength concrete is determined by the water-cement ratio, the strength increasing as the latter decreases (Figure 6.12). According to Johnson[17], when the water-cement ratio is higher than about 0.7 the gel contains microchannels which may remain open after curing so that the concrete will be highly permeable. Such a concrete would be unsuitable for most structural purposes. In practical terms, therefore, for high strength and low permeability the water-cement ratio should be low. Unfortunately, if a concrete mix is to be easily transported and properly worked into the mould cavity it must possess adequate workability: since the easiest way of increasing workability is to add water, there is a conflict between good strength and good workability (Table 6.9). Good workability is especially required when reinforced concrete must contain a high density of reinforcement.

It is possible to measure workability in terms of the work required to achieve adequate compaction and where the requirements of strength

constituents coarse aggregate. For some purposes, concretes are made without the coarse aggregate: mixes of this sort are properly called mortars but may occasionally be incorrectly referred to as cements (as in ferro-cement).

Concrete is required to be strong, free from excessive volume changes and resistant to penetration of water. In special circumstances it may

TABLE 6.9 Strengths of concretes with crushed aggregate on bases of equal workability

Proportions	Quantity of cement (kg/m³ of concrete)	Water/cement ratio	28-day strength MPa	ksi
1:1½:3	359	0.4	60.5	8.8
		0.5	47.0	6.8
1:2:4	289	0.4	60.5	8.8
		0.5	47.0	6.8
		0.6	36.0	5.2
1:3:6	197	0.6	36.0	5.2
		0.7	28.0	4.1

Data from Orchard.[19]

dictate the use of low water-cement ratios mechanical compacting or vibrating equipment must be used: consistency of properties often requires also the addition of chemical plasticizers to assist workability. Although the aggregate has no direct effect upon strength in normal concrete, it does influence workability. The traditional method of employing standard mixes ranging from 1:1:2 to 1:4:8 of cement:sand:aggregate is hardly adequate in modern times. Proper control of workability requires the coarse and fine aggregates to be combined in proportions determined by their individual particle-size gradings: a 1:2:4 mix would then be described as a 1:6 mix by weight. It is the total surface area of the aggregate which is important in determining workability. The procedure for mix design is first to determine the water–cement ratio that will give the required strength and then determine the actual water and cement additions from considerations of workability and nature of aggregate. The modern British method of mix design is described in a Government publication[18] and this, as well as older methods, is discussed by Orchard.[19]

In high-strength concretes the water–cement ratio is no longer the sole factor influencing strength: the aggregate–cement ratio and the nature of the aggregate must be taken into account.

The strength of concrete develops over a period of time (Figure 6.12) and data are only meaningful when related to the time which has elapsed after casting. The 28-day strength is often employed as a standard parameter.

Apart from concretes for special purposes, it is possible to visualize three main types of concrete:

(1) bulk concrete used for foundation work, dams, retaining walls, etc.; strength is not of very great consequence in these applications – impermeability and freedom from cracking are more important;
(2) normal or general-purpose concrete having a 28-day crushing strength from 20 to 40 MPa (3–6 ksi);
(3) high-strength concrete used for prestressed work. The 28-day crushing strength varies

from 40 MPa (6 ksi) upwards. With ordinary cement it is difficult to obtain crushing strengths much greater than 70 MPa (10 ksi) but Gerwick, referring to marine structures,[20] considers that for special purposes 100 MPa (15 ksi) is attainable.

Concrete is much less strong in tension than it is in compression – Orchard[19] states that reinforcement is indicated if tensile stresses greater than 0.7 MPa (0.1 ksi) are generated – and strength is therefore commonly described by subjecting a cylindrical specimen to a compressive axial load and recording the crushing strength. However, concrete for walkways and decking is used in the form of beams and for such applications it is often tested in flexure: the outer fibre stress calculated from simple beam theory gives the modulus of rupture which is about 50% higher than the true tensile strength but, of course, much lower than the compressive strength; perhaps only a tenth.

Reinforced concrete

Since the tensile strength of concrete is normally regarded as negligible, concrete used for structural members which are subjected to tensile stresses such as beams, must be reinforced so that the whole of the tensile load is taken by the reinforcement. It is possible to distinguish several concrete-based materials according to the nature of the reinforcement.

In conventional reinforced concrete the reinforcement consists of fairly widely-spaced bars of steel of diameter 16–40 mm (⅝–1½ in). As the concrete sets it shrinks and grips the steel so that the direct tensile strength of the composite can be put equal to that of the reinforcement in the direction of the load, providing the latter is surrounded by a sufficient thickness of concrete. The cover should not be less than 12 mm (½ in). In the early days, the reinforcement was of plain bars of mild steel. It is now more common to use higher-strength steels, deformed so that surface protrusions aid the interfacial bonding between steel and concrete. In conventional reinforced concrete this bond is established when the steel is unstressed and there is therefore an upper limit to the useful strength of the reinforcement. This is because under load, assuming 'no-slip' at the interface, the concrete is compelled to extend to the same extent as the steel, which it cannot do without fracturing because the fracture strain of concrete is less than that of steel. Extensive cracking of the concrete therefore occurs in the vicinity of the concrete/steel interface. The higher the stress in the steel the wider are the cracks, but since these must not be allowed to extend to the outside surface of the concrete the stress in the steel must be limited.

The effectiveness of reinforced concrete as a structural material derives from two fortunate circumstances: one, that steel and concrete have very similar coefficients of thermal expansion, and two, that the passivating effect of the concrete environment surrounding the steel inhibits corrosion of the latter.

Prestressed concrete

The effectiveness of the reinforcement can be improved by prestressing methods. In these techniques the concrete, prior to entering service, is subjected to a permanent compression which is applied by tendons of high tensile steel (BS 5896 or ASSTS 416) loaded in tension and anchored to the concrete. The magnitude of the precompression should be similar to that of the anticipated tensile stress. Then when the composite is subjected to tension in service the effect is to unload the precompression and if the design is good the concrete itself is never put into tension. Prestressed concrete is normally given a lower factor of safety than normal concrete, but since the tensile properties of the cold drawn steel are greatly lowered by heating, fire resistance has to be ensured by a suitable level of thermal insulation.

Prestressed concrete can be produced in two ways. In pre-tensioning methods the tendons are stressed before the concrete is cast. In posttensioning methods, the concrete is cast with preformed ducts through which the tendons are threaded and loaded when the concrete is strong enough. The concrete used should be of the highest quality, viz a high compressive strength, low shrinkage, high modulus of rupture. Values of 40 MPa for the compressive strength and 4.5 MPa for the modulus may be specified as minima.

Ferro-cement

This is a form of reinforced mortar which is sometimes used for the manufacture of ships' or yachts' hulls. The reinforcement commonly, though not exclusively, consists of one or more layers of fine steel mesh. In large structures additional strength may be provided by rods but the character of the reinforcement is determined by the mesh, the fineness of which controls the size of the microcracks that form under tension. The mortar is applied and finished off by hand methods identical to those used by plasterers in house construction. The method is thus highly labour-intensive and considerable skill is required since the cover of the mortar over the reinforcement may be as little as 2 mm (0.08 in).[21] Because of this small cover, the reinforcement may need to be galvanized.

Glass-reinforced cement (GRC)

Incorporating glass fibres into a cement and sand mix can increase the tensile strength of ordinary cement. The alkaline content of cement can cause surface defects on ordinary glass, resulting in fracture of the fibres, which meant alkaliresistant glass had to be developed. The weight fraction of glass fibre used is typically 3–6% with fibres between 10 and 40 mm long. GRC can be cast for such applications as lamp standards and drainage pipes or sprayed, and it has been developed for a structural panel system.

'Smart' concrete

With the large amount of concrete used in the infrastructure of modern communities, it is important that its condition should be monitored.

There is currently much interest in the use of optical fibres and carbon fibres in concrete to determine changes in stress patterns.

6.5 The strength of wood

Although the stems of many plants can be described as 'woody', wood as a constructional material is obtained solely from the main structures of large trees. It is therefore a natural material with the economic disadvantages of requiring time and space to grow and having to be harvested, but needing little energy input other than natural sunlight. Woods can be divided into hardwoods and softwoods. Softwoods grow quickly, hardwoods grow slowly. This is fortunate for the building construction industry, which makes much use of deal (Scots pine, spruce) but hardwoods are often in short supply. There is, however, a developing possibility of genetic control to produce woods which combine some of the hardwood properties with more rapid growth. The hard/soft distinction is, in fact, botanical (angiosperm and gymnosperm) and, for example, even though balsa seems soft, it is classed botanically as a hardwood. In physical terms, hardness is related to the larger proportion of space filled by material – most usually associated with the botanical hardwoods.

The most noteworthy feature of the strength of wood is its anisotropy. This is because one of its functions in nature is to conduct fluid vertically upwards, and much of the structure of wood therefore consists of stacks of vertically elongated, roughly parallel, hollow cells called fibres, fracheids, parenchyma or vessels. The structure is thus essentially a cellular composite with the majority of cells, typically ~1 mm long and 100 μm diameter, aligned. A further degree of anisotropy is introduced by the varying growth density associated with seasonal changes in climatic conditions – the annular rings. Less usefully now than when natural 'crucks' were employed for building timbers, flaws or 'knots' occur in the aligned growth due to branches. The cells themselves have composite walls made of sheets of another composite – cellulose fibres in a hemicellulose matrix. Ultimately the strength derives from the cellulose fibres. Sometimes wood tissues may store crystals of calcium oxalate or even silica with adverse effects on the working qualities.[22]

Because of the voids in the structure, wood is easily crushed normal to the grain. The fibres are fairly easily separated so that tensile strength normal to the grain is also low. The fibres of some trees, notably ash, are straight, having grown in an orderly, parallel fashion, with the result that cleavage occurs very readily. Other types, such as the common elm, exhibit a more convoluted grain and these are less easily split. The best strength of wood is developed in tension, parallel to the grain, but the resistance to longitudinal compression is rather poor since the cells then fail by buckling at a stress which is possibly only one-third of the longitudinal tensile strength. In fact, the applications of wood rarely utilize its tensile strength directly. As a structural member in building etc., it is most frequently stressed in a bending mode in the axial plane, or compressed normal to the grain. Wood is, in fact, one of the few materials where $\sigma_{comp} < \sigma_{tensile}$, and where high toughness is available with high strength.

Although woods vary widely in density, from balsa (0.1 tonnes/m^3) to lignum vitae (1.25 tonnes/m^3), the basic wood substance has a density, constant for all species, of about 1.5 tonnes/m^3. The variation is accounted for by the amount of space held within the hollow cells. The mechanical properties of woods are roughly in proportion to density as indicated in the table comparing the strengths of timber, Table 6.10 (Dinwoodie[23]). This applies both in the green and dried state, and the shrinkage which accompanies drying is associated with increase in strength.[23] The horizontal portion of the curve shown in Figure 6.13 equates to the amount of water before shrinkage commences.

Whilst wood is an important structural material, its extreme anisotropy makes it a difficult material to use for general engineering purposes; it is also subject to fungal attack and the

TABLE 6.10 Strength and stiffness of certain timbers (Dinwoodie[23])

Timber	Density tonnes/m^3	Bending (modulus of rupture)		Compression (maximum strength parallel to grain)		Impact (maximum drop of hammer)	Young's modulus	
		MPa	ksi	MPa	ksi	m	GPa	10^3 ksi
Softwoods								
Western red cedar	0.37	65	9.4	35.0	5.1	0.58	7.0	1.0
Scots pine	0.51	89	13	47.4	6.9	0.71	10.0	1.5
European larch	0.56	92	13	46.7	6.8	0.76	9.9	1.4
Hardwoods								
Balsa	0.18	23	3	15.5	2.2	–	3.2	0.5
Obeche	0.35	54	8	28.2	4.1	0.48	5.5	0.8
Beech	0.69	108	16	51.8	7.5	0.99	10.1	1.5
Greenheart	0.99	181	26	89.9	13.0	1.35	21.0	3.0

Figure 6.13 Relationship of longitudinal compressive strength to moisture content of the timber. (Taken from Dinwoodie.[23])

hemicellulose matrix is sensitive to water, leading to shrinking, swelling and warping. On the other hand cutting is easy, it is readily joined mechanically or by adhesives, and fire resistance is better than might be expected since it holds strength while charring. Hornbeam and applewood have been used historically for gearwheel cogs and screws in watermills; lignum vitae has self-lubricating properties which suit it for use as propellor shaft bearings in the stern tubes of ships. Wood is also used for ladders – hemlock for the stiles, oak for the rungs – but is rapidly being superseded by aluminium alloys for this application. Even in developed countries wood is still used for structural purposes in domestic buildings. However, its use in major structures such as bridges and ships is now limited to more primitive locations.

In plywood, planar anisotropy is reduced by gluing together thin sheets of wood with the grain alternately at right angles. This produces an excellent sheeting material. The world production of timber and timber products is about 10^9 tonnes, surprisingly similar to the annual production of iron and steel. In 1974 the value of timber and timber products (excluding pulp and paper) imported into the UK was £850 million and constituted 3.7% of total imports.[24]

6.6 Materials selection criteria for static strength

Strength is an extensible property. This means that the strength of a component can be increased simply by making it larger – this is not true of other properties such as toughness or weldability. Thus, when different load paths in a compound structure, such as a building truss, are subjected to different loads, it is often more convenient to cope with the higher loads by increasing the cross-sectional area of the members involved, so as to avoid the problems of identification which arise when materials of different strength are incorporated in the same design. In some injection-moulded thermoplastic applications the need for increased strength may be accommodated by local thickening of sections and a similar principle can be applied in fibre-reinforced plastic construction when the laminations making up the thickness of a member can vary from say, three to forty plies. Clearly, then, the materials engineer must not automatically reach for a higher-strength material when designing for higher loads.

Nevertheless, a large part of the explosion in materials science research which occurred during the first half of the present century was devoted to producing materials of increased strength and it is necessary to enquire into the purpose of this effort.

There are three reasons for seeking to employ materials of increased strength: (a) decreased volume, (b) decreased weight, and (c) decreased cost. The first of these will apply when stowage space is critical, as when, for example, engineering parts are required to fit into a very small space. It may be noted that the converse sometimes applies because there are many applications when the requirement is that the object shall be of a given shape and size which is not at a minimum. For example, pressure-activated devices may need to operate at low pressure and emergency stop-pushes must be large enough to avoid damage to operators' hands or feet. The dimensions of a toy, or even a lathe bed, are functions of shape and size requirement. In this sense, increased volume implies low-strength materials. Decreased weight is important especially in applications in which transport is involved – so frequently referred to in aerospace application but significant also in transport by water and by road. Cost is always important because with the possible exception of some extreme performance-oriented applications it is

the basis on which all final decisions will be taken.

The structural efficiency of a structural member can be taken as the ratio of the load which the member will support and its weight. Thus, consider a simple tension member, i.e. a tie, of cross-sectional area A and length L. Then structural efficiency = (Load/Weight) = $(P/LA\rho)$ where ρ = density. Now $P = \sigma_{DS}A$ where σ_{DS} is the allowable design stress, given by the yield stress σ_{YS} divided by whatever load factor, f, is permitted by considerations of safety or relevant Code of Practice. Thus

$$\text{Structural efficiency} = \frac{P}{LA\rho} = \frac{\sigma_{DS}A}{LA\,\rho}$$

$$= \frac{\sigma_{YS}}{\rho}\cdot\frac{1}{fL}$$

Since f and L are design constants the materials selection criterion for maximum structural efficiency is σ_{YS}/ρ, sometimes called the strength-weight ratio, since it refers to the condition of minimum weight. If the cost of the material per unit mass is C_R then $C_R\,\rho/\sigma_{YS}$ is the cost of unit length of a bar having a cross-sectional area sufficient to support unit load. Although derived for a simple tie-bar these expressions apply generally to any member in which dimensional or geometric design changes do not alter the efficiency with which the material is bearing its load. This will apply to any member whose cross-sectional area is required to sustain a uniformly distributed load. Thus, consider a thin-walled pressure-vessel of diameter, D, and thickness, t, required to sustain an internal pressure, p. The largest membrane stress will be the hoop stress, σ_H, given by $\sigma_H = pD/2t$. As before,

$$\sigma_H = \frac{\sigma_{YS}}{f} \qquad \therefore t = \frac{pDf}{2\sigma_{YS}}$$

But the volume of material in the vessel =

$$\pi DtL = \frac{\pi D^2 Lpf}{2\sigma_{YS}}$$

and its weight is

$$\frac{\pi D^2 L p f \rho}{2\sigma_{YS}}$$

The only variables in this expression are σ_{YS} and ρ: reasoning as before produces the same materials selection criteria.

Different criteria apply when the geometry of loading is more complex.

Consider the case of a cantilever beam of length L and rectangular cross-section of breadth b and depth d, required to support at its end a load P. The structural loading efficiency of a beam can be defined as the ratio of sustainable load to weight of beam. Structural loading efficiency =

$$\frac{\text{Load}}{\text{Weight}} = \frac{P}{Lbd\rho} = \frac{M}{L^2bd\rho}$$

where $M = PL$ is the maximum bending moment.

The maximum fibre stress $\sigma = M/Z$ where Z is the section modulus of the beam, $bd^2/6$. Thus $M = \sigma Z = \sigma bd^2/6$ and therefore

$$\frac{\text{Load}}{\text{Weight}} = \frac{M}{L^2bd\rho} = \frac{1}{6}\frac{\sigma}{\rho}\frac{d}{L^2}$$

But

$$d = \left(\frac{6M}{b\sigma}\right)^{1/2} = \left(\frac{6PL}{\sigma b}\right)^{1/2}$$

Therefore

$$\frac{\text{Load}}{\text{Weight}} = \left(\frac{1}{6b}\right)^{1/2}\frac{\sigma^{1/2}}{\rho}\left(\frac{P}{L^3}\right)^{1/2}$$

Thus, if the designer exercises his option to hold b constant and treat only d as a variable, this gives $\sigma^{1/2}/\rho$ as the materials selection criterion rather than σ/ρ as previously. The term $(P/L^3)^{1/2}$ is termed the structural loading index.

The effect of the stress exponent is to bring materials of different strength closer together and make it easier for the designer to consider

materials of unfavourable σ_{YS}/ρ ratio if they possess other properties of value.

If $C_R \rho/\sigma_{YS}$ is compared for a selection of materials it will be found that if strength were the only property of interest everything would be made either of structural steel or reinforced concrete, since these are the materials that can supply strength at the lowest cost. The reason so many other materials are in constant use is that other properties are important such as corrosion resistance (copper, aluminium, polymers), temperature resistance (titanium, tungsten, nickel), ease of fabrication (zinc, polymers) and so on. The nature of a given design will indicate to the designer whether he should consider material of low, medium or high strength. Having established the area of interest the initial selection will be made not on the basis of strength alone but rather on that combination of strength with whatever other properties are deemed important, and which can be obtained at least cost. The final arbiter will be least total cost of the completed component.

References

1. *Copper Data*: Publication TN20. Copper Development Association.
2. J. E. TRUMAN: *Metallurg. Mater. Technol.*, February 1980.
3. E. J. DULIS: in *High Alloy Steels*. Iron and Steel Publication No. 86.
4. A. M. HALL: ASTM Spec. Tech. Pub. No. 498, 1971.
5. A. G. HAYNES: *J. Roy. Aeronaut. Soc.* August 1966, **70**, 766.
6. P. C. POWELL: *Selection and Use of Thermoplastics*. Engineering Design Guide No. 19. Design Council/OUP, 1977.
7. R. M. OGORKIEWICZ: *The Engineering Properties of Plastics*. Engineering Design Guide No. 17. Design Council/OUP, 1977.
8. P. I. VINCENT: in *Thermoplastics – Properties and Design* (ed.: R.M–Ogorkiewicz). John Wiley, 1974.
9. M. C. HOUGH and R. DOLBEY: *The Plastics Compendium, Volume 1*. Rapra Technology Ltd, 1995.

10. C. S. SMITH: in Proc. Conf. 'Composites-Standards, testing and design'. NPL, Teddington, April 1974.

11. B. HARRIS: *Metallurg. Mater. Technol.*, February 1981.

12. J. SUMMERSCALES and D. SHORT: *Composites*, July 1978; **9**, 157.

13. J. D. BIRCHALL, A. J. HOWARD and K. KENDALL: *Chem. Br.*, 1982; **18**, 86.

14. J. D. CURRY: *Proc. Roy. Soc., London*, 1977; **B196**, 443.

15. J. D. BIRCHALL, A. J. HOWARD and K. KENDALL: *Metallurg. Mater. Technol.*, 1983; **15**, 35.

16. J. D. BIRCHALL, A. J. HOWARD and K. KENDALL: *J. Mater. Sci. Lett.* 1982; **1**, 125.

17. R. P. JOHNSON: *Structural Concrete*. McGraw-Hill, 1967.

18. DEPARTMENT OF THE ENVIRONMENT: *Design of Normal Concrete Mixes. HMSO*, 1975.

19. D. F. ORCHARD: *Concrete Technology*, Vol. 1: *Properties of Materials*. Applied Science Publishers, 1979.

20. B. C. GERWICK in Proc. Conf. 'Concrete Afloat.' Concrete Society/RINA, 1977.

21. R. G. MORGAN: Proc. Conf. 'Concrete Afloat', Concrete Society/RINA, 1977.

22. H. E. DESCH – revised J. M. Dinwoodie: *Timber, its Structure, Properties and Utilization*. Macmillan, 1981, p. 60.

23. J. M. DINWOODIE: Timber – a review of structure–mechanical property relationships. *J. Microscopy*, May 1975; **104** (1), 14.

24. J. M. ILLSTON, J. M. DINWOODIE and A. A. SMITH: *Concrete, Timber and Metals*. Van Nostrand Reinhold, 1979, p. 106.

7
Toughness

Although strength is traditionally the principal parameter of design it is not the only, or even the most, important property. Hardly ever is a material suitable for use if it possesses just one desirable property – usually it must exhibit a suitable *combination* of properties. In the case of engineering structures it is important that strength be combined with toughness. This is because experience has shown that most service failures at temperatures below the creep range occur not as a result of general plastic distortion but because of fracture at nominal stresses lower than those for general yield. Early design procedures did not take toughness into account to any considerable extent, certainly not in the numerate part of the design procedure, because there was no adequate theoretical basis for doing so. As a result, failures of large structures in the past were not uncommon. That they have not occurred even more frequently is because:

(1) it was common to employ high factors of safety with the result that design stresses were very low;
(2) joining methods were mechanical, such as bolting and riveting, so that failure of one part of a large structure did not necessarily develop into failure of the whole structure;
(3) the use of fairly thin material allowed a chance for stress concentrations to be unloaded by localized plastic deformation.

This situation has changed progressively over the years. The increasing use of precise computer designs has been accompanied by lower factors of safety; welding is now the most important method of joining; material tends to be thicker. For example, the pressure vessel of the American pressurized water nuclear reactor is manufac-

tured by welding plates up to 304 mm (12 in) in thickness.

In fact it is possible to provide the required toughness because material that is tough at all service temperatures is available. The problem is to know

(1) what level of toughness is required to ensure satisfactory performance at reasonable cost, and
(2) how to specify it in a numerate manner. At a given strength level increased toughness means increased cost and the designer does not wish to specify more toughness than is required.

It is only recently that good progress has been made in devising reliably numerate methods for assessing toughness.

7.1 The meaning of toughness

Toughness is resistance to fracture. Absence of toughness is denoted by the term brittle and when a material can be induced to fracture with the expenditure of little effort it is so described. The effort expended can be thought of in terms of stress or energy giving different but equally valid ways of looking at the fracture problem, as in Table 7.1. This table shows that fracture can also be categorized in terms of the speed with which it propagates.

Fracture occurs by the advance of a crack, and the micromechanisms of crack advancement are many and varied. The most important micromechanisms of fracture are (1) cleavage, (2) microvoid coalescence, (3) stress-corrosion, (4) fatigue, (5) creep rupture.

TABLE 7.1

	Brittle	Tough
Stress	Fracture occurs at a level of stress below that required to produce yielding across the whole cross-section	Fracture occurs at a level of stress which corresponds to that required to produce yielding across the whole cross-section
Energy	Fracture is a low-energy process	Fracture is a high-energy process
Speed	Fracture is fast	Fracture is slow

Cleavage

This is fracture occurring by separation at crystallographic planes of low indices. It can be demonstrated in body-centred cubic and close-packed hexagonal metals but it is most important when it occurs spontaneously in steels, when it produces the classic form of brittle fracture.

Microvoid coalescence

This is crack advancement by the coalescence of voids produced by the tearing away of second-phase particles from the matrix in which they are set. This form of fracture can occur in all types of materials, and in low-strength materials produces the classic ductile (i.e. non-brittle) type of fracture. However, in high-strength materials it is a low-energy process, which can justifiably be described as brittle fracture even though it is quite different in nature from the classic form of brittle fracture involving cleavage as described above.

Stress-corrosion cracking

This is a form of failure in which much of the energy for crack growth is provided by chemical corrosion reactions occurring at the crack tip. It is a very insidious and damaging form of failure and unfortunately, as yet, is not very well understood.

Fatigue

Fatigue is crack growth induced by cyclic or fluctuating stresses. Although now comparatively well understood, this is still the most important form of fracture and is dealt with in Chapter 9.

TABLE 7.2

Micromechanism	Nature of fracture	
	Brittle	Tough
Cleavage	Likely in steels	Unlikely
Microvoid coalescence	Possible in high-strength materials	Common

The two important static low-temperature mechanisms can be classified as shown in Table 7.2.

Before failure by fracture can occur there must be a defect from which it can be initiated. Since, as described above, ductile fracture initiates from voids produced by separation at high-energy interfaces around non-metallic inclusions or other types of second-phase particles, it follows that resistance to ductile fracture is increased if the volume fraction of second-phase particles is decreased in a given type of material. It follows that the material is more fracture-resistant when it is cleaner. The resistance to fatigue in high-strength materials is also increased by making the material cleaner.

In the case of brittle fracture the initial flaw must be larger than a certain critical size before fast propagation can occur at a given level of stress. The flaw may be a pre-existing defect such as a quenching crack or a welding crack, or it may be produced by microvoid coalescence occurring in an initially ductile fracture. If the flaw is initially smaller than the critical size for fast fracture it may grow to critical size as the result of some process that supplies external energy, such as fatigue or stress corrosion or plastic deformation. A flaw of critical size may be formed by the coalescence of several smaller flaws.

Clearly, it is desirable to ensure as far as possible that there are no pre-existing defects capable of initiating fracture, but this is difficult to do, and under cyclic or fluctuating loading conditions it is in any case fairly certain that even if a defect is not already present when the component enters service one will be produced sooner or later. It is therefore always wise to make the assumption that a flaw of some sort is already present when a component goes into service, and ensure that if it is able to grow in service it can be detected before it attains a size capable of initiating fast fracture at the design stress. The situation is at its simplest when it is possible to use materials of low strength. These are usually inherently tough, and fracture, if it should occur in a static mode, is likely to be ductile, or high energy, slow and soon arrested, except in steels at low temperatures.

The problem is more difficult with high-strength materials, which are likely to be of lower toughness so that fracture is fast and possibly catastrophic. This is especially the case in structures of high stored energy such as pressure vessels or pipe systems containing pressurized, compressible fluids. The stored energy of the system may then be so high that fracture, once initiated, is able to propagate even in a shear mode.

These considerations suggest that the 'useful' strength of structural materials may never exceed some limiting value – perhaps 3000 MPa (435 ksi) since at this level of strength elastic strains may exceed 2% and thus provide a very large reservoir of elastic strain energy for unstable crack propagation.

Many tests have been invented to assess the toughness of a material, e.g. the notched tensile test, the Charpy impact test, the drop weight tear test (DWTT), etc. The early tests were highly arbitrary and merely attempted to imitate the conditions of service that were known to decrease toughness. The toughess of steel is decreased by

(1) decreasing temperature;
(2) increasing strain rate; and
(3) increasing plastic constraint.

Accordingly, the Charpy test is carried out over a range of temperatures at high strain rate using a notched specimen. But the conditions of test are quite arbitrary – only one strain rate is available, the Charpy specimen is too small to provide information directly applicable to thick plate, and the machined notch cannot simulate the conditions that exist in the vicinity of a real flaw. Thus the results obtained from the test do not have any absolute significance and their inter-pretation depends upon correlation with pre-vious service failures. This is unsatisfactory because:

(1) it is not possible to extrapolate from one type of design situation to another; and
(2) the absence of failures does not tell how close to failure a series of apparently successful designs has been.

Despite these limitations, the Charpy test and others serving a similar purpose are still widely used, both for materials evaluation and quality control.

However, what is required is a test which allows the determination of a fundamental mate-rial property which can be incorporated into design calculations. This has recently been ach-ieved with the development of fracture mechan-ics which now provides the basis for the most technologically advanced methods of toughness evaluation.

7.2 The assessment of toughness

There are two main ways of assessing materials for resistance to fracture:

(1) the transition temperature approach and
(2) the fracture mechanics approach.

The first of these methods can be applied only to steels in the lower range of strength. In these materials there is a fairly sharp transition from brittle behaviour at low temperatures to tough behaviour at high temperatures. This transition is generally measured by some form of impact test such as the Charpy test and the test data may be shown either by plotting the energy absorbed in fracture or the appearance of the fracture surface against temperature, giving the diagram shown in Figure 7.1.

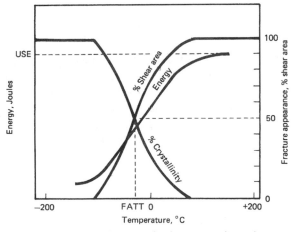

Figure 7.1 Charpy diagram for low-strength steel.

Fracture appearance is described according to the proportions of the fracture surface which appear crystalline or fibrous. The latter is termed the shear area (SA). Thus: % crystallinity + % SA = 100. Specimens fractured at temperatures close to the transition temperature will show a mixed fracture appearance. Fracture of a brittle type is initiated in the region of high triaxial stresses near the root of the notch, and because the propagation of this brittle fracture is accompanied by very little plastic deformation, cleavage in the individual grains of varying orientation causes this part of the fracture surface to have a sugary or crystalline appearance, with bright reflecting facets. As the crack grows outwards towards the boundaries of the specimen it enters regions in which there is less plastic constraint and the mechanism changes to ductile

fracture, i.e. microvoid coalescence, and further propagation of the crack is accompanied by considerable plastic deformation. This part of the fractured surface therefore exhibits a fibrous, silky appearance.

If the transition from brittle to tough behaviour is quite distinct then it is possible to specify a single value of temperature to represent the transition. This transition temperature then gives, for the steel under test, the lower limit of permissible temperature in service. That is to say, if a given steel is to be a candidate material for a given application its measured transition temperature must be lower than the temperature of intended service.

It is a very simple matter to determine the transition temperature of very low-strength steels since the transition between the brittle and tough regions is nearly vertical. However, as the strength of steel increases the transition becomes less sharp and it becomes possible to define the transition temperature in a variety of ways. One way is to determine the temperature corresponding to a given value of energy absorbed. For example, in the early investigations of the notorious 'Liberty' ship failures it was found that satisfactory performance correlated well with an energy value of 15 ft.lb. A disadvantage of this method is that transition energy values are influenced by the yield strength of the material concerned so that as steels are strengthened the transition occurs at higher energy values. It is thus difficult to compare the transition energies of steels having different yield stresses. This disadvantage does not apply to definitions of transition temperature based on fracture appearance. In Figure 7.1 the fracture appearance transition temperature (FATT) is based on 50% SA (shear area). Unfortunately, it is more difficult to obtain an accurate numerical assessment of fracture appearance and it is quite common for specifications to combine both energy values and fracture appearance in their requirements.

The transition approach fails with high-strength materials because the transition becomes so indistinct as to be almost indeterminate. In terms of energies the upper shelf energy (USE) is not much greater than that at the lower, and the

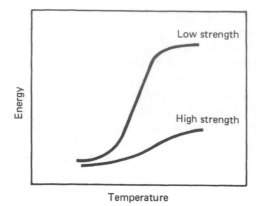

Figure 7.2 Comparison of Charpy energies for low- and high-strength steels.

Figure 7.3 Centre-cracked plate loaded in tension.

difference is spread over such a wide range of temperatures that the concept of transition temperature has hardly any meaning (Figure 7.2).

It is therefore especially with materials of higher strength (non-ferrous as well as ferrous) that the toughness concepts of fracture mechanics have proved so valuable.

7.3 Fracture mechanics

Designing to prevent initiation of fracture requires a knowledge of the relationship between the applied stress and the critical size of the crack to cause unstable propagation. The means of doing this had its origin in the work of Griffith, who proposed that the condition for unstable crack growth in plane stress was given by

$$\frac{d}{dc}\left(-\frac{\sigma^2 \pi c^2}{E} + 4c\gamma_s\right) = 0$$

σ is the stress acting on the end of a plate of unit thickness into which a central crack of length $2c$ is cut so as to lie normal to the direction of σ (Figure 7.3). γ_s is the surface energy per unit length of the crack, and E is Young's modulus.

The first term in the brackets represents the loss of elastic strain energy in the plate due to unloading of material close to the surfaces of the crack. The second term represents the increase in energy required by the crack surfaces. When the rate of increase of the first term is equal to or greater than the rate of loss of the second term then conditions are energetically favourable for propagation of the crack. This gives the critical applied stress for unstable fracture as

$$\sigma_F = \sqrt{\frac{2E\gamma_s}{\pi c}}$$

In the form $\sigma_F \sqrt{c}$ = constant, this relation has been found to hold well for brittle and semi-brittle materials, but for reasonable crack lengths the derived values of γ_s are not realistic for metals. It was suggested, therefore, that a plastic work term γ_p be added to the surface energy term, giving the Orowan-Irwin equation:

$$\sigma_F = \sqrt{\frac{(2\gamma_s + \gamma_p)E}{\pi c}}$$

γ_p is the plastic work done per unit area of crack extension and since it may be 10^3 greater than γ_s it represents the major deterrent to crack propagation. It varies greatly in different materials and is also sensitive to temperature.

The Orowan-Irwin equation may be written

$$\sigma_F^2 \, \pi c = EG$$

The greater is G the higher is the allowable applied stress for a given crack length, and G is thus a measure of fracture toughness. Although it has been called the crack extension force per unit crack length, it really represents the energy released when advance of the crack tip produces unit area of new crack surface. Thus, G can have the dimensions of force per unit length although it may equally be regarded as energy per unit area. The value of G at the point of fracture propagation is denoted by G_C. Neither G nor G_C is a material constant since the value is a function of the plastic zone size at the tip of the crack, which is determined by the conditions of test, principally geometry. G_C is, for example, a function of thickness. When, however, a crack propagates under conditions of plane strain G_C is at a minimum, denoted by G_{IC}, and it may then be regarded as a material constant.

An alternative approach to the assessment of fracture toughness is based on the idea that the effect of a crack is to produce a localized perturbation of the stress field. A concentration of stress is produced at the tip of the crack. The stress field around the tip of the crack (Figure 7.4) can be written:

$$\sigma_x = \frac{\sigma\sqrt{\pi c}}{\sqrt{2\pi r}}\cos\frac{\theta}{2}\left(1 - \sin\frac{\theta}{2}\sin\frac{3\theta}{2}\cdots\right)$$

$$\sigma_y = \frac{\sigma\sqrt{\pi c}}{\sqrt{2\pi r}}\cos\frac{\theta}{2}\left(1 + \sin\frac{\theta}{2}\sin\frac{3\theta}{2}\cdots\right)$$

$$\sigma_{xy} = \frac{\sigma\sqrt{\pi c}}{\sqrt{2\pi r}}\sin\frac{\theta}{2}\cos\frac{\theta}{2}\cos\frac{3\theta}{2}$$

When $\theta = 0$ and $x \to 0$, σ_y is the stress tending to open up the crack. Thus

$$_y = \frac{\sigma\sqrt{\pi c}}{\sqrt{2\pi x}}$$

Figure 7.4 Notation for stress field at tip of crack.

The factor $\sigma\sqrt{\pi c}$ is written K, termed the elastic stress intensity factor. The value of K at propagation of fracture is denoted K_C and is related to G_C, the crack extension force, as follows:

$$\sqrt{\pi c} = K_C \quad \text{and} \quad \sigma^2\pi c = EG_C$$

$$\therefore K_C^2 = EG_C \text{ (plane stress)}$$

For plane strain:

$$K_{IC}^2 = \frac{EG_{IC}}{1 - v^2}$$

K_C is not a material constant since it is determined by geometry, but when the thickness of the specimen is large enough to ensure plane strain conditions K_C has a minimum value for the given material. This is denoted K_{IC}, since crack advancement is then in the opening mode. K_{IC} is therefore a material constant (Figure 7.5). The dimensions of K are stress$\sqrt{\text{length}}$ and are expressed in MPa m$^{1/2}$ or similar.

Figure 7.5 Variation of fracture toughness with thickness.

The application of fracture mechanics to design problems is in principle simple. The equation

$$K_{IC} = \sigma_F\sqrt{\pi c}$$

gives the relationship between three variables, K_{IC}, applied stress at fracture and length of crack. A knowledge of any two of these factors then allows calculation of the third. K_{IC} for a given material is first determined by inserting a crack of known length into a suitable specimen of the material and measuring the applied stress at fracture. Knowing K_{IC}, it is then possible to ensure in service either that the service stress is well below the fracture stress determined for such cracks as may unavoidably be present, or that no cracks occur that are long enough to cause failure at the service stress.

In practice, the problem is not so simple. First, in dealing with specimens and structures of finite size the equation for K_{IC} must be written in the form

$$K_{IC} = \alpha \, \sigma \sqrt{\pi c}$$

where α is a correction factor for the geometry of the flaws and structure. The determination of α in practical cases requires complex mathematical methods. Second, the accurate determination of flaw size is not easy in real structures. Third, the actual level of stress in real structures is not always accurately known. In materials that are too thin to develop plane strain a K_C value appropriate to the given thickness may be employed. The use of centre-cracked panels of thin sheet or plate may enable α to be put equal to unity for the purposes of preliminary evaluation.

The stress field will only adequately be described by linear elastic theory so long as plastic phenomena are negligible in relation to the dimensions of the specimen, i.e. crack length and residual cross-section. But real engineering materials are hardly ever totally brittle – there is usually a zone of plastic material at the crack tip as it advances (Figure 7.6).

If this zone is considered to have a radius r_p then the opening stress σ_y when $x = r_p$ can be put equal to σ_{YS}. But $\sigma_y = K_{IC} / \sqrt{(2\pi x)}$ and when $x \simeq r_p$ then $\sigma \simeq \sigma_{YS}$ and

$$r_p = \frac{1}{2\pi} \left[\frac{K_{IC}}{\sigma_{YS}} \right]^2 .$$

The parameter $[K_{IC}/\sigma_{YS}]^2$ is thus proportional to the plastic zone size. If the linear elastic basis of fracture mechanics is to be applied to toughness testing without unacceptable error then the presence of a plastic zone will impose restrictions on the procedures used. That is, if a measurement of K_{IC} is to be valid, the plastic zone size must be small in relation to the cross-section of the specimen. Generally, it is not known prior to testing whether a given specimen design will provide sufficient constraint to maintain a plane strain fracture mode. The usual practice, therefore, is to perform the test and consider the result provisional, say K_Q. The ASTM then recommend that if it turns out that the value of the expression $2.5 [K_Q/\sigma_{YS}]^2$ is equal to or smaller than the thickness of the specimen then K_Q can be taken as a valid value of K_{IC}.

The expression $2.5 [K_Q/\sigma_{YS}]^2$ is empirical and very conservative, and for tough low-strength materials may demand specimens that are impractically large. The figures shown in Table 7.3 are calculated for a steel of yield stress 350 MPa (50 ksi) and show the minimum thicknesses of a test piece required to give valid data for the K_{IC} values shown.

The inability to obtain valid K_{IC} data is of no consequence if the material is to be used in

Figure 7.6 Plastic zone at the tip of a crack.

TABLE 7.3

K_{IC}		t	
MPa m$^{1/2}$	ksi in$^{1/2}$	mm	in
30	27	18	0.7
60	54	73	3.0
90	82	165	6.5
120	109	294	12.0

thicknesses small enough to permit full-thickness test specimens. For example, many aluminium alloys employed in aircraft are used in thicknesses so much less than the thickness required to give plane strain that K_{IC} determinations would not be of value. Instead, K_C values are obtained from centre-cracked panels of the thickness to be employed in service and used in the same way as K_{IC} values. However, the K_C values thus obtained relate only to material of that thickness.

The problem of obtaining valid data is more serious when the material is to be used in thicknesses so great that specimens cannot be large enough either to provide valid K_{IC} data or match the thickness of the service material. Parts of some nuclear pressure vessels approach 304 mm (12 in) in thickness, and Cowan[1] has described recommendations within the ASME Code for the use of Charpy V and drop weight test data, obtained on standard pressure vessel steels, to obtain a 'reference fracture toughness' K_{IR} curve. This avoids the need for directly measured K_{IC} values but the procedure is, of course, conservative. The minimum thickness for valid K_{IC} measurement decreases as the yield stress of the material increases.

If the fracture toughness equation is written in the form

$$c = \frac{K_{IC}^2}{\alpha^2 \, \sigma^2 \, \pi}$$

it is seen that the crack resistance of a material is proportional to K_{IC}^2. However, this is only true so long as there is no restriction on the value of σ. More generally, the design stress is limited to some fixed fraction of the yield stress, say σ_{YS}/f where f is a design factor imposed by the relevant Code of Practice. Then

$$c = \frac{f^2}{\pi\alpha^2} \left[\frac{K_{IC}}{\sigma_{YS}} \right]^2,$$

so that the crack resistance is proportional to $[K_{IC}/\sigma_{YS}]^2$. This is the same as saying that the crack resistance of a material increases with plastic zone size.

7.4 General yielding fracture mechanics

Whilst the development of linear elastic fracture mechanics allowed enormous strides to be made in the understanding of toughness, to the extent that this property can now be unambiguously measured in many materials, it is unfortunately true that the great majority of structural and pressure vessel steels are not amenable to this discipline. This is because in most cross-sections of these generally rather tough materials there is too much plasticity at the crack tip for the assumption of linear elasticity to be sound and since, as previously discussed, the ASTM criterion then demands excessively large section thicknesses for K_{IC} testing to be valid, there is a need for the theory to be extended into the elastic-plastic regime.

In the presence of gross plastic yielding it is possible to say, following a suggestion by Wells[2], that the parameter which determines crack advancement is the level of strain at the crack tip, and further that for a given material this level of strain will be the same in specimens of different size and therefore varying plasticity. This means that relatively small laboratory specimens, which may yield extensively, can be used to predict the behaviour of large structures in which plasticity is more restricted. The strain at the crack tip is assessed by measuring the displacement, δ, between opposite crack surfaces at the crack tip (COD). The critical value at fracture, δ_c, can be regarded as a material constant. The higher is δ_c the greater is the toughness of the material.

In cases where the crack tip plasticity in the application is not too extensive a quantitative treatment is available in which the fracture criterion is given by[3]

$$\delta_c \geqslant \frac{\beta K_1^2}{E\sigma_{YS}}$$

where β is a constant taking values between $\frac{1}{2}$ and 1, and σ_{YS} is the yield stress.

The unambiguous determination of δ_c presents difficulties of experimental technique and interpretation, and the J–integral criterion introduced by Rice[4] may overtake crack-opening-displacement as a means of assessing toughness in yielding bodies. The J–integral is strictly a path-independent energy line integral which surrounds a crack tip but it is more conveniently interpreted as measuring the rate of change of potential energy with respect to crack length. This must be determined from load-displacement measurements. Fracture occurs when the value of J equals or exceeds the critical value J_c. In an elastic body J becomes equal to G, the energy release rate. The determination of J can require rather tedious graphical integrations but it avoids the difficulties of interpretation associated with the measurement of crack opening displacement.

7.5 Toughness in polymers and adhesives

An important aspect promoting the use of polymers in engineering is their toughness. In the past, the fracture behaviour of polymers and adhesives has usually been assessed by a range of largely unconnected empirical tests such as impact and falling weight. Whilst such tests have their uses in a selection context for ranking, they do not provide quantitative information with which to compare materials or enable the prediction of long-term performance.

As polymeric materials have become increasingly used in engineering as, for example, in the gas pipes described on page 162 or in the polymer matrix composites and adhesives used in aircraft, it has become important to understand fracture behaviour in both failure analysis and design contexts and to be able to select for required levels of toughness on a quantitative basis.

There are special physical characteristics of polymers which cause difficulties in applying the previously-described linear elastic fracture mechanics or general yielding fracture mechanics, as employed for ceramics and metals. Polymers differ in the degree of visco-elasticity exhibited; where this is low, then LEFM (Linear Elastic Fracture Mechanics) can be acceptably applied to produce definitions of toughness in terms of G (p. 80) and K_C (p. 81). Where there is visco-elasticity, however, G and K_C will be time-dependent and may be expected as functions of crack speed in order to predict long-term behaviour[5].

As regards general yielding fracture mechanics there are also particular difficulties in its application to polymeric materials. As would be expected, the yield criteria relating to shear may need modification for time-dependence but, additionally, the shear strain at the crack tip will not be the only mode of deformation and energy dissipation involved.

At a molecular level all polymers are anisotropic, often complex structures and failure may occur through either chain slippage or chain scission at very different levels of stress. Failure will, therefore, be much influenced by chain orientation, degree of mutual entanglement and cross linking and, additionally, by the detailed organization of the structure – spherulites, microfibrils, crystal lamellae, defects and voids.

In particular, in many glassy and semi-crystalline polymers (e.g. PC, PS, PMMA and PE, PP), large numbers of straight silvery zones traversed by fibrillar matter, which are called crazes[6], develop at stresses well below the breaking stress. At higher strains many polymers (e.g. PVC, PE, PP) have a tendency to deform by opening up small voids, appearing white, throughout the whole volume (stress-whitening, intrinsic crazing). Crazing may be initiated by surface defects and enhanced by active environments where a 'crazing agent' may be able to diffuse into the polymer, increasing chain mobility and thus the initiation of crazing. The impact strengths of polymers such as polystyrene have been improved by a number of methods including: the use of high molecular weight grades, the use of plasticizers, the addition of reinforcing fillers, copolymerization and addition of elastomers.

It should be noted that rubbers are often both highly elastic and very non-linear, but providing G is evaluated for the non-linear system, standard fracture mechanics can be applied. There is, in fact, a well-developed body of knowledge of fracture in relation to the use of rubber for tyres and springs.

7.6 Materials selection for toughness

It was formerly traditional in engineering practice to design initially on the basis of strength or stiffness and to disregard toughness unless it presented itself as a problem. Unfortunately, when it did this it was generally in the form of a large engineering failure: modern practice therefore is to regard toughness as having an importance at least equal to, and possibly greater than, other mechanical properties. As always, the selection problem has two aspects – that of the design as a whole and that of the material.

When considering the design as a whole, it is generally true that large constructions and thick sections are less tough than small parts and thin sections. This is because of the greater plastic constraints which apply in the former cases. It may, for example, be necessary in large ships to use higher-strength steels because ordinary mild steel would require sections so thick (designed on a strength basis) that the resulting plastic constraints would be sufficient to spoil its normal toughness. But within a given class of materials there is generally an inverse relationship between strength and toughness (Figure 7.7), so that unless the toughness is properly matched to the design the use of a high-strength material may introduce problems due to inadequate *material* toughness.

To some extent, selection procedures are easier with high-strength materials because, provided the property data are available, fracture mechanics calculations can be incorporated into the design process. With low-strength material, until COD or J–integral measurements become better established, it is necessary to employ ranking parameters of toughness obtained from, for example, Charpy or drop-weight tear tests (DWTT). Since these cannot be assimilated into calculations involving design stresses the

Figure 7.7 Variation of fracture toughness with yield strength for various classes of high-strength steels. (From J. F. Knott.)[8]

designer must rely upon his experience and judgement, together with a judicious reading of specifications.

The concern with low-strength materials is confined mainly to structural steels, since low-strength non-ferrous materials are not readily susceptible to any of the forms of brittle fracture, and therefore specifications for structural steels offer grades with varying levels of toughness expressed in terms of the standard tests. The difficulty is to know what level to specify for a given application. It has been stated[7] that the required Charpy value for a structure is proportional to the square of the yield stress and to the section thickness. Standard specifications, however, cannot offer for each steel a complete range of tabulated Charpy data for all likely ambient temperatures and the preferred approach is to specify minimum values at arbitrarily selected temperatures which are low enough for these values to be discriminating as between one grade and another, but need have no direct relationship with the actual service temperature envisaged. Selection for toughness must then be based upon the designer's assessment of the relative severity of the duty that the material has to perform (meaning mainly temperature and thickness). The level of numeracy in this procedure is unsatisfactory and needs to be improved.

Selection in terms of fracture toughness data can be carried out with more precision, and there is now a considerable amount of such data available. Although this may be reported in terms of G or K the latter is more convenient because specimens of different materials having the same crack length will exhibit the same failure stress if they have the same K_{IC} but not if they have the same G_{IC} (Young's modulus enters into the calculation).

Table 7.4 shows comparative data for various materials. In practice, there is a good deal of

TABLE 7.4. Typical values of plane strain fracture toughness K_{IC}

	Strength		Fracture toughness	
	MPa	ksi	MPa m$^{\frac{1}{2}}$	ksi in$^{\frac{1}{2}}$
Steels				
Medium-carbon steel	260	37.7	54	49
Pressure-vessel steel A533B Q and T	500	72.5	200	182
Maraging steel	1390	202	110–176	100–160
	1700	247	99–165	90–150
	1930	280	85–143	80–130
300M	2033	295	66	60
	1895	275	83	75
AISI 4340	1758	255	77	70
	1930	280	61	55
AISI 4340				
Tempered 200°C, commercial purity	1650	239	40	36
Tempered 200°C, high purity	1630	236	80	73
Aluminium alloys				
2024–T4	346	50	55	50
2024–T851	414	60	24	22
7075–T6	463	67	38–66	35–60
7075–T651	482	70	31	28
7178–T6	560	81	23	21
7178–T651	517	75	24	22
Titanium alloys				
Ti–6Al–4V	830	120	50–60	45–55
Ti–6Al–5Zr–0.5Mo–0.2Si	877	127	60–70	55–65
Ti–4Al–4Mo–2Sn–0.5Si	960	139	40–50	36–45
Ti–11Sn–5Zr–2.25Al–1Mo–0.2Si	970	141	35–45	32–41
Ti–4Al–4Mo–4Sn–0.5Si	1095	159	30–40	27–36
Thermoplastics				
PMMA	30	4	1	0.9
GP Polystyrene	–	–	1	0.9
Acrylic sheet	–	–	2	1.8
Polycarbonate	–	–	2.2	2.0
Others				
Concrete	–	–	0.3–1.3	0.3–1.2
Glass	–	–	0.3–0.6	0.3–0.5
Dense alumina			2.5–3	2.3–2.7
Zirconia toughened alumina			6–12	5.5–11.0
Partially stabilized zirconia			8–16	7.3–14.6
Douglas fir	–	–	0.3	0.3

variation in the reported data which must be attributed to variations within specification ranges for nominally identical standard materials, as well as experimental scatter. It must also be remembered that the usual inverse relationship between strength and toughness[8] means that identical materials heat-treated to different strengths will exhibit different values of K_{IC}. K_{IC} values for specific materials should not be quoted without accompanying values of yield strength, but in general, a K_{IC} value of around 60 MPa m$^{1/2}$ (55 ksi in$^{1/2}$) represents a reasonably tough material. K_{IC} can be used to calculate either fracture stress or crack tolerance but although K_{IC} is a direct measure of toughness provided the design stress can vary freely, it is not an appropriate parameter for selection purposes where the design stress is limited by a code of practice to some fraction of the yield stress. In such cases, as was shown earlier, the critical crack length is given by

$$\frac{f^2}{\pi}\left[\frac{K_{IC}}{\sigma_{YS}}\right]^2.$$

The value of this expression can be illustrated using data given in a paper by Hodge[9], Table 7.5. The final column gives the critical crack length for a stress equal to half the yield stress (i.e. $f = 2$).

Of the three steels A212 has the lowest fracture toughness because of its unfavourable metallographic structure, which consisted of a very coarse-grained ferrite and pearlite. Nevertheless, the steel exhibits a respectable crack tolerance because its low yield stress determines that the permissible design stress will be low. Steel A543

has the best fracture toughness and this is the result of a refined metallographic structure in which the carbide particles are fine and randomly distributed. It also has the best crack tolerance because the increase in K_{IC} is sufficient to accommodate the high permissible design stress. The uniform microstructure of this steel therefore results in a good combination of strength and toughness, a result which is to be expected from a steel designed to have an adequate hardenability. However, some steels – for example, the semi-bainitic types – may have insufficient hardenability when cooled in thick sections, and this often produces objectionable mixed micro-structures which yield an unsatisfactory combination of strength and toughness. Steel A533, when transformed to give the properties shown, has the worst crack tolerance, because although the absolute value of K_{IC} is relatively high it is unable to support the high level of permissible design stress. The greater the section thickness in which a steel must be used, the more difficult it is to provide the required hardenability at acceptable cost.

An example of the use of fracture mechanics applied to fairly thin sheet has been provided by Davis and Quist[10]. They consider aluminium alloy sheets of thickness 6 mm (¼ in) in the form of centre-cracked panels for which $K_C = 1.12\,\sigma\sqrt{\pi c}$ where the crack length is $2c$. The materials data are given in Table 7.6.

It is now necessary to decide upon the smallest length of crack which can be detected with absolute certainty by available non-destructive testing equipment. If it is supposed that this is 30 mm (1.1 in) so that $C = 0.015\,m$, then the

TABLE 7.5

Steel	Structure	σ_{YS}		K_{IC}		$4/\pi[K_{IC}/\sigma_{YS}]^2$	
		MPa	ksi	MPa m½	ksi in½	mm	in
A212	Ferrite-pearlite	283	41	77	70	94	3.7
A533	Mixed transformation products	427	62	95	86	63	2.5
A543	Lower bainite	586	85	181	165	121	4.8

Toughness

TABLE 7.6

Alloy	σ_{YS}		K_{IC}	
	MPa	ksi	MPa m½	ksi in½
2024–T3	276	40	121	110
7178	448	65	41	37.5
2024–T81	400	58	70	64
7075–T73	386	56	82	75

fracture stress is given by $K_C/0.24$. If it is further taken that the design stress is the yield stress divided by the design factor of 1.6 the values for fracture stress and design stress given in Table 7.7 are obtained.

TABLE 7.7

Alloy	σ_{DS}		σ_F		σ_F
	MPa	ksi	MPa	ksi	σ_{DS}
2024–T3	172.5	25	504	73	2.9
7178	280	41	171	25	0.6
2024–T81	250	36	292	43	1.2
7075–T73	241	35	342	50	1.4

Clearly 2024-T3 is safe but would be uneconomic since the fracture stress offers too great a margin of safety over the design stress. 7178 is even more unacceptable since with the given crack length failure would be certain.

2024-T81 might be regarded as provisionally satisfactory although it could be argued that the design factor for fracture should not be lower than that for yield. In this respect 7075-T73 is more satisfactory and might be preferred on this account.

However, this preliminary evaluation is too simple to provide a final selection. It has achieved its purpose in eliminating the first two alloys. The other two alloys remain for more detailed evaluation, which must take into account a wider range of properties including resistance to corrosion, fatigue crack propagation rates, cost, etc.

It should be emphasized that a K_C figure for thin sheet can relate only to material of the thickness on which that figure was determined.

The inverse relationship between strength and toughness found with most materials ensures that if the yield strength of a material is pushed upwards by normal metallurgical methods the critical crack length eventually becomes unacceptably small; i.e. the material is too brittle for

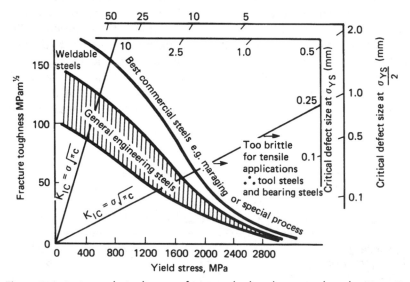

Figure 7.8 Ratio analysis diagram for quenched and tempered steels. (From T. J. Baker[11].)

its purpose if this involves tensile stress. Clearly, a crack length which is too small in one application might be quite acceptable in another, but very small critical crack lengths cannot be allowed in tension-loaded structures. If, for example, the critical crack length for a pressure vessel is less than the vessel thickness then fast fracture is possible; but if the critical crack length is greater than the vessel thickness fast fracture cannot occur because there is insufficient material to grow a crack of that length. In general, therefore, the material chosen for a given application must have such a combination of K_{IC} and yield stress that the critical crack length is appropriate for that application. This length may be only a few millimetres for small-scale engineering applications but many tens of millimetres for large structures. Baker[11] has shown how a conventional ratio analysis diagram (Figure 7.8) can be divided up to show different fields of application according to appropriate crack length.

References

1. A. COWAN: in *Developments in Pressure Vessel Technology.* (ed. R. W. Nicholls). Vol. 1: *Flaw Analysis.* Applied Science Publishers, 1979.
2. A. A. WELLS: *Cranfield Crack Propagation Symposium*, Vol. 1, 1961.
3. R. A. SMITH: *Mater. Eng. Appl.*, 1978; **1**, 121.
4. J. R. RICE: *J. Appl. Mech.*, 1968; **35**, 379.
5. J. G. WILLIAMS: in *Fracture of Non-Metallic Materials*, ISPRA, 1985, eds K. P. Hermann and L. H. Larsson, D. Reidel Publishing, 1987, pp. 227–255
6. H. H. KAUSCH and B. STALDER: ibid., pp. 291–300.
7. A. A. WELLS: *Design in High Strength Structural Steels.* Iron and Steel Institute, Publ. No. 122.
8. J. F. KNOTT: *The Welder*, **41**, No. 202.
9. J. M. HODGE: United States Steel International, Paper No. 26.
10. R. A. DAVIS and W. E. QUIST: *Mater. Des. Eng.*, November 1965.
11. T. J. BAKER: Private communication.

8
Stiffness

Stiffness is the ability of a material to maintain its shape when acted upon by a load. The concept of stiffness in metals is usually approached through Hooke's Law, which is concerned with the relationship between stress and strain (although Hooke's actual terms were load and extension). When a metal is loaded, the stress–strain curve is at first approximately linear and its slope is a measure of the stiffness of the metal. If the loading is in tension or compression the value of the slope is known as Young's modulus, or the modulus of elasticity, denoted by E in the engineering literature; when the loading is in shear it is known as the modulus of rigidity, or shear modulus, denoted by G. These two elastic constants are related through Poisson's ratio, v, as follows:

$$G = \frac{E}{2(1 + v)} \qquad (8.1)$$

Of course, the stress–strain relationship of materials in general is not always linear, and then stiffness must be measured by alternative parameters such as the tangent modulus or secant modulus. This also applies to metals as they start to enter the plastic range.

8.1 The importance of stiffness

There are three reasons why stiffness is important. One is concerned with stable deflections, another with absorption of energy and the third with failure by instability.

Deflections

Deflections increase as stiffness decreases. Consider, for example, the end-deflection, δ, of a cantilever of length l, subjected to an end load P, (Figure 8.1). The deflection δ is given by

$$\delta = \frac{Pl^3}{3EI} \qquad (8.2)$$

where I is the second moment of area of the cross-section of the cantilever. It follows that if two cantilevers – one of aluminium, the other of steel – are constructed to have identical second moments of area, the deflection in the former will be three times as great as that in the latter since Young's modulus for aluminium is only one-third of that for steel. It is not possible to produce any significant improvement in the performance of aluminium, or any other metal, by alloying because Young's modulus is a structure-insensitive property, and microstructural or compositional variation cannot produce more than about 10% variation in either direction. This inability to control Young's modulus within a given material means that if, for some reason, the designer is compelled to use a material of low stiffness he must compensate for this by increasing the stiffness of his structure, i.e. by increasing its second moment of area. The ways in which this might be done,

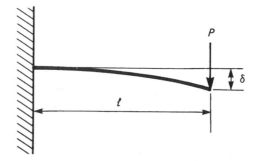

Figure 8.1 A cantilever subjected to an end load.

and the associated implications, are discussed later in this chapter.

Although there is a well-established prejudice against large deflections in massive structures such as ships, bridges and buildings, it is not at all clear that movement of the structure as *a whole* is necessarily harmful. A tall building, subject to windload at the top, can be regarded as a cantilever and the John Hancock building in Chicago, for example, which is 102 stories high, displays a windsway of 40 cm (15.7 in) but there is no suggestion that its overall integrity is thereby threatened. Similarly, it is difficult to see why large deflections in a road or rail bridge, as traffic passes over it, should not be readily accommodated by competent design although the failure by aerodynamic oscillation of the Tacoma Narrows Bridge in 1940 as a result of inadequate torsional stiffness in the bridge deck points to the consequences of failing to take proper account of all possible environmental hazards.

Clearly, where relative motion between adjacent parts in an assembly must be provided then low material stiffness can make design much more difficult or even impossible. Gordon[1] quotes the example of the underground passenger train which was designed to be manufactured in a plastics material. The design study showed that although in the unloaded state operation was satisfactory, when the train was loaded with passengers the sliding doors could not close due to excessive deflection of the main structure. An equally important, though less dramatic, example is presented by long lengths of rotating shafting – correct alignment of the bearings is difficult to maintain if the structure on which the bearings are mounted is of low stiffness.

Problems also arise in complex assemblies which incorporate materials of differing stiffness because there is then the danger that incompatibilities of deformation can lead to local concentrations of stress and ultimately some form of localized failure. Presumably, recently reported occurrences of cladding blocks, and even whole window frames, falling out of tall buildings are related to this sort of effect.

Attempts to save weight by using high-strength materials are also liable to affect stiffness adversely, since although the Young's modulus of the material is not significantly affected by the metallurgical strengthening methods employed, the higher strength allows smaller cross-sections to be employed with a consequent reduction in I, the geometric stiffness. Thin-walled members such as boxes need extra stiffening if they are to carry the full stresses made possible by the use of high-strength structural steels, and this offsets to some extent the savings that are possible.

Energy absorption

When a material is strained it gains elastic strain energy. The energy per unit volume is then equal to the area under the stress–strain curve (Figure 8.2). Energy per unit volume = $\frac{1}{2}\sigma\epsilon = \sigma^2/2E$, where ϵ = strain.

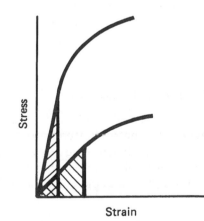

Figure 8.2 Stress–strain curves for two materials of differing stiffness.

Considering a crash barrier required to absorb the kinetic energy of a moving vehicle leaving a roadway, if Young's modulus of the barrier material were reduced by a factor n, then the maximum retarding stress would be reduced by $\sqrt{n}$ which would be good for the occupants of the vehicle (except that elastic strain energy is

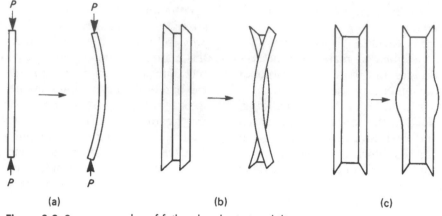

Figure 8.3 Some examples of failure by elastic instability.

recoverable, so that a highly compliant elastic crash barrier would tend to behave like a catapult).*

In less dramatic circumstances the increased deflections encountered in compliant structures are often disadvantageous. In transport vehicles, for example, the soft ride given by excessively compliant shock absorbers can result in more discomfort than the hard ride encountered with stiff ones.

Failure by elastic instability

The simpler methods of stress analysis assume that the overall geometry of a body under load does not change sufficiently to invalidate the analysis. For example, simple beam theory makes the assumption that plane sections remain

---

* Although not strictly germane to the present chapter, Figure 8.2 also shows that the tangent modulus in the elastic-plastic regime is much lower than Young's modulus. A crash barrier which deforms plastically on impact can be designed to provide a still smaller retarding force and the deformation is not recoverable. Some piers and jetties subject to impact by large ships during berthing employ crash stops, in which the energy of impact is taken up by the plastic torsion of steel bars. Unlike elastic stops, the retarding force is nearly constant, thereby minimizing damage to both ship and jetty. After a certain amount of deformation the steel bars must be replaced.

plane. However, it may happen, and this applies particularly to thin, slender bodies or those incorporating cross-sections of high aspect ratio, that twisting or buckling of the stressed body occurs with the result that failure occurs at loads much lower than those predicted by simple theory. Failure by elastic instability can be general or localized, and some examples are shown in Figure 8.3.

8.2 The stiffness of materials

The elastic moduli of materials cover a wide range from diamond, the stiffest material known with a tensile modulus of 1000 GPa (145 × 10^6 psi) to rubbers and plastics at around 0.01 GPa (1.45 ksi). Steel has a tensile modulus of 200 GPa (29 × 10^6 psi) which makes it a very useful structural material but the modulus of aluminium, at 70 GPa (10 × 10^6 psi), is low enough to present problems and nylon, with 3 GPa (0.4 × 10^6 psi), can never find major structural use.

Table 8.1 gives data for the tension moduli of some important materials. The figures commonly quoted for Young's modulus generally refer to normal room temperature: E decreases somewhat with increasing temperature (Figure 8.4a). Young's modulus in metals at room temperature is not time-dependent and is therefore

TABLE 8.1

	Stiffness (GPa)	Density (tonnes/m³)	Materials selection criteria Minimum weight	
			$\dfrac{E^{1/2}}{\rho}$	$\dfrac{E^{1/3}}{\rho}$
	E	ρ		
Concrete (in compression)	27.0	2.40	2.16 [7]	1.25 [7]
Oak: parallel to grain	9.5	0.60	5.10 [2]	3.53 [1]
HM Carbon fibres	400.0	1.95	10.20	3.78
Aluminium N8 alloy	70.0	2.70	3.10 [4]	1.53 [5]
Steel	207.0	7.80	1.84 [8]	0.76 [9]
Glass fibre-reinforced concrete	25.0	2.40	2.10 [7]	1.22 [7]
Glass fibre: 70% resin reinforced plastic mat	10.0	1.50	2.10 [7]	1.44 [6]
Glass fibre: 50% resin reinforced plastic cloth	14.0	1.70	2.20 [7]	1.42 [6]
Glass fibre 20% resin reinforced plastic undirectional	48.0	2.00	3.46 [3]	1.82 [3]
Nylon 33% g.f.	3.5	1.20	1.56 [9]	1.27 [7]
Titanium	116.0	4.50	2.39 [6]	1.08 [8]
Unidirectional graphite-epoxy	137.0	1.50	7.80 [1]	3.40 [2]
45° cross-ply graphite-epoxy	15.0	1.50	2.58 [5]	1.64 [4]
Polypropylene	0.36	0.90	0.67 [10]	0.79 [9]

[1]–[10] = order of merit

not influenced by changes in strain-rate. Young's modulus under creep conditions is strain-rate-dependent.

The behaviour of polymeric materials is very different from that of metals. Not only are they much less stiff, but in consequence of their viscoelastic nature their properties are strongly time-dependent. Stress–strain curves are strain-rate-dependent and stiffness moduli increase as strain-rate increases. In another sense, this

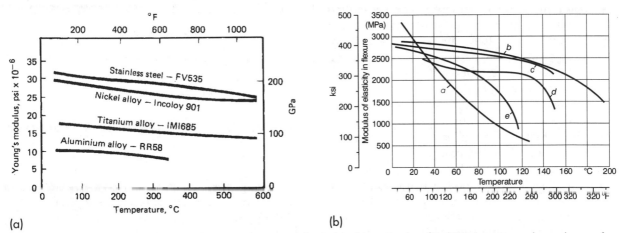

Figure 8.4 (a) The effect of temperature on Young's modulus (taken from Gresham[3]). (b) Temperature dependence of the flexural modulus of elasticity of various engineering thermoplastics[4] a polyacetal, b polysulphone, c polyphenylene ether, d polycarbonate, e ABS, high temperature grade.

93

means that plastics under load become less stiff as time goes by. The stress–strain relationship indicates that plastics also become less stiff with increasing strain (Figure 6.5). However, the effect of temperature on the flexural modulus for a number of polymers is shown in Figure 8.4b.

Temperature has a strong effect on the properties of plastics; so much so that only specialized plastics are used at temperatures above 100°C e.g. polyether ether ketone, polyethersulphone, polysulphone and liquid crystal polymers. The behaviour at temperatures even as low as room temperature of commodity plastics such as polyethylene and polypropylene is therefore best described as creep, and test data should preferably be provided in terms of the creep tests laid down in British Standard 4618 (or ASTM D2990, ISO 899 or DIN 53444). Nevertheless, much data relating to short-term tests are to be found in the literature and these should be used with caution.

One characteristic of polymers is the fact that the strain caused by the applied load is recoverable when the load is removed provided sufficient time is allowed (and provided also that the application of the load did not cause irretrievable damage). This feature must be taken into account when designing for intermittent loading (see Figure 6.4).

Unlike metals, the stiffness of plastics is not independent of microstructure. The crystalline thermoplastics such as polyethylene and the nylons can vary in their degree of crystallinity depending upon the nature of the processing they have received. Higher stiffness is associated with increased crystallinity.

Although Young's modulus for single crystals of metals may be strongly anisotropic, a polycrystalline metal or alloy having its individual grains arranged in a generally random manner will exhibit properties that are approximately isotropic. Plastics, on the other hand, commonly exhibit significant anisotropy on account of the possibility of preferential orientation of the long-chained polymer molecules. Stiffness moduli are higher measured along the direction of orientation than across it.

The stiffness of composites

As shown above, the elastic properties of plastics are inherently inferior to those of metals. On a density-compensated basis, the situation is improved as plastic materials are generally of low density. Reinforcing plastics with strong, stiff fibres such as glass or carbon will improve the mechanical properties, the principle behind composites. The axial stiffness of a composite containing continuous, aligned fibres can be calculated from the properties of its constituent parts using the 'Rule of Mixtures'. The derivation of this equation is given in Figure 8.5. This clearly demonstrates that the composite modulus is dependent on the stiffness of the fibres and the matrix and the relative volumes of each. The tensile modulus of polyester resin and glass fibre composites increases from 5% to 25% that of steel as the fibre content increases from 25% to 80% w/w. Despite the fact that the fibres used are generally more dense than the plastic matrices, the specific properties (specific modulus and specific tensile strength) are still very high.

$$\text{fibre strain} = \text{matrix strain} = \varepsilon_c$$

$$\therefore \varepsilon_c = \frac{\sigma_f}{E_f} = \frac{\sigma_m}{E_m}$$

$$\text{but} \quad \sigma_c = V_f \sigma_f + V_m \sigma_m \quad (V_f + V_m = 1)$$

$$\therefore E_c = \frac{\sigma}{\varepsilon} = \frac{V_f \sigma_f + V_m \sigma_m}{\sigma_f / E_f}$$

$$= E_f \left(V_f + \frac{V_m \sigma_m}{\sigma_f} \right)$$

$$= E_f \left(V_f + \frac{V_m E_m}{E_f} \right)$$

$$\underline{E_c = V_f E_f + V_m E_m}$$

ε	= strain
σ	= stress
E	= Young's modulus
V	= volume fraction

Subscript m	= matrix
Subscript f	= fibre
Subscript c	= composite

Figure 8.5 Composite 'Rule of Mixtures'.

8.3 The stiffness of sections

The most important structural component subjected to bending is the beam.

As a typical example of a beam, consider the cantilever shown in Figure 8.1. Equation 8.2 shows that, as in most equations of this type, E is accompanied by I, the second moment of area of the cross-section.

Now E is a material property whereas I is a geometric property of the design: it is important to distinguish between material properties and design properties because they may be varied independently. We may define a stiff *material* as one with a high value of E, whereas a stiff *design* is one with a high value of I. Thus, if it is desired to use a material with a low value of E because of some other especially favourable property then the designer has the option of overcoming the disadvantage of low material stiffness by increasing the stiffness of the design, i.e. increasing I.

Figure 8.6 shows three sections of equal area:

(a) a square cross-section;
(b) a rectangular cross-section of aspect ratio 3 to 1; and
(c) a hot-rolled steel section from British Standard 1449 chosen also to have an aspect ratio of 3 to 1.

The efficient disposition of material in the hot-rolled section has increased the second moment of area to more than 50 times that for the square and more than 17 times that for the solid rectangle.

Similar considerations apply in relation to flooring or decking which is to be laid on joists or purlins. The simplest method of flooring over joists would be to lay an assembly of simple planks of rectangular cross-section, as shown in Figure 8.7(a). This would not be very efficient, and the joists would need to be quite closely spaced. For the same cross-sectional area, the second moment of area can be doubled if the planks are modified as shown in Figure 8.7(b). However, parts with such large variations in thicknesses are not only difficult to manufacture but may respond to stresses in service in ways that are difficult to predict, and a better solution to the problem is shown in Figure 8.7(c) in which less of the material lies on the centreline of the cross-section and the thickness of the section is more uniform.

An even more efficient type of section, frequently used for roof decking[2] is shown in Figure 8.7(d). It is easy to find dimensions such that the second moment of area of the decking section is twice that of the plain rectangular plate shown at (a), whilst simultaneously reducing the cross-sectional area by a factor of 8.

The resistance to bending of a section can be increased for a given weight by making it hollow. Consider, for example, two hollow sections – one square, the other circular – constructed to have

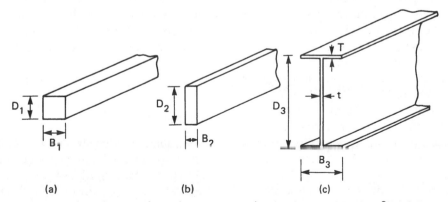

Figure 8.6 Three beams of equal cross-sectional areas (area = 31.40 cm^2): (a) D_1 = B_1, I_1 = 82.16 cm^4; (b) D_2 = 3B_2, I_2 = 246.5 cm^4 (c) D_3 = 3B_2, I_3 = 4381 cm^4

Figure 8.7 (a), (b) and (c): Planks of equal cross-sectional area; (d) decking material.

areas and depths equal to those of the rectangular section in Figure 8.6(b). The required wall thicknesses are one-eleventh and one-eighth, respectively, of the depth, whilst the second moments of area are increased by factors of 1.66 and 1.17 respectively, as compared with that of the solid rectangular section. The square shows the higher increase because more of the cross-section is further from the plane of bending.

It is generally true that the stiffness of a section can be increased by placing as much as possible of the material as far as possible from the axis of bending. The extent to which this has been achieved can be measured by the radius of gyration. This is defined by putting the second moment of area of the section, I, equal to Ak^2 where A is the area of the cross-section and k is the radius of gyration. Thus, although stiffness is increased by both A and k, the square term means that the latter is more effective. Further, the constant need for economy of weight means that it is desirable to hold A constant, or even have it reduced.

There is, however, a limit to which k can be increased with A held constant or reduced, because of the consequent reduction in the thickness of the material. This, carried to extremes, increases the likelihood of failure by instability.

Suppose a cylindrical tube of the type discussed above is to have its second moment of area increased by a factor ρ while the area remains constant. Then, since $I = \pi r^3 t = Ak^2 = 2\pi r t(r^2/2)$ this requires that r be increased by a factor of $\sqrt{\rho}$ whilst simultaneously the thickness of the tube must be reduced by a factor of $\sqrt{\rho}$. However, if I for the tube were to have its second moment of area increased a hundred-fold its wall thickness would be reduced to a tenth of what it was. Such a thin-walled tube, when subject to stress, would be extremely vulnerable to localized buckling.

Failure of a strut

When long slender structural members are subjected to uniaxial compressive loads they are known as struts and failure occurs by overall flexural buckling (Figure 8.3. (a)). The longer and more slender the struts are, the smaller is the failure load.

The standard formula for the failure load of a strut was developed by Euler and can be expressed as follows:

$$\text{Euler buckling load, } P_E = \frac{\pi^2 E I}{l^2} \qquad (8.3)$$

where E = Young's modulus; I = second moment of area; l = length of strut. If I, the second moment area, is written as Ak^2 where A is the area of the cross-section and k is the radius of gyration, Euler's equation can be put into terms of stress:

$$\text{Euler buckling stress, } \sigma_E = \frac{\pi^2 E}{(l/k)^2}$$

The ratio l/k is known as the slenderness ratio of the strut and Euler's equation only agrees with the measured failure load (or stress) of a strut when the slenderness ratio is rather high. When it is very low, i.e. when the strut is short and stubby, Euler's buckling stress becomes greater than the yield stress in compression of the material of which the strut is made, and it is obvious that failure will then occur by crushing in simple compression rather than by buckling. The relationship between the Euler buckling stress, σ_E and the slenderness ratio is shown in Figure 8.8. σ_{YS} is the yield stress.

For steel, Young's modulus E is taken as $200\,\text{GN/m}^2$ and yield stress σ_{YS} as $250\,\text{MN/m}^2$. Then $E/\sigma_{YS} = 800$ and

$$\left(\frac{l}{k}\right)^2 = \frac{800\pi^2}{(\sigma_E/\sigma_{YS})}$$

For aluminium, E is taken as $68\,\text{GN/m}^2$ and σ_{YS} as $230\,\text{MN/m}^2$ so that

$$\frac{E}{\sigma_{YS}} = 300 \quad \text{and} \quad \left(\frac{l}{k}\right)^2 = \frac{300\pi^2}{\sigma_E/\sigma_{YS}}$$

These results indicate (Figure 8.8) that whereas steel starts to buckle rather than yield at a slenderness ratio of 89, aluminium buckles at a corresponding figure of 54, but this idealized relationship overestimates the buckling stresses actually measured in practice. The discrepancy is due to manufacturing imperfections, the fact that no strut is ever perfectly straight, and the difficulty of obtaining precise alignment between the direction of the compressive load and the axis of the strut. Contributions to the theory of imperfections in struts by Perry, Robertson and Dutheil have culminated in the expression for buckling stress given in British Standard 449, from which the curves in Figure 8.9 were calculated for two grades of structural steel. Experimental results lie within the shaded areas

Figure 8.8 Idealized buckling behaviour of steel and aluminium struts. Solid line = steel; dotted line = aluminium.

Figure 8.9 Buckling behaviour of two grades of structural steels.

and the lower bound curves were calculated from

$$\sigma_{BS} = \frac{\sigma_{YS} + (\eta + 1)\sigma_E}{2}$$
$$- \left\{ \left[\frac{\sigma_{YS} + (\eta + 1)\sigma_E}{2} \right]^2 - \sigma_{YS}\sigma_E \right\}^{1/2}$$

In this expression η is an imperfection parameter equal to $0.3\ (l/100k)^2$.

This expression for buckling stress is empirical and will no doubt be modified further as the theory is developed. For present purposes, it serves to illustrate two points: (1) when the slenderness ratio exceeds a certain value characteristic of the given material there is little benefit to be obtained from increased yield strength, and (2) if aluminium alloys are to be made more competitive with rolled-steel joists it is necessary to use thinner, more open sections, with higher radii of gyration. Because of their thinness such sections need to be strengthened against other types of instability failure such as torsional failure. This can be done by localized thickening in the outer parts of the sections, giving rise to lipped and bulbed sections (Figure 8.10).

Figure 8.10 Bulbed and lipped sections.

Although these designs are available there has not been sufficient incentive for sections of this sort to be put into large-scale commercial production.

Although frames for building construction, television and transmission towers and the like necessarily contain struts, there is not a wide range of applications. In contrast, end-loaded panels are much more widely used and it must be noted that these may also fail by a kind of buckling mechanism.

Buckling of a panel

When a plate is subjected to an end load P which lies in the plane of the plate it is described as a panel. If t is the thickness of a panel of width b, then the vertical stress sustained by the panel is given by P/bt. If the panel is thick enough, failure will occur by plastic crushing when the applied end stress attains the yield stress of the material. Thinner panels, however, fail by buckling at a lower value of stress given by

$$\sigma_B = \frac{\pi^2 E}{3\ (1 - v^2)\ (b/t)^2} \tag{8.4}$$

This equation is similar to the Euler equation for buckling of a strut with the thinness ratio of the panel taking the place of the slenderness ratio of the strut. The two types of behaviour are, however, rather different because whereas Eulerian buckling is the result of an overall instability, panel buckling is a form of local instability. In practical terms this means that whereas the strength of a strut disappears virtually to zero immediately buckling is initiated, a buckled panel will continue to support a significant, although much lower, load and it would be wasteful not to allow for this residual strength. Although the stress analysis of a buckled panel is rather complex it is accepted that the distribution of stress across the width of an end-loaded buckled plate is not uniform, varying from a minimum at the centre of width to maxima at the two edges. It follows from this that better utilization of material is achieved if a given panel area is divided into a number of panels of lesser width. This can

Figure 8.11 A panel subjected to an in-plane end load.

be effected by providing longitudinal stiffeners at suitable spacings (Figure 8.11). In equation (8.4)b is the distance between stiffeners.

8.4 Materials selection criteria for stiffness

Deflection of a beam

As shown in Figure 8.1, the deflection of a cantilever beam, δ, is given by $\delta = Pl^3/3EI$. If the cross-section of the beam is square, of breadth b, then the stiffness

$$\frac{P}{\delta} = \frac{Eb^4}{4l^3} \quad \text{and} \quad b = \left[\frac{4l^3}{E} \cdot \frac{P}{\delta}\right]^{1/4}$$

The weight of the beam is

$$lb^2\rho = l\rho\left[\frac{4l^3}{E} \cdot \frac{P}{\delta}\right]^{1/2} = 2l^{5/2}\left(\frac{P}{\delta}\right)^{1/2}\frac{\rho}{E^{1/2}}$$

where ρ is the density.

Therefore, for a given stiffness P/δ, the weight of the beam is minimized when $E^{1/2}/p$ is maximized. $E^{1/2}/\rho$ is therefore the materials selection criterion.

However, the designer can increase the geometric stiffness of the beam by control of the aspect ratio of the cross-section. If he replaces the square cross-section of the beam with a rectangular section of depth d and breadth b, it is sensible to hold b constant and allow d to vary. In this case it turns out that the weight of the beam is given by

$$(4l^6b^2)^{1/3} \cdot \left(\frac{P}{\delta}\right)^{1/3}\frac{\rho}{E^{1/3}}$$

and the materials selection criterion becomes $E^{1/3}/\rho$.

Buckling of a strut

The Euler buckling load, $P_E = (\pi^2EI)/l^2$. Since a strut is free to buckle in any lateral direction there is no point in considering other than axisymmetric sections. Assume, therefore, that the strut is a round rod of diameter d, for which the second moment of area is

$$I = \frac{\pi d^4}{64} \quad \therefore d^4 = \frac{64.P_El^2}{\pi^3E}$$

Structural efficiency

$$\frac{\text{Load}}{\text{Weight}} = \frac{\pi^2E}{l^2} \cdot \frac{\pi d^4}{64} \cdot \frac{4}{\pi d^2} \cdot \frac{1}{p} = \frac{\pi^2Ed^2}{16l^3p}$$

$$= \frac{\pi^2E}{16l^3p} \cdot \frac{8P_E^{1/2}l}{\pi^{3/2}E^{1/2}}$$

$$= \frac{\pi^{1/2}}{2}\left(\frac{P_E}{l^4}\right)^{1/2}\frac{E^{1/2}}{p}$$

$\therefore$ The materials selection criterion is $E^{1/2}/p$.

Buckling of a panel

The buckling stress of a solid panel in compression is

$$\sigma_B = \frac{\pi^2}{3(1-v^2)} \cdot E \cdot \left[\frac{t}{b}\right]^2$$

which, taking $v = 0.3$, becomes $3.62\,E\,(t/b)^2$.

Buckling load $P = 3.62\,E\,(t^2/b^2).\,tb$

$$\therefore t^3 = \frac{1}{3.62}\frac{Pb}{E}$$

Structural efficiency $=$

$$\frac{\text{Load}}{\text{Weight}} = 3.62\frac{Et^3}{b} \cdot \frac{1}{tbp}$$

$$= 3.62\frac{E}{p} \cdot \frac{t^2}{b^2} = \frac{3.62}{b^2} \cdot \frac{E}{p}\left(\frac{Pb}{3.62E}\right)^{2/3}$$

$$= 1.54\frac{E^{1/3}}{p}\left(\frac{P}{b^2}\right)^{2/3}$$

The materials selection criterion is thus $E^{1/3}/p$.

8.5 Comparison of materials selection criteria

We are now in a position to examine the performance of several constructional materials in terms of the criteria that have been developed (Table 8.1). The orders of merit revealed in the table demonstrate above all the importance of density in weight-sensitive applications. Steel, which in absolute terms has the highest Young's modulus of all the materials considered, ranks bottom, equal with polypropylene, in terms of $(E^{1/3}/p)$. The two best materials, wood and carbon-fibre-reinforced-plastic (CFRP), are both materials of low density and, further, when the GRP reinforced with 30% glass fibre in the form of chopped strand mat is compared with the version containing 50% woven roving it is seen that although the increased glass content has raised the absolute value of stiffness significantly, this has been offset by the concurrent increase in density.

The effect of anisotropy is worth noting. The best values are produced by the most anisotropic materials – oak, unidirectional GRP and unidirectional CFRP. Wood in the form of plywood panels, and CFRP and GRP as cross-ply or random laminations, are much less competitive.

Although polymers in general are highly compliant materials, their low densities enable them to find wide use in small-scale applications, because whether injection-moulded or laminated, it is a simple matter to provide additional stiffness where it is most needed by local thickening of cross-sections without undue increase in weight. It is common practice to stiffen plastic sections with the use of ribs, which can be incorporated into the injection moulding process relatively easily. As already mentioned, additional stiffening is frequently applied to metallic structures by welding on (sometimes riveting), stiffeners, but this procedure is less convenient for several reasons. Being an extra operation, it introduces additional cost; since metals are dense, the increase in weight is not insignificant; where welding is involved, careful consideration at the design stage is necessary because of the propensity for welds to introduce defects and reduce the fatigue resistance of the structure.

The theory of beam stiffening is put into good practice with the sandwich construction of composite panels, where stiff composite laminates are bonded each side of a light, rigid, polymeric foam such as PVC, or similar materials e.g. balsa. This construction is used for ocean-going racing yachts, for example. The sandwich is essentially acting as an I-beam in terms of the stiffness increase.

It is necessary to retain a proper perspective regarding the validity of the materials selection criteria. Although wood and CFRP appear at the top of the order of merit, for example, the former is not used in modern aircraft structures although the use of CFRP is expanding steadily (see Chapter 15). Again, although the parameter for aluminium is twice that for steel, it has failed to establish itself in large-scale structures although a few small bridges and ships have been built in aluminium alloy. Such materials selection criteria are of value in the early stages of materials selection, especially for novel applications, since they provide the best means of ensuring that all possible contenders, even apparently outlandish ones, are properly considered. The final stages of materials selection will involve a much wider range of more detailed considerations and, as always, cost will be the final arbiter.

References

1. J. E. GORDON: *The New Science of Strong Materials*. Penguin Books, 1968.
2. A. C. WALKER (ed.): *Design and Analysis of Cold Formed Sections*. International Textbook Co.
3. H. E. GRESHAM: *Met. and Mater.*, November, 1969.
4. H. DOMININGHAUS: *Plastics for Engineers*. Carl Hauser Verlag, 1993.

9
Fatigue

Fatigue is a dangerous form of fracture which occurs in materials when they are subjected to cyclic or otherwise fluctuating loads. It occurs by the development and progressive growth of a crack and the two characteristic and equally unfortunate features of fatigue fracture are, first, that it can occur at loads much lower than those required to produce failure by static loading, and second, that during the more or less lengthy period of time that is required for fracture to propagate to the point of final failure there may be no obvious external indication that fracture is occurring.

Although fatigue failure is most familiar when it occurs in metals, probably no material is immune to this form of failure and other materials in which it is known to occur include concrete and polymers, and even living matter.

However, fatigue failure was first diagnosed in metals and most of the research carried out to elucidate the nature of fatigue has been performed on metallic materials. The first recorded observations related to the axles of railway wagons in the 19th century. Nevertheless, it is only recently that significant understanding of the micromechanistic processes involved, and the rate at which they occur, has advanced sufficiently to enable the design engineer to take some account of fatigue in a numerate manner.

The ultimate aim must be to prevent fatigue fracture occurring altogether, but final solution of the fatigue problem does not seem to be a realistic prospect for the foreseeable future. The major problem is the fact that fatigue behaviour is dominated by details of design. Thus, although it is possible to assess the inherent fatigue resistance of a material, and even find ways of increasing it, these efforts usually produce a rather inconspicuous improvement in the behaviour of many engineering components.

This is not just a matter of defective design (although many fatigue failures have been directly caused by shortcomings in design): it is rather that many features harmful to fatigue resistance are difficult to avoid in practical machine parts. Fortunately some of these harmful features can be ameliorated by competent design. For example, the effects of stress concentrations at geometric irregularities such as keyways, oil-holes and changes in cross-section are serious but they are now well documented[1] and the careful designer can do much to avoid repeating the mistakes of the past. But undoubtedly the most damaging feature of engineering design from the point of view of fatigue is the joint. Unfortunately, the presence of a joint, whether bolted, riveted, adhesively bonded or welded, can render the fatigue behaviour of large-scale jointed structures almost totally insensitive to materials development. This means that significant improvements in the fatigue resistance of jointed structures are extremely hard to come by.

Joints of one sort or another are very common in engineering structures, and the materials engineer faces some difficult problems. However, there are areas in which positive contributions can be made.

First, there are many engineering applications – helicopter rotor blades and ball races are examples – which because of their simplicity of form *do* respond to improvements in materials properties. Second, if the cost-benefit analysis for the use of an advanced material is not favourable, the materials engineer, with the design team, must make sure that the design and manufacture is entirely consistent with the materials to be utilized. Third, it is necessary to continue with research into ways of increasing the materials component of fatigue behaviour – the improved performance of aircraft alloys and of powder metallurgy products, and the recent advances in adhesive bonding, shows that it can be worthwhile.

Materials selection in relation to fatigue must be based on an understanding of the major features of fatigue failure and these are dealt with in the following sections.

9.1 Micromechanisms of fatigue in metals

Fatigue failure in metals starts with the initiation of a crack. The crack then propagates across the cross-section of the part until the residual ligament is unable to support the load and final failure occurs by a static mechanism. There are thus two quite distinct fatigue processes involved – initiation and propagation.

Initiation of fatigue cracks is due to crystallographic slip and mostly starts at, or very close to, the surface of the part. This is because engineering metals are generally polycrystalline so that grains at the surface of the part, being incompletely surrounded by other grains, are freer to deform than those within the bulk of the material: favourably oriented grains at the surface therefore start to slip locally at stresses that are lower than the stress required to produce general yielding. Grains within the bulk of the specimen, even if favourably oriented, cannot deform at low loads because of the support and constraint provided by the surrounding material. Fatigue properties therefore vary with surface finish, so care is required in design and manufacture to avoid stress concentrators such as a poorly machined surface or sharp sectional transitions.

Initiation processes

Initiation of a crack can occur in two main ways:

(1) By formation of slip bands, due to crystallographic slip in a surface grain, followed by development of a crevice which eventually deepens into a crack lying in the favoured crystallographic plane. This process is only important in parts made of soft ductile materials. It is greatly accelerated by the presence of geometric stress concentrations.

(2) As a result of severe strain incompatibility across inclusions or hard second-phase particles. This process tends to occur in metallurgically hardened alloys in which the matrix is resistant to the crystallographic slip required to form a slip band crack.

The mechanism of initiation which occurs in any given case is thus the result of competition between the ability of the material to sustain the imposed strain discontinuities across the various inhomogeneities within the material and the resistance of the matrix to crystallographic slip.

Crack growth in smooth ductile specimens

When crack initiation occurs by crevice formation at a slip band there are two distinct stages in the subsequent growth of the crack. Stage I occurs whilst the crack is confined to the slip plane on which it was initiated. In specimens without stress concentrations Stage I growth can occupy up to 90% of the total life of the specimen. Eventually, however, as the crack lengthens, the plastic zone at its tip becomes large enough to be independent of the crystallographic nature of the material and the crack then grows as if it were in a continuum. This is known as Stage II crack growth and as the transition occurs the direction of crack growth changes so as to maximize the crack opening displacement during subsequent growth. This will generally correspond to a direction that is about normal to the maximum principal tensile stress in the region of the crack tip. Because Stage II growth is faster than Stage I growth, it generally produces the largest area of a fatigue fracture surface. It is Stage II growth that produces the beach markings commonly seen on fatigue fracture surfaces. Changes in orientation of beach markings can often be correlated with changes in loading conditions. If the applied stress is low the beach markings may be too fine to be seen with the naked eye.

A Stage I crack is sometimes known as a micro-crack and a Stage II crack as a macrocrack.

Stage I and Stage II crack growth are fundamentally different. The former is nucleated by reversed crystallographic slip on a particular slip band and so long as it remains a Stage I crack it can only propagate in the direction of that band. The ease with which a Stage I crack nucleates and grows is therefore dependent upon the strength of the matrix in which the slip band forms. If, in any particular instance, Stage I growth is a significant part of the whole fatigue process, then the overall fatigue resistance of an engineering component can be increased by increasing the strength of the material by normal metallurgical methods.

In contrast, Stage II growth, when it occurs in metals of good toughness, is a continuum mode of growth that is not greatly influenced by conventional methods of metallurgical strengthening and reduction of the rate of Stage II crack growth is not easily achieved.

Crack growth in smooth, hard specimens

In an unnotched low-strength material the fatigue strength will increase with matrix strength irrespective of whether the strengthening is achieved by cold working, alloying or heat treatment. However, as the matrix becomes progressively harder and Stage I nucleation becomes correspondingly more difficult, the stage is eventually reached when some additional factor must operate to bring about nucleation. This is available at the stress concentrations produced by second-phase particles. Thus, the methods used to improve the fatigue resistance of a high-strength material must be different from those applicable to a low-strength material because in the former case the cyclic stress required to initiate a crack will depend not on the hardness of the matrix but on the size, shape and distribution of non-metallic inclusions and other second-phase particles. To improve the fatigue performance of a high-strength material it is necessary to make it cleaner. This may necessitate the use of expensive processes such

as electroslag refining for steels or the use of higher-purity base material for aluminium alloys.

The well-known correlation between fatigue strength and tensile strength, known as the fatigue ratio,* fails as the strength increases above a certain level, (Figure 9.1). It fails because the stress required to initiate cracks at second-phase particles bears no direct relationship to the stress required to produce crystallographic slip in a surface grain and hence is not related to the tensile strength of the material. For example, increasing the tensile strength of air-melted steels beyond about 1000 MPa (145 ksi) produces little, if any, further increase in fatigue strength; the corresponding figure for aluminium alloys is 300 MPa (43.5 ksi). However, titanium alloys seem to be much better in this respect; perhaps because they are intrinsically cleaner materials.

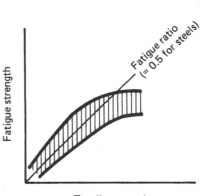

Figure 9.1 Relationship between fatigue strength and tensile strength.

Crack growth in notched specimens

Whereas in unnotched specimens Stage I crack growth may occupy 90% or more of the total fatigue life, with Stage II growth taking up the remaining 10%, in specimens containing stress concentrations these figures can easily be reversed, with Stage II growth accounting for

* Fatigue ratio, $FR = \dfrac{\text{Fatigue strength}}{\text{Tensile strength}}$

more than 90% of the total life. Since Stage II growth is much faster than Stage I growth this means, unfortunately, that the total life is greatly reduced. Thus, a high stress concentration in a machine part can cause quite disastrous effects on fatigue performance. Therefore, wherever there are unavoidable features such as fillets, changes in cross-section and engineering details such as oil-holes, keyways and especially joints, strenuous efforts must be made to minimize the inevitable stress concentrations.

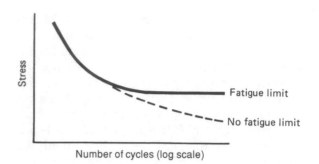

Figure 9.2 Typical stress-life (S–N) curves.

9.2 The assessment of fatigue resistance

There are two distinct lines of approach. One way is to use stress-life relationships, generally known as S–N curves, in which S is the applied stress and N is the total fatigue life measured in cycles of stress. The other way is to use fracture mechanics data to estimate rates of fatigue crack propagation (FCP).

The S–N method is the older of the two and is widely used in all branches of engineering. The S–N curve for a material is determined by taking specimens of that material and subjecting each one to a different cyclic stress until it fails. For each specimen, the number of cycles to failure is noted and each value of N is plotted against the corresponding stress amplitude (Figure 9.2). The practice in regard to the definition of fatigue stress is frequently confusing. The preferred practice is to define S as stress amplitude, and indeed this is essential in any definition of fatigue ratio. Frequently, however, it is plotted as maximum stress, a practice which usually refers to a specific type of loading in which the stress varies from zero to some maximum value. The maximum stress is then equal to double the stress amplitude when the stress varies sinusoidally. Sometimes, as in the literature relating to concrete, the stress is plotted indirectly as a fraction of static strength.

Cyclic loading generally produces failure, however low the stress may be. However, with some materials the S–N curve levels off (Figure 9.2),

suggesting that for these materials a limit of stress can be specified – known as the fatigue limit – below which infinite life can be expected. The fatigue limit is thought to be associated with the phenomenon of strain-ageing. Thus, all ferritic steels of tensile strength not exceeding 1100 MN/m^2 (160 ksi) may be expected to show a fatigue limit: titanium alloys also show a fatigue limit. Aluminium alloys and non-ferrous metals in general do not, although the non-heat-treatable aluminium-magnesium alloys may.

It must be understood that the bulk of S–N data is subject to severe limitations. There are three principal restrictions: (1) configuration of stress, (2) mean stress, and (3) stress concentrations.

Configuration of stress

Because of the greater simplicity of loading, S–N curves are mostly determined under conditions of uniaxial stress. Tests employing complex stress systems are difficult to devise and perform, and the few experimental programmes that have been carried out provide data not for direct design purposes but rather to test the validity of various theoretical procedures which employ uniaxial data to solve problems involving complex stress using the Levy-Mises or Tresca yield criteria. These experimental data relating to the behaviour of materials under complex stress are insufficiently comprehensive to be used for the purposes of materials selection.

Mean stress

When a cyclic stress is fully reversed the mean stress is equal to zero, (Figure 9.3a) and much fatigue data has been obtained in this way. However, the mean stress σ_m, is often not equal to zero (Figure 9.3b,c) and it is then necessary to take account of the observed fact that if the mean stress is increased in the tensile direction then the stress amplitude must be reduced in order to maintain the same fatigue life. This is usually done by means of constant-life diagrams based on the original ideas of Goodman, Gerber and Soderberg. Figure 9.4 shows a conventional constant-life diagram in which mean stresses are plotted as abscissae and stress amplitudes as ordinates. The conditions of stress which lead to failure after a life of 10^7 cycles are given by line

ABC. A is the static tensile strength of the material and B is its fatigue strength for a mean stress of zero. The line extends towards C into a region of compressive mean stress. Other lines on the diagram, such as ADE, relate to different values of life. When the constant-life lines are not obtained directly from experimental data they are commonly approximated by one or other of the lines shown in Figure 9.5.

It is rather difficult to visualize how fatigue crack growth can proceed when the loading cycle is entirely compressive, but it has been suggested that plastic flow at the crack tip can produce residual tensile stresses which are momentarily sufficient to cause crack growth. It would seem sensible to suppose that fatigue performance would always be improved by the presence of a compressive mean stress but this

Figure 9.3 Mean stress, σ_m, for different stress cycles (NB Stress ratio, $R = \sigma_{min}/\sigma_{max}$).

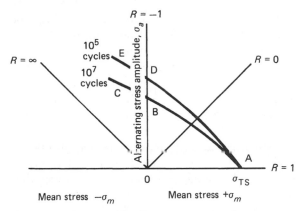

Figure 9.4 Constant-life diagram in terms of stress amplitude, σ_a and mean stress, σ_m.

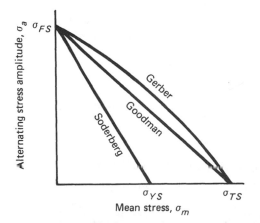

Figure 9.5 Constant-life diagram showing predictions of Goodman, Gerber and Soderberg.

does not necessarily seem to be so. Some materials do show a continuous enhancement of fatigue strength as the mean stress becomes increasingly compressive, but with other materials it seems merely to remain constant.

Stress concentrations

The great majority of machine parts and structural members contain notches and stress concentrations of one sort or another. One way of dealing with these is to obtain appropriate values of the elastic stress concentration factor, K_T (calculated, or assessed, by analogue methods such as photoelasticity), and use the local values of stress thereby obtained with $S–N$ or constant-life curves actually obtained on smooth specimens. Usually this procedure is conservative because materials vary in their sensitivities to notches and it is often found that the actual reduction in fatigue strength caused by a stress concentration is less than the amount suggested by the elastic stress concentration factor. This is probably due to unloading of the stress concentration by local plastic deformation; this mechanism would be expected to be more effective in low- rather than high-strength materials, and in confirmation of this view it is found that the notch sensitivity of all materials rises with increasing strength.

Attempts have been made to quantify these effects by defining a notch sensitivity factor, q, in terms of the local stress concentration factor K_T, and the fatigue strength reduction factor, K_f.

$$q = \frac{K_f - 1}{K_T - 1}$$

where

$$K_T = \frac{\text{local stress}}{\text{nominal stress}}$$

and

$$K_f = \frac{\text{fatigue strength of unnotched specimen at } N \text{ cycles}}{\text{fatigue strength of notched specimen at } N \text{ cycles}}$$

Thus, when $q = 0$ the notch has no effect and when $q = 1$ the notch exerts its full effect. From the point of view of materials selection it would be convenient if q were a true material constant but, unfortunately, this is not found to be the case: q seems to vary with different types of notch and loading. Although the design engineer often uses estimated q-values in his design calculations, there is insufficient reliable notch sensitivity data for the method to be used for the purposes of quantitative materials selection. It should be noted, moreover, that since notch sensitivity increases with strength, any expectations of improving the fatigue strength of a material by boosting the tensile strength are doomed to be increasingly disappointed as the strength rises, and this is in line with the previous discussion.

In difficult cases there is no more reliable way of assessing fatigue performance in the presence of a stress concentration than by testing a full-scale prototype such as a whole aeroplane wing or fuselage cabin. This could hardly be a general procedure but may be vital where innovative design is associated with safety-critical parts.

Cumulative damage and the Palmgren–Miner rule

Fatigue damage accumulates more rapidly at high stress amplitudes than at low, so that when the stress amplitude applied to a specimen fluctuates widely, any method of predicting total fatigue life must take account of the varying rates at which fatigue damage accumulates.

The Palmgren–Miner rule proposes that if cyclic stressing occurs at a series of stress amplitudes S_1, S_2, S_3 ... each of which would correspond to a failure life of N_{f_1}, N_{f_2}, or N_{f_3} if applied singly, then the fraction of total life used at each stress amplitude is the actual number of cycles, N_i divided by the lifetime at that amplitude N_{f_i}. The damage then accumulates such that

$$\frac{N_1}{N_{f_1}} + \frac{N_2}{N_{f_2}} + \frac{N_3}{N_{f_3}} + \ldots \frac{N_i}{N_{f_i}} \ldots = 1$$

The Palmgren–Miner rule takes no account of loading sequence, i.e. high-low or low-high. Neither does it consider the effect of mean stress.

Although for many applications use of the rule is mandatory, it may give conservative or unsafe predictions by factors of 5–10 on endurance. Boulton[2] refers to experiments which have given Palmgren-Miner summations ranging from 0.033 to 30. Fortunately, there is a good deal of experience relating to the use of the rule so that the worst of the errors in prediction can be avoided. However, there is no doubt that the rule may often be misleading, and it continues to be used partly because of its simplicity (although the calculations may be lengthy and tedious) but mainly because there is no alternative.

Fracture mechanics and fatigue

It is clear that in many engineering situations it is realistic to neglect any contribution to fatigue life from Stage I crack growth, either because an existing defect makes it non-existent or because stress concentrations make it vanishingly small. Since, in reasonably ductile materials, the rate of advancement of a Stage II crack is only weakly dependent on the microstructure through which the crack grows, the best that can be done is to estimate that rate and then arrange to take the cracked part out of service safely before failure occurs. This has been made possible by the work of Paris and Erdogan,[3] who demonstrated that in the mid-range of behaviour there exists a simple power relationship between crack growth rate and range of stress intensity factor during the loading cycle.

$$\frac{da}{dN} = A\,(\Delta K)^m \qquad \text{Paris-Erdogan Law}$$

A and m are constants which must be determined experimentally.

If the crack propagation law for the material is known it is possible to calculate by integration the number of cycles required for the crack to grow from one length to another.

Further, if the fracture toughness, K_C, is known for the material of interest then this parameter can be used, together with the value of the maximum design stress, to calculate the critical value of crack length, a_f, at which fast fracture will occur and hence, by integration of the Paris law, the total life in cycles of the cracked part.

During fatigue crack growth, $\Delta K = \Delta\sigma a^{1/2}\alpha$, where α is the compliance factor for the given geometry. The crack growth law gives $da/dN = A(\Delta K)^m = A(\Delta\sigma\,\alpha a^{1/2})^m$. Integrating this expression for the number of cycles required for the crack to grow from an initial length a_0 to the critical length for fast fracture gives

$$\int_{a_0}^{a_f} a^{(-m/2)}\,da = A\,\alpha^m\,\Delta\sigma^m \int_{N=0}^{N=Nf} dN$$

$$= \frac{2}{2-m}\left[a_0^{\,(1-m/2)} - a_f^{\,(1-m/2)} \right]$$

$$= A\alpha^m\,\Delta\sigma^m\,N_f$$

(Note that this expression fails when $m = 2$).

The early literature reported values of m mostly lying between 2 and 4, although values much higher than this have been encountered. It is becoming clear that for the purpose of the above calculation appropriate values of m lie between 2 and 3, the higher values being found in materials of low toughness. Thus, when $m = 3$

$$N_f = \frac{2}{A\alpha^3\,(\Delta\sigma)^3}\left(\frac{1}{\sqrt{a_0}} - \frac{1}{\sqrt{a_f}} \right)$$

Although integrating for total life is a useful procedure more information is obtained if the calculation is performed in a stepwise manner. This demonstrates that growth of the crack accelerates with respect to life, as shown in Figure 9.6. In fact, the real situation is worse than this because it is now known that the Paris-Erdogan law applies only over the middle range of crack growth rates, i.e. between rates of about 10^{-6} and 10^{-4} mm/cycle. A plot of log da/dN against log ΔK shows three regimes of behaviour, and the central regime in which the

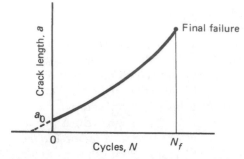

Figure 9.6 Fatigue crack length as function of life measured in cycles of stress.

Paris-Erdogan law applies is preceded and followed by the regimes in which m varies and takes much higher values (Figure 9.7).

It is interesting to note that Region A shows that there exists a value of ΔK below which the crack is non-propagating, i.e. it merely opens and closes without growing forward. This is called the threshold for fatigue crack growth, ΔK_{TH}. The rate of crack growth in the threshold region is much slower than calculation from the Paris-Erdogan law would predict. Nevertheless, because understanding of threshold effects is currently rather poor it is usual at present to err on the side of safety and neglect this region in life calculations by assuming that crack growth according to the Paris-Erdogan law occurs down to the lowest values of ΔK. However, it is worth noting that high-strength brittle materials show lower thresholds than low-strength tough materials, and that thresholds are lower as the mean stress is increased in the tensile direction.

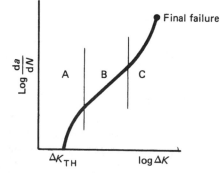

Figure 9.7 Fatigue crack growth rate as function of stress intensity range.

In region C, as K_{max} approaches the limiting fracture toughness of the material, K_{IC} or K_C, the Paris-Erdogan law underestimates the fatigue crack propagation rate. This acceleration of the logarithmic growth rate seems to be associated with the presence of non-continuum fracture modes such as cleavage, intergranular and fibrous fracture (which are activated at high levels of K). There is also a marked sensitivity to mean stress.

Empirical expressions to account for the accelerating region are not lacking:

$$\frac{da}{dN} = \frac{A\Delta K^m}{(1-R)K_{IC} - K_{max}} \qquad \text{Ref. 4}$$

$$\frac{da}{dN} = \text{const}\left[\frac{\Delta K^4}{\sigma_1^2(K_{IC}^2 - K_{max}^2)}\right]^n$$

where

$$\sigma_1 = \frac{\sigma_{YS} + \sigma_{TS}}{2} \text{ and } n \cong \frac{3}{4}, \qquad \text{Ref. 5}$$

but in critical cases it is simpler and safer effectively to eliminate the accelerating region by imposing an upper limit on K_{max} of $0.7\,K_{IC}$.

Within region B there is a good understanding of propagation behaviour and a considerable data-bank now exists for fatigue crack growth rates. It has been found that within given classes of materials these rates are not greatly affected by the usual metallurgical variations. Mogford[6] has plotted bands of data from various workers for a wide range of constructional steels to show that they are very narrow and overlap. He gives the following expressions for the upper bounds to the growth data:

$$\frac{da}{dN} = 9 \times 10^{-12}.\Delta K^3 \quad \text{for ferrite/perlite steels,}$$

$$\frac{da}{dN} = 1.7 \times 10^{-11}.\Delta K^{2.25} \text{ for martensitic steels,}$$

$$\frac{da}{dN} = 1 \times 10^{-11}.\Delta K^3 \qquad \text{including weldments,}$$

and states that a satisfactory approximation for general use in the absence of specific data would be

$$\frac{da}{dN} = 10^{-11} \cdot (\Delta K)^3$$

with $\frac{da}{dN}$ in mm/cycle and ΔK in MPam$^{1/2}$

Saxena[7] has shown that in weldments of ASTM A514A steel the fatigue crack growth behaviour of weld metal and parent metal were very similar and could be characterized by the single expression

$$\frac{da}{dN} = 1.6 \times 10^{-11} \cdot (\Delta K)^3$$

It seems that a truly rational and satisfying procedure for selecting materials to maximize fatigue resistance will be limited to the most highly developed and fatigue-sensitive applications because in run-of-the-mill applications the potential benefits will be too small to justify the expenses that would be incurred.

It is important to note that design procedures which are based solely on Stage II fatigue crack propagation rates must be highly conservative since no account is taken of any contribution to fatigue life from Stage I growth. This is probably only justified in certain high-technology applications.

Low-cycle fatigue

When an engineering component or structural member subjected to fatigue loading need only withstand 10^4 or 10^5 cycles, or less, during a normal lifetime it is possible for the component to operate at stress levels much higher than the conventional fatigue limit. This is called low-cycle or low-endurance fatigue.

As the fatigue life decreases down towards $N = \frac{1}{4}$ (corresponding to the static tensile strength) the $S-N$ curve flattens out so that small variations in stress produce large changes in endur-

ance. Low-cycle fatigue is therefore generally discussed in terms of applied strain rather than applied stress. The situation is confused by the fact that some workers have measured total applied strain (i.e. elastic + plastic), whereas others have measured plastic strain only, on the basis that only plastic deformation can lead to fatigue damage.

When total strain range, ϵ_t, is plotted against endurance, N, on a log-log basis, the scatter-band conforms reasonably well to the relationship

$$\epsilon_t + N^m = C$$

where m and C are constants. Most materials give results that lie close to a single line and it is only as the endurance approaches the high-cycle regime that different materials peel off to give different fatigue strengths (Figure 9.8).

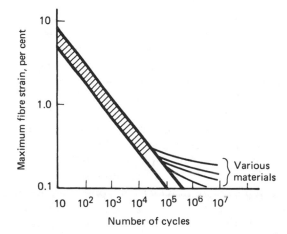

Figure 9.8 Relationship between maximum strain and endurance for several different materials. (After Low,[8] by permission of the Council of the Institution of Mechanical Engineers.)

Coffin, measuring plastic strains, proposed the relationship $\epsilon_p N^{1/2} = C$ and also postulated that C could be put equal to $\epsilon_f/2$ where ϵ_f is the fracture ductility, $\ln A_0/A_f$.

Manson, on the other hand, obtained a good correlation with widely differing materials for the relationship $\epsilon_p = (\epsilon_f/N)^{0.6}$.

It seems likely that the exponents in both Coffin's and Manson's laws cannot be true constants. Unfortunately, no analysis of data has yet provided a basis for reliably assigning particular values of the constants to particular materials for the purposes of materials selection.

9.3 Factors influencing fatigue of metals

Fatigue in jointed members

Mechanical joints

A mechanical joint may be pinned, riveted or bolted, but whichever method is used the fatigue strength of the assembly as a whole is reduced to a small fraction of the plain fatigue strengths of the component members. For example, by determining S–N curves for pin joints incorporating sliding-fit steel pins through high-strength aluminium alloy and steel lugs, Heywood[9] found that, at 10^7 cycles, S was equal to only about 4% of the tensile strength of the steels and 2½% of the tensile strength of the aluminium alloys.

There are two important factors influencing the fatigue resistance of mechanical joints. One is the stress concentration introduced by the joint. The other is fretting between contacting surfaces. Fretting may occur on the cylindrical surface of the hole through which the pin, rivet or bolt passes or, if the joint is riveted or bolted, it may occur between the faying surfaces of the plates, in which case the final failure may occur away from the holes. There are various practical methods for increasing the fatigue resistance of a mechanical joint, such as interference fits between mating parts and minimizing relative motion at faying surfaces – it may even be possible to ensure that all of the load on the joint is borne by frictional forces acting between the plate surfaces. It is also desirable to incorporate anti-fretting compounds into mechanical joints. However, none of these methods takes any account of the basic properties of the materials making up the joints, and it seems to be true that within a given class of materials variations in

composition, or in the metallurgical nature of the component materials in a mechanical joint, have little influence on the final fatigue resistance of joints required for long endurances. For short endurances, say up to 2×10^6 cycles, increasing the tensile strength of the bolt and plate materials in a bolted joint will allow a greater clamping pressure to be applied to the plates and thereby delay the onset of fretting[10] but at longer endurances when failure eventually occurs due to fretting fatigue performance is not greatly influenced by the joint materials. Differences in fatigue resistance will, of course, be observed in moving to a completely different class of materials. Heat-treated aluminium alloys, for example, are notoriously poor in fatigue as compared with steels.

Welded joints

The fatigue strength of a welded joint is always lower than the plain fatigue strength of the unwelded material even though the static strength of a welded joint may be equal to that of the parent material.

The fatigue strength of a given welded joint is determined by (1) the size and distribution of the defects within the deposited weld metal, (2) the magnitude of the stress concentration factor at the junction of the weld metal and the parent plate and (3) in the case of steel, the decarburization at the surface of the weld metal and heat-affected zone. None of these factors gives much scope for materials selection and it seems that the fatigue strength at long endurances of joints made from alloys within a given class is not greatly affected by variations in the alloys concerned. That is to say the high cycle fatigue strength of a welded high-strength steel is no greater than that of a welded low-strength steel. A similar remark could be made about joints in aluminium alloys and presumably other materials as well. This insensitivity of fatigue performance to basic material strength is found with all types of welded joints and welding processes.

This distressing feature of weld behaviour seems at first sight to destroy any incentive to employ high-strength materials for structural purposes involving fatigue, since welds are almost invariably a feature of these applications. In fact, high-strength materials are frequently used and this can be justified in three ways. First, the above remarks apply only to long endurances and as interest centres on lives decreasing below 10^6 cycles the benefits to be derived from materials of high static strength progressively increase (Figure 9.9). Second, the use of high-strength materials provides protection against static overload and occasional peak stresses in cases where the loading spectrum is fluctuating rather than simply cyclic. Third, where the fatigue spectrum incorporates a high value of mean stress it is necessary, even where the stress range is low, to employ high-strength materials so as to avoid yielding across the net section due to the high value of maximum stress.

Figure 9.9 *S–N curves for various welded steels (after Frost and Denton[11]).*

Most fatigue failures in welded structures start from the weld toe and many of the measures taken to reduce the undesirable effects of welding relate to the treatment of this part of the weld. Its geometry may be improved by grinding or weld toe remelt using TIG or plasma, to reduce stress concentration. As described for the more general case below, the fatigue performance of a weld may be improved by hammer or shot peening of the surface.

It is known that all fusion-welded structures, prior to stress relief, contain residual stresses of yield stress magnitude. Therefore, welded parts in which the stress is nominally compressive will, if not given a precautionary stress relief, inevitably be subject to tensile fluctuations. Such stress relief may be applied locally or may be part of an overall heat treatment cycle for the whole component to obtain optimized microstructures.

Adhesive joints

There is much interest in the use of adhesively bonded metal-to-metal joints, particularly the joining of aluminium body panels in the automotive industry. The stress distribution across an adhesive joint will be more uniform than that obtained with mechanical fasteners. Residual stresses and weld defects are also avoided, so that the fatigue characteristics can be attractive. However, this remains a relatively new area, and further experience of these joints is required (see Chapter 18).

Surface processing

The fatigue strength of a metallic specimen will be increased if the surface of the specimen is hardened. Conversely, softening of the surface has the effect of decreasing the fatigue strength.

Cold working

Cold working of a surface, as for example in shot peening, induces compressive residual stresses which act to increase endurance by inhibiting the opening of the fatigue crack. In some materials plastic strain may increase fatigue strength by as much as 30% but in others the effect is much smaller. It does not necessarily persist to long endurances. A similar but more precisely controlled effect can be produced by surface rolling. This is especially effective in notched components: the fatigue performance of crankshafts can be significantly improved by cold-rolling the journal and web fillets. The same effect is found in cold-rolled screw threads.[11]

Case hardening

Surface hardening by flame or induction methods, by carburizing or by nitriding all increase the fatigue limit of steels, especially in bending and torsion (the effect in direct stressing may be small or non-existent).[11] These processes are especially valuable in pieces containing stress concentrations: in some cases the fatigue strength of a notched specimen may be more than doubled in this way.

However, the fatigue limit of a case-hardened specimen cannot be predicted.[11] The initiation and growth of a fatigue crack is greatly influenced by the hardened layer. If the crack is initiated at the surface of the piece its rate of growth may be slow as it passes through the hardened layer but will accelerate as it grows into the core. However, crack initiation more commonly occurs just below the hardened layer.

Surface softening

Soft surface layers, which are always harmful to fatigue performance, may be produced by decarburization in steel or by cladding processes in high-strength aluminium alloys. Decarburization of steel parts unprotected during heat treatment or welding may be harmful and, although some subsequent improvement may be accomplished by machining or less, satisfactorily, shot peening, it is better to avoid the problem in the first place.

Cladding high-strength aluminium alloys with pure aluminium or aluminium-zinc alloy so as to improve corrosion resistance is in a different category as a deliberate process: the 25% or more reduction in plain fatigue strength which results must be accepted if service conditions are such that this is the only way to control the corrosion problem. However, attempts are being made to develop corrosion-resistant claddings which are capable of developing strengths comparable to those of the core material.

Plated coatings

Steel parts are frequently provided with protective plated coatings of metals such as nickel, chromium, cadmium and zinc. Although the details vary from one case to another, the generalized picture seems to be that these treatments are harmful to fatigue resistance, the deleterious effect becoming worse as the strength of the steel increases. Two factors are operative in producing this undesirable result. First, the flawed nature of the plated coating provides ready-made sites from which cracks can propagate; second, the residual stresses existing within the coatings are frequently tensile so that propagation of any crack, once formed, is assisted. Any process which will modify the residual stresses and if possible render them compressive will therefore enhance the fatigue performance.

It should be emphasized that where corrosion fatigue is involved, plating with zinc or cadmium can be beneficial.

9.4 Fatigue of non-metallic materials

Although fatigue was first documented as a serious problem in metallic materials, it also occurs in non-metallic materials and many of the characteristics are similar to those observed in metals. Non-metallic S–N curves have the same form as those obtained from metallic materials, and it is also frequently possible to observe striations on the fracture surfaces. However, in view of the enormous variations in structures and properties encountered in moving from metals to, say, polymers or concrete or the various composites, it is clear that significant differences in fatigue behaviour must be expected to occur.

The first point to be established is whether any significant contribution to fatigue life can be anticipated from fatigue crack initiation processes. If the material, by its nature, contains discontinuities or interfaces favourable for the inception of a macrocrack, then there will be no benefit to fatigue life from initiation processes. This appears to be established in the case of concrete in which fatigue cracks are known to originate within the cement paste or near the interface between the cement paste and the aggregate. A similar situation must surely exist in composites in which the interface between fibre and matrix offers an excellent location for crack growth. For the purposes of materials selection in the present state of knowledge it is clearly prudent to assume that in any non-metallic material fatigue life is no more than the number of cycles required for a crack to grow from some starting size to the size required for final failure. There is probably no other useful generalization to be made since the many non-metallic materials vary so widely in their make-up and nature.

Fatigue of polymers

Although fatigue in polymers is superficially similar to fatigue in metals it is not as well understood. This is partly because of the need to take account of certain factors which are not important in metals (or at any rate much less so).[12] These include molecular weight, degree of cross-linking, crystallinity, transition effects and thermal effects (internal heating).

The importance of thermal effects means that fatigue in polymers is very sensitive to the frequency of cyclic stressing. If conditions are not isothermal, the hysteritic heating effect generated during each cycle of stress causes the elastic modulus to decrease, so that eventually the specimen is unable to support the load and therefore fails prematurely. To take account of this, ASTM Standard D671–71 defines the failure life under conditions of thermal fatigue as the number of cycles at a given stress required to cause the apparent modulus to decay to 70% of

the original value.[13] The temperature effect is also a potent reason for limiting the stress. The mechanical loss for thermoplastics varies from about 0.1 to 1 whereas for metals it is about 0.0001. The flexural fatigue curves for a number of thermoplastics are shown in Figure 9.10.

Key:-
a PA6 (23°C; f = 15 Hz)
b PA66 (23°C; f = 15 Hz)
c SAN (injection moulded)
d Acetal copolymer
e ETFE (23°C; f = 15 Hz)
f PMMA (extruded sheet)
g Polysulphone
h PC (23°C; f = 30 Hz)
i PVC alloy (23°C; f = 10 Hz)
j Styrene butadiene (20°C)

Figure 9.10 Flexural fatigue strength of thermoplastics, data from H Dominghaus[17].

One of the most important factors influencing the fatigue resistance of plastics is degree of crystallinity: the lowest fatigue crack growth rates are found in crystalline polymers such as nylon and polyacetal (Figure 9.11). In contrast, polymers exhibiting a high degree of cross-linking exhibit very high crack growth rates. It is unfortunate that the high strength and modulus of these highly cross-linked polymers which make them so attractive for engineering purposes are associated with the low ductility and toughness which presumably account for their poor fatigue performance. The molecular weight of a polymer also appears to be important in

Figure 9.11 Fatigue crack growth data for several polymeric solids. (From Hertzberg and Manson.[13]). PA = polyacetal; PS = polystyrene; PPO = polyphenylene oxide; PVDF = polyvinylidene fluoride; PC = polycarbonate; PSF = polysulphone; PMMA = polymethylmethacrylate. Note superior fatigue resistance in crystalline polymers (Nylon 66, Polyacetal, PVDF, ST801).

relation to fatigue resistance: according to Hertzberg and Manson[12] resistance to fatigue crack propagation resistance at a given stress level is dramatically improved by increasing the molecular weight or by the addition of a high-molecular weight fraction to the matrix. They also comment that fatigue resistance is enhanced by the addition of second-phase particles that lead to enhancement of toughness.

Fatigue of fibre-reinforced plastics

Some brittle non-metallic materials fail by cracking after a more or less extended period of time under a steady load. This behaviour is sometimes referred to as static fatigue. It appears to be generally agreed that failure of fibre-reinforced plastics occurs more rapidly under repeated loading than it does under a steady load.

A fibre-reinforced plastics material is a composite and if the fibrous component is to be used effectively it must be stressed more highly than

the plastics matrix. Thus, although the plastics matrix must carry a proportion of the applied load its more important function is to transmit load to the fibres. It follows that conditions at the interfaces between the fibres and the matrix are critical and this is where failure is most likely to be initiated. Initial breakdown of the resin occurs by crazing and probably occurs fairly generally. This must reduce the load-carrying ability of the material as a whole but the processes that lead to failure originate at the fibre-matrix interface and the strength of the resin-fibre bond is thus very important. In glass-reinforced plastics fracture occurs by fibre-resin debonding and break-up of the resin matrix but in plastics reinforced with carbon or boron fibres debonding occurs much less readily and the $S-N$ curves fall less rapidly with increasing number of cycles.[14] It is therefore not surprising to find that the fatigue properties of GRP are rather poor, with the fatigue strength at long endurance being as little as 20% of the static tensile strength,[15] whereas the corresponding figure for CFRP may be 70%.[16] Hertzberg and Manson[13] consider that the superiority of carbon (and boron) fibres may be due partly to their higher thermal conductivities which give relief from hysteritic heating.

With a given type of reinforced plastic the factors that would be expected to influence fatigue performance are proportion of reinforcing fibres, morphology of reinforcement, i.e. random chopped mat, unidirectional filament, woven cross-ply roving, etc., and the nature of the resin. These are also factors that control the static strength and indeed within a given group there is frequently a close correspondence between static strength and fatigue strength. Thus Owen and Morris[16] considered that the fatigue properties of their CFRP specimens depended only on the strength of the composites as determined by fibre content.

Boller,[15] working on specimens of GRP, found that although the fatigue performance of a chopped mat reinforcement was significantly poorer than that of woven cross-ply fabric there was little difference between the performance of different types of woven fabric (Figure 9.12). Since it is known that the strength of GRP

Figure 9.12 *S–N* curves of notched specimens of polyester reinforced with various glass fabrics and mat, showing effect of notch. Test conditions: 0° to warp, 73°F and 50 per cent RH and zero mean stress (taken from Boller[15].

Figure 9.13 (a) *S–N* curves of unnotched specimens of heat-resistant polyester reinforced with glass fabric (zero mean stress). (Taken from Boller.[15]). (b) Temperature dependence of the tensile/compressive fatigue limit of UP resin laminates (60% glass cloth; loading in the warp direction of the weave; mean stress $\sigma_m = 0$; taken from Domininghaus[17]).

reinforced with fabric is smaller when measured at an angle of 45° to the warp than when measured parallel to the warp it is not surprising to find that fatigue properties are influenced similarly (Figure 9.13). The effect of temperature is seen in Figure 9.13(b). A review by Harris[14] is of interest (Figure 9.14).

The nature of the resin is significant: Boller[15] found the fatigue strength of GRP at 10^7 cycles expressed as a percentage of static tensile strength to be 20–25% for polyester resin as against 40% for epoxy.

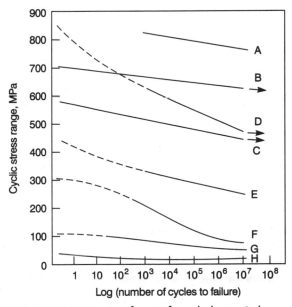

Figure 9.14 *S–N* curves for reinforced plastics (taken from Harris[14]). Points at extreme left represent static strength data.

A = Boron/epoxy laminate: 10 ply: 0 ± 45°: axial tension.
B = Carbon (type 1) polyester: unidirectional: V_f = 0.40: axial tension.
C = As B: repeated flexure cycling.
D = Carbon/epoxy laminate 18 ply: 0 ± 30°: axial tension.
E = As D: axial compression cycling.
F = Glass/polyester: high-strength laminate from warp cloth: axial tension/compression.
G = Glass/polyester: composite: chopped strand mat laminate: tension/compression.
H = Polyester dough moulding compound: V_f = 0.12: tension/compression.

115

It is worth emphasizing that fibre-reinforced plastics exhibit a considerable degree of scatter in their results both for static strength and fatigue strength, and this is true for CFRP as well as GRP. Whilst the effect of a notch is to reduce properties somewhat (Figure 9.12), it also has the effect of reducing the scatter.

Welded joints in plastics

The significance of the presence of welds in metals in determining the fatigue strength of assemblies has been discussed previously (p. 110). In the case of welded plastics, assuming that the weld itself is of high integrity there could still be the effect of stress concentrations introduced by the geometry of the join and the upset or protrusions produced. An important point is that local deformation associated with stress concentration could give rise to a local increase in temperature in view of the high damping capacity of the material. This could lead to increased deformation and accelerated failure. For amorphous polymers, molecular flow when the parts are pressed together, results in alignment of the molecules parallel to the weld line, which can reduce the fatigue strength of the weld. If high cooling rates are used with semi-crystalline polymers, an amorphous structure will develop at the weld. The thermal history is also important in determining residual stress.

Fatigue of ceramics

As we have seen, the study of metal fatigue is relatively mature, but by comparison little is known about the cyclic fatigue failure of ceramics and ceramic matrix composites. Indeed, none of these materials have ever really been designed with their microstructure optimized for fatigue failure resistance. Yet ceramic materials are proposed for structural applications such as gas turbine and engine components and artificial joint prostheses, all likely to experience fatigue loading. It is increasingly realized that ceramics are susceptible to fatigue failure. While current

technologies such as transformation toughening and fibre reinforcement have resulted in fracture toughnesses an order of magnitude greater than those attainable 20 years ago, the limited fatigue data shows that fatigue thresholds are as low as 50% of the fracture toughness[18]. Also, these data follow a Paris power-law relationship of the form:

$$\frac{\mathrm{d}a}{\mathrm{d}N} = C(\Delta K)^m$$

where C and m are constants. The value of m for ceramics varies between 10 and 100 for ceramics, much higher than the values typically found for metals (in region B of Figure 9.7). Following the equations developed in page 107, these high values imply that the number of cycles to failure will be proportional to the reciprocal of the applied stress raised to a large power. Hence for ceramics a two-fold increase in the applied stress reduces the projected life by 3–30 orders of magnitude. It is also worth noting that the fatigue and fracture surfaces of advanced ceramics are very similar in appearance.

In order to try and guarantee acceptable component life it is important to proof test ceramic components or resort to scanning electron microscopy to ascertain the initial defect population characteristics.

Fatigue of concrete

The fact that concrete will fail in fatigue has long been known. The results of fatigue tests on concrete exhibit a great deal of scatter and interpretation of the results is also complicated by the effects of age, moisture gradients within the concrete and strain rate. These factors have been investigated at the Transport and Road Research Laboratories by Galloway and his co-workers.[19] At a given stress level the cycles to failure increase as the concrete ages. This is presumably associated with the fact that the static strength of the concrete is also increasing with age so that the given stress represents a progressively smaller fraction of the static strength.

Raithby[20] suggests that since variations in fatigue behaviour due to changes in the major manufacturing and testing conditions such as mix proportions, aggregate type, moisture condition, and age follow closely the variations in static strength due to similar changes in the same conditions, it is reasonable to estimate the fatigue performance of a particular type of concrete for various ages from a 'characteristic' design curve obtained by normalizing the cyclic stress in relation to the quasi-static flexural strength.

If the concrete is reinforced or prestressed with steel then the fatigue performance of the concrete assumes a different significance since the fatigue properties of the steel then control the performance of the structure as a whole. Since the purpose of reinforced concrete design is to ensure that tensile loads are carried by the steel whilst compressive loads are carried by the concrete, the presence of even quite extensive cracking in the concrete should not be too serious from the point of view of mechanical failure. However, steel is a material that corrodes very readily and its efficacy as a reinforcement in concrete depends upon the high alkalinity of the surrounding cement which has a passivating effect. In marine structures in particular the 'cover' of concrete plays an important part in insulating the steel from the aquatic environment and excessive cracking would be expected to lead to depassivation of the steel and consequent corrosion and loss of strength.

9.5 Materials selection for fatigue resistance

The selection of a material for use in a fatigue-dominated application is probably the most difficult task faced by the materials engineer.

It is useful at the outset to recognize that engineering design has had, and often still has, an excessive attachment to the provision of static strength. It cannot be too strongly emphasized that what is required in all cases is the best possible *combination* of properties and if, in order to obtain improved fatigue resistance, it is

necessary to sacrifice a degree of static strength then this must be accepted. The metallurgical treatments currently applied to advanced alloys commonly sacrifice a degree of static strength so as to optimize overall properties including toughness and fatigue resistance.

In relation to metallic materials, the first important point to establish in any given case is how much of the total life can be provided by Stage I crack growth. In the case of a joined component this can be virtually disregarded: the total fatigue life will be essentially determined by the rate of propagation of a macrocrack. In the case of a machine part such as a crankshaft which is unjointed but contains stress concentrations, then Stage I endurance may well be short; however, there is scope for optimization and it is worth considering how this might be done. As previously discussed; fatigue cracks in ductile materials initiate in narrow bands of intense slip and this process will be inhibited by increase in yield strength such as may be produced by stabilized dislocation networks and by increased amount of a fine dispersion of particles. On the other hand, it will probably be accelerated by increasing amounts of large hard particles, especially when these are set in a hard matrix (fatigue cracks, when initiated at fractured intermetallics, may grow for short distances as Stage I cracks before changing direction). The crack initiation process will also be accelerated as the magnitude of any local stress concentration factor increases.

Clearly, the first priority is to design out all stress concentrations as far as practicable. Since complete removal of stress concentrations is often impossible, consideration must then be given to minimizing as far as possible the effects of those that are left, by grinding and polishing of all contours and surfaces. Further improvements can be produced by surface hardening of critical fillets and contours by surface rolling, flame hardening or nitriding.

If the residual maximum elastic stress concentration factor is then not too high it is possible to envisage the use of a material of high static strength. However, the higher the strength, the greater is the need to eliminate coarse second-phase particles and to ensure that the matrix is refined and homogeneous, and this will be expensive. Thus, in steels, a ferrite-pearlite structure is generally undesirable; on the other hand, whilst a hardened and tempered structure would be capable of giving the best results, this can only be achieved with sufficient hardenability to deliver a uniform, refined microstructure. In very large artefacts this would require high contents of alloying elements and if it happened that a rather large wrought component had its origin in an ingot which was not a great deal larger, then metallurgical homogeneity would be hard to achieve. Should the material under consideration be an aluminium alloy, then strengthening of the matrix will necessarily involve precipitation hardening and because of the intrinsic nature of aluminium alloys the production of a refined, homogeneous microstructure is even more difficult. The avoidance of coarse inter-metallic compounds, solute-depleted zones, coarse grain-boundary precipitates – all of which are potent sites for crack initiation – requires close control of alloy purity and highly developed techniques of heat treatment.

Nevertheless, all of this can be done, and the technical expertise is available, but the materials are expensive and the technology critical: it then becomes a matter of judgement as to whether this is the best way to proceed. In high-technology applications characterized by an unavoidable need for maximum fatigue resistance then it may indeed be necessary to squeeze the maximum of Stage I endurance out of the material, and this seems to be the case in aircraft materials. In other applications it may be preferable to neglect any possible contribution from initiation processes and centre attention on Stage II crack growth. One or other of two philosophies may then be adopted. The low-technology approach is to aim for slow Stage II crack growth by keeping the design stresses low. This will permit the use of a ductile material of low static strength and in such a material it is likely that the major stress concentrations present will be at least partially unloaded by plastic deformation. Even if this does not occur and a crack is fairly rapidly initiated at a stress concentration it will tend to grow out of the high-stress region and as it enters a field of lower stress may even become non-propagating. Of course, low design stresses imply heavy, inefficient structures. The alternative approach is to recognize that the presence of a crack, even quite a large crack, does not necessarily prevent an engineering component from performing quite satisfactorily. There is a documented example of a turbine shaft which initiated a crack in the first year of its service life and survived in service for a further 19 years before final failure. The immanent requirements of any procedure for dealing with a flawed structure are first to know that the crack is there – that is to say, to be able to detect the crack by some reliable method of observation and monitor its progress – and second, to be able to calculate its rate of advancement by means of the appropriate fracture mechanics law. This allows the use of strong materials and efficient structures but if catastrophic failure is to be avoided it has the necessary consequence that life between inspections may be short and down-time for periodic repairs may be large. This is the 'safe-life' approach to fatigue control. In applications of extreme criticality, as in Class I aircraft components, the 'fail-safe' approach may supervene and the need for some controlled structural redundancy then detracts from the overall structural efficiency. There are also, strict regulations for the inspection and the non-destructive evaluation of cracks.

References

1. R. E. PETERSON: *Stress Concentration Design Factors.* John Wiley & Sons, New York, 1953.
2. C. F. BOULTON: *Welding Inst. Res. Bull.*, December 1975.
3. P. C. PARIS and F. ERDOGAN: *J. Bas. Eng., Trans. ASME Series D*, 1963; **85**, 528.
4. R. G. FORMAN, V. E. KEARNEY and R. M. ENGLE: *J. Bas Eng. Trans. ASME Series D*, 1967; **89**, 459.
5. P. T. HEALD, T. C. LINDLEY and C. E. RICHARDS: *Mater. Sci. Eng.*, 1972; **10**, 235.

6. I. L. MOGFORD: in *Developments in Pressure Vessel Technology*, Vol. 1: *Flaw Analysis* (ed. R. W. Nicholls). Applied Science Publishers, 1979.

7. A. SAXENA: Proc. Conf. 'Fatigue '81': Society of Environmental Engineers, University of Warwick, March 1981.

8. A. C. LOW: *International Conference on Fatigue*. Institution of Mechanical Engineers, 1956, p. 206.

9. R. B. HEYWOOD: *Designing against Fatigue*. Chapman and Hall, 1962.

10. P. C. BIRKMOE, D. F. MEINHEIT and W. H. MUNSE: *Proc. Amer. Soc. Civ. Eng., Struct. Div.*, 1969; **95**, 2011.

11. N. E. FROST, K. J. MARSH and L. P. POOK: *Metal Fatigue*. Oxford University Press, 1974.

12. R. W. HERTZBERG and J. A. MANSON: Proc. Conf. 'Fatigue '81', Society of Environmental Engineers, University of Warwick, 1981

13. R. W. HERTZBERG and J. A. MANSON: *Fatigue of Engineering Plastics*. Academic Press, 1980.

14. B. HARRIS: *Composites*, 1977; **8**, 214.

15. K. H. BOLLER: *Mod. Plastics*, 1957; **34**, 163.

16. M. J. OWEN and S. MORRIS: *Mod. Plastics*, 1970; **47**, 158.

17. H. DOMININGHAUS: *Plastics for Engineers*. Carl Hauser Verlag, 1993.

18. R. O. RITCHIE and R. H. DAUSKARDT in *Encylopedia of Advanced Materials'*, eds. D. BLOOR, R. J. BROOK, M. C. FLEMINGS and S. J. MAHAJAN, Pergamon Press 1994.

19. J. W. GALLOWAY, H. M. HARDING and K. D. RAITHBY: Transport and Road Research Laboratory Report LR 865, 1979.

20. K. D. RAITHBY: in *Developments in Concrete Technology* (ed. F. D. Lydon). Applied Science Publishers, 1979.

10
Creep and temperature resistance

Creep is deformation that occurs over a period of time. Under certain conditions it will, if allowed to do so, culminate in fracture. Generally, creep is the result of an externally applied load but can also occur as the result of self-weight. Lead sheet, when used on an inclined roof or vertical face, will, after a period of years, be thicker at the bottom than it is at the top; not necessarily a serious matter.

After extensive creep, however, the lead will often exhibit cracks, which is more serious. There are thus two aspects to the creep phenomenon, one being concerned with deformation; the other with fracture or creep rupture. A typical deformation-limited situation is that of a blade in a steam turbine which must not lengthen in service to the point at which it fouls the casing. An example of a rupture-limited situation is the tungsten filament of an electric light bulb. Although the windings may sag due to progressive creep strain, thereby decreasing the output of light, the lamp does not actually fail until the coil breaks. In many applications it is necessary to consider both forms of failure. In aircraft engines, for example, it is deformation within prescribed limits which must form the initial basis of design, but it is also recognized that during emergencies the service conditions will for a short time be severely exacerbated and the designer then needs to know for how long a given part will operate under these extreme conditions without fracture. The only circumstances in which the possibility of creep rupture may safely be neglected are those in which the service condition involves stress relaxation. The simplest example here is that of a screwed fastener. When two articles are clamped together by a bolt and nut, the clamping force is provided by the elastic extension of the shank of the bolt as the nut is tightened

down. If the conditions of service are such that creep occurs the stress in the bolt is progressively relaxed and as it does so the danger of fracture recedes. Of course, the clamping force simultaneously decreases and bolts on equipment such as pressure vessels which operate under creep conditions must regularly be retightened, and if this is done often enough, rupture again becomes a hazard.

10.1 The evaluation of creep

All materials creep under load at all temperatures, but a very wide range of creep behaviour is revealed when comparisons are made in terms of the three important parameters that describe the creep process – namely, stress, temperature and time. The generalities of creep behaviour are well understood, i.e. the higher the temperature and the higher the stress the greater is the creep rate and the shorter is the time to fracture, but the complete quantitative description of the creep behaviour of engineering materials, particularly of complex heat-resisting materials, is often lengthy and complex.

Creep behaviour is described by the conventional creep curve (Figure 10.1), made up of three successive stages, viz: primary (or transient), secondary (or steady state) and tertiary. This behaviour is exhibited by all simple materials, whether metals, plastics or ceramics, but complex materials may show considerable variations. An engineering part should spend the majority of its service life in the steady-state range of creep since once the tertiary stage is entered the creep strain accelerates rapidly to fracture.

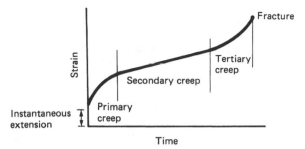

Figure 10.1 Conventional creep curve, showing the stages of creep.

The strain rate during the steady-state regime is often described as follows:

$$\dot{\epsilon}_{ss} = A\sigma^n \exp\left(\frac{-Q}{RT}\right)$$

where σ is stress, T is temperature (°K), A is a constant, n is the stress exponent, and Q is the activation energy for creep (J/mole). These are material constants which must be determined experimentally; n typically takes a value between 3 and 8, but may be higher.

The creep curve is difficult to describe in mathematical terms. There have been numerous attempts to do so, most of limited usefulness. There are also many apparently gross, but often useful, simplifications: for example, it is commonly assumed that time to rupture is inversely proportional to the creep rate, leading to equations such as

$$l_r = A \exp\left(\frac{Q}{RT}\right)$$

Whatever its form, the whole creep curve is the only comprehensive way of describing what is a basic material property, just as the stress-strain curve is the only complete way of describing short-term plastic behaviour. Unfortunately, creep curves are difficult and expensive to determine and although many are available (e.g. Figure 10.2), it is not practicable to obtain such a curve for every possible combination of stress and temperature. Even with established materials, information useful for engineering parts required to give service for periods of 20 years or more is often not available. Therefore, although codes of practice worldwide specify design stresses in the creep range in terms of creep strength or creep rupture strength at 100,000 hours (11.4 years), the scarcity of suitable data means that less appropriate criteria based on, say, creep rate at 1000 hours (6 weeks) have to be used.

For design purposes, however, it is not necessary to refer to the complete creep curve and there are various ways of providing the necessary information (Figure 10.3a,b). In general, there is more information available relating to creep rupture than to creep strain. Just as in

Figure 10.2 Creep curves for Nimonic 90 showing the effect of stress: (a) at 700°C (1292°F), (b) at 750°C (1382°F). (Reproduced by courtesy of Wiggin Alloys Ltd.[4])

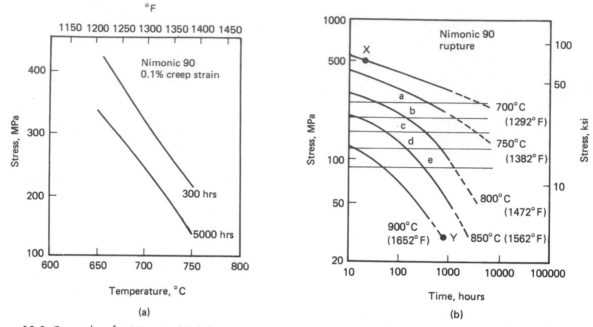

Figure 10.3 Creep data for Nimonic 90; (a) stress to give 0.1% creep strain; (b) stress for creep rupture. (Reproduced by courtesy of Wiggin Alloys Ltd.[4])

short-term testing it is easier to determine tensile strength rather than a proof stress, so in creep testing it is much easier to test for rupture than for a given strain. The standard creep-rupture test is carried out at constant stress and temperature, and measures the time to rupture with no account being taken of deformation, except for total elongation of the fractured test piece to provide a criterion for rupture ductility. This information can be useful for quality control and the comparison of materials. For many design purposes, however, it is inadequate and the data for materials are often incomplete. This is the nub of the problem. A great deal of effort has been devoted to finding ways in which meagre data can be supplemented or extended by interpolation or, more critically (and hazardously), extrapolation. Up to a score of equations or correlation parameters have been proposed for these purposes. Some, such as the Dorn and Larson–Miller parameters, are based on physical principles; others, e.g. the Manson–Haferd parameter, are purely empirical, but the aim is first to find a way of plotting data such that it lies on simple curves or, preferably, straight lines, and then to obtain a function which incorporates either time and temperature or time and stress into a single correlation parameter.

One of the oldest ideas is that creep rupture data are linear when the temperature of test is plotted against the logarithm of time to rupture. Manson and Haferd[1] extended this idea by assuming that lines of constant stress, plotted separately, would converge at a point, which defines the two constants T_c and log t_c in the Manson–Haferd correlation parameter:

$$\text{MHCP} = \frac{T - T_c}{\log t_r - \log t_c} \quad \text{(see Figure 10.4)}$$

Larson and Miller[2] on the other hand, taking as their starting point the Arrhenius relationship

$$\frac{d\epsilon}{dt} = A \exp\left(\frac{-Q}{RT}\right)$$

in which Q is the creep activation energy and R the universal gas constant, proposed that plots at

Figure 10.4 The Manson–Haferd method for correlating creep data.

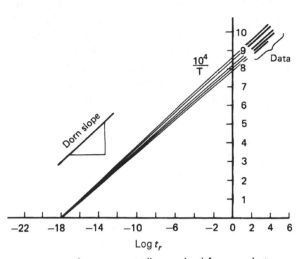

Figure 10.5 The Larson–Miller method for correlating creep data.

constant stress of $1/T$ against $\log t_r$ would intersect at a single point having the coordinates $0, -C$ (Figure 10.5). Their correlation parameter is given by $\text{LMCP} = T(C + \log_{10} t_r)$.

The Dorn[3] correlation parameter, which is written

$$\text{DCP} = t_r \exp\left(\frac{-Q}{RT}\right),$$

is also evaluated from plots of $1/T$ against $\log t_r$ but in this case the separate lines of constant stress are predicted to be parallel, of slope $2.3R/Q$ (Figure 10.5). This parameter was intended for the correlation of creep strains against stress but has also been used for rupture.

These, and other, correlation parameters are used to fill out scarce data, and provided this is only done over a carefully restricted range this can be very valuable. For the purposes of materials selection there are two possible uses for correlation parameters: one is to provide for the extrapolation of short-term data to long times, the other to provide a basis for the comparison of a wide span of creep-resisting materials. Neither of these aims is met particularly well by the available methods. It is

instructive to apply the three parameters mentioned to the data in Figure 10.3b. Figures 10.4 and 10.5 show stresses from this graph at *a* to *e* reworked in the manner of Manson and Haferd and Larson and Miller. The linear extrapolations do not truly intersect cleanly in the manner

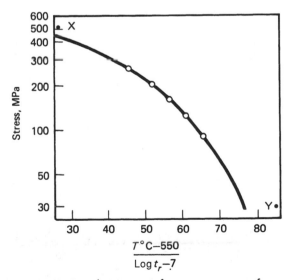

Figure 10.6 Correlation curve for creep rupture of Nimonic 90 using the Manson–Haferd correlation parameter.

shown – 'interpretation' is needed to give reasonable values for the required constants. Corresponding values of the Dorn parameter were also calculated. The three resulting correlations are shown in Figures 10.6–10.8. The value of the extrapolations can be judged from points X and Y which represent values taken from extreme positions on the original curves. It is seen that

the extrapolated curves are often in error. It is clear that extrapolation of creep data is hazardous and should only be carried out in the light of considerable experience of the materials under consideration.

Originally, Larson and Miller proposed that the value of C in their parameter should be 20 for all materials (t is then in hours). However it is found that in practice C can vary from 17 to 23, but nevertheless the correlation parameter $T(20 + \log t)$ is commonly used as a basis for comparisons within collections of widely disparate materials (Figure 10.9). Such comparisons are useful, if of limited accuracy.

Figure 10.7 Correlation curve for creep rupture of Nimonic 90 using the Larson–Miller correlation parameter.

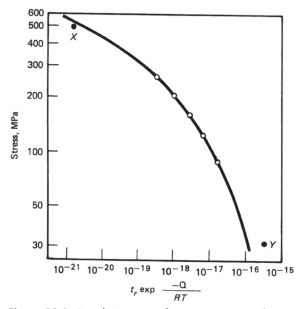

Figure 10.8 Correlation curve for creep rupture of Nimonic 90 using the Dorn correlation parameter.

10.2 The nature of creep

The temperature of service determines whether or not creep must be considered as a possible mode of failure for a given material component, and the critical level of temperature depends upon the anticipated service life. For example, the aluminium alloy 2618 (RR58), containing copper, magnesium, nickel and iron, was used for the forged impellers in the Whittle jet engine for the Gloster Meteor aircraft. In this application, the temperature of operation was up to 200°C (400°F) and the stress was high enough to limit the life to a few hundreds of hours. Clearly, 200°C is a creep-producing temperature. The same alloy is used for the main skin of Concorde and over most of this structure the temperature does not exceed 120°C (250°F). However, the Concorde airframe was designed for a life in service of 20,000–30,000 hours, a long enough period for 120°C to constitute a possible hazard but has not proved to be so. Creep is important in both applications even though the temperatures are very different.

If we wish to obtain a first approximation for the maximum operating temperature of a metallic material it is often useful to consider its melting point. This is because in metals the useful operating temperature is limited by the rate at which internal microstructural changes take place. Now at a given temperature, the

Figure 10.9 Creep resistance of various materials using the Larson-Miller correlation parameter. (Data by courtesy of Wiggin Alloys Ltd. (nickel alloys); The Institute of Materials (titanium alloys); and Fulmer Research Institute (ferrous materials).) N.B. The curves for the ferrous materials lie at the centres of scatter bands which have been omitted for the sake of clarity.

energy of thermal fluctuations in a metal lattice is given by kT, where T is the absolute temperature and k is Boltzmann's constant; any process which is thermally activated will proceed at a rate which is proportional to exp (Q/kT), where Q is an energy barrier, called the activation energy for the process in question, which must be surmounted by the applied thermal energy. Secondary creep in metals proceeds by processes which are dependent upon the diffusion of vacancies, and there is therefore very good correspondence between the activation energy for recovery creep Q_{CR} and that for self-diffusion Q_D. A high value of Q_D is therefore essential for high creep resistance and since Q_D is approximately proportional to the melting point

of the metal, T_m, on the Kelvin scale, it follows that the first requirement of a creep-resisting alloy is a high melting point. In agreement with this, experience shows that, in any alloy, it is difficult to produce useful mechanical properties at temperatures higher than $2/3\ T_m$ for that material and sometimes the actual achievement is much less (see Table 10.1).

The best creep-resisting materials owe their development to what might be described as inspired pragmatism, but with hindsight it is possible to understand the scientific basis of what has been achieved. Creep resistance is a form of strength and, apart from the influence of grain boundaries, and thus grain size, the mechanisms which are employed to enhance creep strength

125

TABLE 10.1.

Metal	Melting point		Potential operating temperature, $2/3\ T_m$		T/T_m actually achieved
	°C	°K	°K	°C	
Al	660	933	620	350	0.56 (RR58 at 250°C)
Cu	1083	1356	900	630	
Ni	1453	1726	1150	880	0.74 (Nimonic 115 at 980°C)
Fe	1536	1809	1200	930	0.47 (ferritic steel at 575°C)
					0.57 (austenitic steel at 750°C)
Ti	1668	1941	1290	1020	~0.4
Zr	1852	2125	1420	1150	
Cr	1900	2173	1450	1180	0.6 could be achieved if Cr could be made sufficiently ductile
Hf	2222	2495	1640	1370	
Nb	2468	2741	1830	1550	0.54 could be achieved if a satisfactory coating could be found
Mo	2620	2893	1923	1650	
Ta	2996	3269	2180	1910	
W	3380	3653	2433	2160	0.76 (electric light filament at 2500°C)

are broadly the same as those used to improve ordinary room temperature strength, i.e. solution-hardening, precipitation-strengthening and dispersion hardening, in these contexts the difference between one creep-resisting material and another depends on the relative persistence of the hardening mechanisms with increase of temperature. At low temperatures – less than $0.25\ T_m$ – thermal softening cannot occur so the creep rate decreases rapidly with time, eventually to zero. This is termed logarithmic creep, and it may be attributed to strain-hardening, i.e. a progressive increase in dislocation density and refinement of the dislocation network. The total strain is very small and it is relatively insensitive to stress and temperature. It is only important in applications where high dimensional stability is critical, e.g. instrumentation.

When the temperature is $1/3\ T_m$, or higher, the strain-hardening which results from crystallographic slip within the grains is not retained within the structure since the temperature is high enough for thermally-activated processes such as dislocation climb and cross-slip which transform the dislocation networks into configurations of lower energy i.e. the formation of a subgrain structure. This results in recovery creep, and the creep rate under these conditions is given by

$$\dot{\epsilon} = A\sigma^n \exp\left(\frac{-Q_{CR}}{RT}\right)$$

where T = temperature; σ = stress; Q_{CR} = activation energy; R = universal gas constant; A, n = constants.

The strong temperature-dependence of recovery creep is clear, as is the stress-dependence, where the stress exponent, n, takes the value 3–8. In recovery creep two separate processes occur simultaneously, i.e. strain-hardening and thermal softening; but provided the temperature is high enough the latter is the rate-controlling process. It is customary, therefore, to divide recovery creep into two stages, primary and secondary. Primary or transient creep (β-creep) occurs early in the creep process when the initially undeformed structure is able to sustain a high rate of strain-hardening, so that although thermal softening is occurring, it is not fast enough to nullify the hardening – the creep rate therefore decreases. Primary creep can be descri-bed by the equation $\epsilon = \epsilon_0 + \beta t^{1/3}$ where t is time and ϵ_0 is the initial extension.

After a period the rate of strain-hardening diminishes to a level at which the thermal softening processes are able to keep pace. The coarsening effect on the dislocation structure produced by dislocation climb and polygoniza-tion then balances the refining effect of the strain-hardening, and if the stress is moderate the creep rate may remain constant for a significant period. This is called secondary, or steady-state creep (κ-creep). At high stresses, the time taken up by secondary creep may be vanishingly small, due to the onset of tertiary creep which is governed by quite different mechanisms.

Neither primary nor secondary creep result in irreversible damage to the material, no matter how long they continue. This is not true of tertiary creep. Whilst the accelerating creep rate which is characteristic of tertiary creep could arise from metallurgical changes such as recrys-tallization, or changes in cross-sectional area due to macroscopic instability, such effects are not the general cause of tertiary creep. This is usually associated with the slow but continuous forma-tion of internal micro-cracks. These represent irreversible damage and final fracture is inter-crystalline, suggesting that the cracks are formed as the result of processes localized at grain boundaries. Two mechanisms have been observed:

(1) *Triple-point cracking* is associated with the triple-point where three grain boundaries meet. In general, the grains at a triple-point will be of differing orientations and in responding to the applied stress they endeavour to deform in directions that are mutually incompatible. The result is to set up a stress concentration which eventually nucleates a crack.
(2) *Cavitation* is a different type of intercrystal-line fracture which commences with the formation of small spheroidal cavities at intervals of a few microns at grain bound-aries that are mainly, though not exclusively, transverse to the applied stress. The cavities

slowly grow, giving the appearance of a string of beads and eventually coalesce to form cracks. The life of the specimen is determined by the time needed for the ligaments between the coalesced voids to neck down to zero cross-section. The mecha-nism is thought to be vacancy diffusion: considering a grain boundary which is per-pendicular to an applied tensile stress, the elongation of the specimen provides a driv-ing force for the diffusion of vacancies from grain boundaries to the walls of the pores, which is equivalent to the deposition of atoms at the grain boundaries, thus lengthen-ing the specimen. The rate of diffusion creep (Herring-Nabarro creep) is given by

$$\dot{\epsilon} = \frac{C\,D\sigma}{d^2}$$

where D = diffusion coefficient; d = grain diameter; σ = stress; C = constant.

Diffusion creep is thus much more weakly stress-dependent than recovery creep, but is strongly favoured by small grain size. It is important only at high temperatures (>0.8 T_m). Cavitation and triple-point cracking may be found in the same specimen but cavitation tends to be found at higher temperatures and lower stresses than triple-point cracking, which is more characteristic of higher stresses and rather lower temperatures. This may be explained by the greater magnitude of the stress concentrations set up at grain corners when the stress is high and the reduced ability of the grains to unload them when the temperature is lower.

10.3 The development of creep-resisting alloys

Melting point

For the reasons previously mentioned it is desirable for a creep-resisting material to have a high melting point, and the crude truth of this statement is demonstrated by the poor creep resistance of metals such as lead and tin. The rate of self-diffusion is lower in metals with a higher melting point. Nevertheless, the two metals that form the basis of the best creep-resistant alloys, iron and nickel, do not have especially high melting points (Table 10.1). Clearly, factors other than melting point are important, and many of the higher melting point metals are inapplicable for practical reasons.

The metals with the highest melting points, all above 2000°C (3630°F), are the refractory metals, niobium, molybdenum, tantalum and tungsten. Their use is limited, first because they are manufactured by the power metallurgy route and therefore cannot easily be produced in large or complex sections, and second because of their low resistance to oxidation.

However, sheet and wire products are readily available; sheet molybdenum is used for furnace heat shields and as already mentioned tungsten is used for electric lamp filaments and electronic value components. These applications employ controlled atmospheres. Attempts have been made to overcome the poor resistance to oxidation by the provision of protective coatings but these have had limited success.

Relative to its melting point the creep resistance of titanium is rather poor – certain alloys are, however, used as aircraft engine alloys, the low density of titanium offering considerable incentive in this application.

Although chromium is an important alloying element in many creep-resisting alloys it is not itself used because of pronounced lack of ductility as so far produced. Zirconium and hafnium are too expensive for general use.

Lattice structure

Metals of face-centred-cubic structure are the basis of the best creep-resisting alloys (e.g. Nimonic alloys, austenitic stainless steels) and this is because of the nature of the defects that can exist in these alloys. The face-centred-cubic lattice is a close-packed structure which is stacked in an ABCABC sequence when the lattice

is perfect. However, real metallic lattices are not normally perfect and to a greater or lesser extent they contain irregularities in the stacking sequence known as stacking faults. Whereas in the simple cubic lattice an edge dislocation can be formed by the insertion of a single half-plane of atoms, the more complex face-centred-cubic structure requires the insertion of two extra half-planes. When these become separated the degree of separation, which varies from metal to metal, defines the magnitude of the resulting stacking fault. The more extended the dislocation the more difficult it is for dislocation climb to occur and the higher, therefore, is the creep resistance.

The equation quoted by Lagneborg[5] for steady-state creep in pure metals:

$$\dot{\epsilon} = AD\gamma^{3.5}\left(\frac{\sigma}{E}\right)^5$$

where D = self-diffusion coefficient; γ = stacking fault energy; E = Young's modulus; A = a constant; σ = stress, suggests that a creep resistant material should have a low diffusivity, a high elastic modulus and a low stacking fault energy. The first two of these are provided by materials of high melting point.

Stacking fault energy accounts, at least in part, for the difference between ferritic and austenitic steels shown in Table 10.1. The BCC structure of ferritic steels is not close packed and cannot therefore contain stacking faults of the same type as FCC structures. Any faults which are present are of high energy. Within the range of face-centred-cubic metals, aluminium with a high stacking fault energy and relatively low melting point is inferior to copper with low stacking fault energy and high melting point (Table 10.2).[6]

If low stacking fault energy is advantageous to pure metals then, considering alloys, it will be equally advantageous to add alloying elements which lower the stacking fault energy, as in brasses, bronzes and nickel-base alloys. One of the reasons for adding cobalt to Nimonic alloys is to reduce stacking fault energy. The effect on stacking fault energy of adding aluminium to

TABLE 10.2.[6]

	Aluminium	Copper	Cu–7.5 Al	BS 304 stainless steel	BS 303 stainless steel
Stacking fault energy (erg/cm^2) = (mJ/m^2)	200	40	2	20	8

copper is shown in Table 10.2. The sensitivity of stacking fault energy to chemical composition is illustrated by the stainless steels data in Table 10.2 where BS 303 contains a sulphur addition for free machining.

Solid solution strengthening

Elements in substitutional solid solution increase creep resistance, and the greater the concentration the lower is the creep rate.

Although there may be different explanations of creep-strengthening by solutes for different alloys there does seem to be a genuine solution-hardening effect.[5] The contribution to creep resistance due to solid solution alloying is useful but not large, and although many high-temperature alloys exist which are quite heavily loaded with elements in solid solution, these alloys are frequently used for applications in which stress is not a major factor: it is then their ability to combine resistance to hot corrosion with good workability and fabricability that is of value. Many of the alloys based on the nickel–chromium binary system with additions of cobalt, molybdenum or tungsten fall into this category. So also do the alloys of iron, nickel and chromium – when these also contain carbon they are very similar to the more highly alloyed austenitic stainless steels.

Elements in interstitial solid solution are in a different category since they may enhance the creep resistance of materials by virtue of the strain-ageing mechanism. Strain-ageing may occur in many materials but is best known for its effects in ferritic steels when caused by nitrogen and to a lesser extent carbon. The usefulness of ferritic carbon steels up to temperatures as high as 350°C (660°F) is largely due to strain-ageing effects; alloy steels operating at temperatures higher than this may also benefit from this mechanism.

Cold work

Creep resistance is greatly increased by prior cold work, but this effect could not be expected to persist for long in the steady state regime at temperatures around or above $0.5T_m$. However, longer-lasting benefit may be expected in the lower-temperature part of the creep range and this is valuable in the silver-bearing coppers used in motors and generators.[7]

Precipitation and dispersion-hardening

The most important method of improving the creep strength of a metal is to incorporate within the grains a fine dispersion of second-phase particles. These particles perform two functions: (1) they impede dislocation glide and so increase work hardening; (2) they inhibit recovery by anchoring the dislocation networks formed by strain-hardening.

Properties are optimized by a certain magnitude of precipitate size and spacing. When the particles are very small they do not present significant obstacles to dislocation climb. When the spacing is large (1) the dislocations may more readily loop between them, and (2) the flexibility of the dislocation line is such that only part of it needs to climb. The result is a minimum in the creep rate (Figure 10.10).

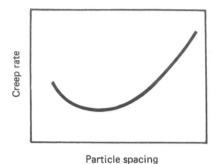

Figure 10.10 Effect of particle spacing on creep rate.

There are two ways of producing particle-hardening: dispersion-hardening and precipitation-hardening. Examples of the former are SAP (sintered aluminium products) in which the hardening particles are alumina, (SAP is no longer available) and TD (thoria dispersed) nickel. Examples of the latter are the precipitation-hardened nickel alloys. Particle-hardened materials exhibit a stronger stress-dependence of the creep rate than do pure metals and solid solutions, and dispersion-hardened materials are more strongly stress-dependent than precipitation-hardened materials.[5] The size of the hard second phase in dispersion-hardened materials is typically 0.3–300 nm; discontinuously-reinforced metal matrix composites represent an extension of these alloys in some respects, although the size of the hard second phase is about 3–300 μm. This size difference means that for MMCs, strengthening by dislocation blocking and bowing decreases, whereas strengthening by dislocations being punched out from the particles increases. Load transfer from the metal matrix to the ceramic reinforcement results in improved creep strength over the metal alloy, as ceramics exhibit higher creep strength due to their high melting point, lower diffusion coefficient and lower dislocation mobility. However, this assumes good interfacial bonding between matrix and reinforcement, and tertiary creep tends to occur relatively quickly in discontinuously reinforced MMCs as cavitation is initiated at the matrix-reinforcement interface, particularly at fibre ends or particulate asperities. The creep behaviour of discontinuously-reinforced MMCs has been described by a power law expression (see p. 121), using the concept of a threshold stress below which creep will not occur. This concept has also been applied to dispersion strengthened alloys where the threshold stress results from the interaction of dislocations with the second phase. The size of the second phase in MMCs is generally too large for such an effect, but many of the MMC studies utilize aluminium alloy material fabricated by powder metallurgy and thus also containing a fine dispersion of aluminium oxide particles from the original powder surface, which may explain the threshold stress phenomenon.

The creep rate of continuous ceramic fibre reinforced metal alloys is much lower than discontinuously-reinforced materials, which is to be expected from the more efficient load transfer. However, these are highly specialized and expensive materials, only produced in experimental volumes. For example, the Space Shuttle fuselage and nose landing gear use a total of 242 unidirectional-boron reinforced aluminium alloy tubes, providing a weight saving of about 45% over the monolithic aluminium alloy.

Unfortunately, precipitation-hardened structures are not thermodynamically stable so that if the temperature of service is not a good deal lower than the ageing temperature, the precipitate coarsens progressively in service, with resultant increase in inter-particle spacing and progressive decrease in creep resistance. Thus, the creep life of a precipitation hardening alloy will depend upon (1) the condition of the alloy when it enters service, and (2) the temperature of service relative to the heat-treatment temperature at which the alloy develops its best properties.

It is highly desirable to minimize particle coarsening and there are two ways of doing this:

(1) make the chemical composition of the precipitate as complex as possible, and
(2) reduce the thermodynamic driving force for coarsening.

As an example of (1) the Nimonic alloys are strengthened by the γ' phase (Ni_3Al containing dissolved titanium). The alloys also contain molybdenum which does not dissolve in γ'. Therefore, if the precipitate is to coarsen, not only must aluminium and titanium diffuse into the precipitate but molybdenum must also diffuse away.

Reducing the driving force for coarsening involves reducing the interfacial energy between precipitate and matrix. The lattice dimensions of the precipitate should be as close as possible to those of the matrix. In order for some particles to grow larger, it is necessary for other particles to dissolve. The time t required for a small particle of radius r to dissolve as a result of growth of a particle of large radius R, can be obtained[8] from the Thomson-Freundlich equation for the variation of solubility with surface curvature assuming $R \gg r$ and taking into account the solute diffusivity.

$$t = \frac{r^4 kT}{DC \gamma R^2}$$

where C = equilibrium solubility of particle; D = diffusivity; and γ = particle/matrix interfacial energy. Thus, coarsening time is increased by employing a particle which has a low solubility in the matrix and a low degree of lattice mismatch.

To extend the high temperature life of a precipitation-hardened material it is possible to use two precipitates in the same alloy, one more sluggish in its response to the service temperature than the other. It is then possible to arrange that as the first precipitate over-ages and ceases to be effective, the second comes into operation.

Glen[9] has represented these ideas graphically, obtaining extra clarity by plotting data as curves of total creep strain against creep rate. Several idealized curves of this sort are shown in Figure 10.11. Curve 1 represents the behaviour of a pure metal and curve 2 that of a simple substitutional solid solution. Curve 3 illustrates the behaviour of an alloy which is subject to strain-ageing whilst curve 4 and 5 typify the responses from

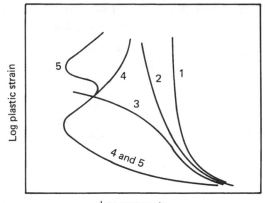

Figure 10.11 Hypothetical creep-rate curves. (After Glen[9].)

two precipitation-hardening alloys, one (curve 4) with a single precipitate, the other (curve 5) with two precipitates one more sluggish than the other.

The Nimonic alloys previously mentioned are some of the most advanced creep-resisting alloys presently available. Nickel, the basis metal, has a fairly high melting point and has the additional advantage of a face-centred-cubic structure. To the nickel is added sufficient chromium to provide oxidation resistance without destroying the face-centred-cubic structure. The first alloy was Nimonic 75, essentially 80/20 Ni/Cr with additions of titanium and carbon for precipitation hardening.

	C	Cr	Ti	Ni
Nimonic 75	0.12	20.0	0.4	bal.

It was then found that a more effective precipitation hardening agent was one based on the FCC phase Ni_3Al, in which titanium can replace some of the aluminium to give $Ni_3(Ti,Al)$, termed gamma-prime, γ'. The first of the turbine blade materials to be strengthened with γ' was Nimonic 80.

	C	Cr	Ti	Al	Ni	Zr	B
Nimonic 80A	0.05	20.0	2.3	1.3	bal.	0.05	0.003

Further improvements were then made by adding cobalt to lower the stacking fault energy of the nickel. The cobalt also provided solid solution strengthening and additions of molybdenum were made for the same purpose.

	C	Cr	Co	Mo	Ti	Al	Ni	Zr	B
Nimonic 115	0.16	15.0	15.0	3.5	4.0	5.0	bal.	0.04	0.014

Precipitates at the grain boundaries are important in controlling creep rupture ductility and impact resistance. If no carbides are present, grain boundary sliding causes premature failure. On the other hand, continuous films provide easy paths for impact failure. The optimum conditions are provided by discrete globular particles.

An important factor in improving the high-temperature performance of the more complex alloys has been the decreased rate of precipitate coarsening obtained by decreasing the lattice mis-match between precipitate and matrix. In Nimonic 80A this was ~0.5% but in Nimonic 115 it is reduced to 0.08%.

An unfortunate result of more complex alloying was that workability was adversely affected and it became necessary to develop improved methods of processing, such as vacuum melting to reduce the level of impurities. The better alloys are now cast, and further improvements have been obtained by directional solidification to give columnar structures and some gas turbine blades are manufactured as single crystals.

There is considerable current interest in the use of nickel-aluminides, Ni_3Al and $NiAl$ in their own right. The Ni_3Al alloys developed thus far, compared with the nickel-base superalloys, have superior fatigue properties, better oxidation resistance, lower density and higher high-temperature strength. Hence, Ni_3Al alloys are under evaluation for use as turbocharger rotors in diesel-engine trucks. Other possible applications include gas turbine blades and high temperature dies and moulds. NiAl alloys exhibit good hot ductility, making hot extrusion or forging attractive manufacturing routes. Compared with the nickel-base superalloys, the NiAl alloys possess better oxidation resistance, lower density, higher melting point and better thermal conductivity, making them candidate materials for use at high temperature, e.g. gas tubine blades. However, improvements in the fracture toughness and impact resistance are necessary.

10.4 The service temperatures of engineering materials

From the previous section it is clear that although temperature does not alone control creep behaviour it is the most important single factor. It is therefore useful to relate the more important materials to the temperature ranges in which they might be expected to give useful service. The temperatures of interest to the materials engineer range from cryogenic regions, say −200°C (−328°F), up to the operating temperature of a tungsten-filament lamp, say 2000°C (3630°F). For present purposes we need only consider the range from room temperature upwards.

Room temperature to 150°C (300°F)

The only engineering metal which cannot be used at any temperature above room temperature is lead: even at room temperature it creeps appreciably and although it is used extensively as a roofing material, and in the past for domestic piping, this is principally because of its ease of fabrication. Copper gives better service in this application but is more difficult to fabricate and use. Super-purity aluminium has no technical disadvantages for this purpose but is prohibitively expensive.

All thermoplastics exhibit creep of engineering significance at room temperature (see, for example Figure 10.12), and few of them are suitable for continuous service at temperatures much in excess of 100°C (212°F). 70°C (150°F) is about the upper limit for low-density polyethylene, polyvinylchloride and GP polystyrene but glass-filled nylon can be used at 150°C (300°F). Thermoplastics behave differently according to their degree of crystallinity. Where this is considerable – as in polyethylene, polyacetal and nylon – the crystalline melting point is available as a criterion of temperature sensitivity (although service temperatures are usually much lower than this) but amorphous polymers such as polystyrene, polycarbonate and the acrylics

Figure 10.12 Tensile creep behaviour at a stress level of 10 MPa for Nylon 6 (–) and polyethersulphone (– – –) as a function of temperature. (The Nylon 6 specimens were in the dry state and were tested in dry air.

TABLE 10.3. Heat deflection temperature at 1.80 MPa, °C (°F)

	°C	°F
Acrylic	70–100	160–212
ABS	80–105	175–220
PTFE	55	130
Nylon 6	80	175
Nylon 6 + 30% glass fibre	200	390
Nylon 6 + 30% carbon fibre	200	390
Nylon 66	70–110	160–230
Nylon 66 + 30% glass fibre	250	480
Nylon 66 + 30% carbon fibre	260	500
Polyacetal copolymer	110	230
Polyacetal copolymer + 30% carbon fibre	160	320
Polycarbonate	140	280
Polycarbonate + 30% glass fibre	150	300
Polycarbonate + 30% carbon fibre	150	300
LDPE	35	95
HDPE	50	120
PET	70	160
PET + 30% glass fibre	210	410
PPO-modified	130	270
Polypropylene	60	140
Polystyrene	80	175
Polystyrene + 30% glass fibre	100	212
High impact polystyrene	75	170
Polysulphone	175	350
Polysulphone + 30% glass fibre	185	365
Polysulphone + 30% carbon fibre	185	365
Polyethersulphone	200	390
Polyethersulphone + 30% glass fibre	215	420
Polyethersulphone + 30% carbon fibre	215	420
uPVC	70	160
SAN	85	185
Polyetherimide	200	390
PEEK	160	320
PEEK + 30% glass fibre	315	600

Data from MC Hough and R Dolbey[10]

soften progressively over a wide temperature range and it is difficult to find a simple criterion to assess temperature sensitivity. Although it is in no sense a true measure of creep resistance, the deflection temperature under load test (ASTM D648) is widely used in this context. Table 10.3 gives some data.[10]

In this test, a specimen is subjected to a standard load and heated at a certain rate. The softening temperature is that at which a standard deformation is attained. Although useful for comparisons between one material and another, the softening temperature cannot be used directly to predict a suitable service temperature. It should be noted that, differently from metals, provided the polymer has not suffered irreversible damage, creep strains are recoverable upon removal of the load. Data is often quoted for the maximum service temperature without mechanical loading in air, both for short term and for continuous use. These are generally based on the Underwriter Laboratories Temperature Index, which gives varying temperature–time limits for a standard range of moulding thicknesses from which a maximum deterioration of 50% of the original mechanical properties are allowed.

The creep properties of copper within this temperature range are of interest in connection with electrical machinery. The possibilities for alloying are limited because of the need to maximize electrical conductivity, and whilst plain tough pitch copper can be used at temperatures around 150–200°C (300–400°F), for applications such as commutator ring segments, silver-bearing coppers are normally employed with the silver content averaging around 0.08%. The creep properties of these materials at temperatures up to 250°C (480°F) have been described by Bowers and Lushey.[7] Alternatively, the use of silver plating on copper alloy components may be employed to ensure current-carrying capacity with strengthened material.

The skin of Concorde also operates at temperatures around or above 100°C (212°F) and for this purpose the Al–Cu–Mg–Ni–Fe alloy RR58, discussed in Chapter 15, is used. Aluminium alloys are also used at higher temperatures (see next section).

150–400°C (300–750°F)

Some thermoplastics may be used at temperatures above 150°C (300°F), notably polyetheretherketone (PEEK) and polyetherketone (PEK) where heat distortion temperatures for unreinforced grades are 160°C (320°F) and 165°C (330°F) respectively and where the addition of 30% glass fibre reinforcement raises these figures to 315°C (600°F) and 340°C (645°F) (see Table 10.3). A maximum temperature for continuous use is estimated to be 250°C (480°F) for both materials, making them suitable for wire and cable insulation in hostile environments and for engine parts. Polyethersulphone (PES) has a continuous use temperature of 190°C (375°F) when reinforced with 30% glass fibre. Self-extinguishing with low smoke emission, it has many applications for aircraft and automobile fittings and components, and also has the advantage that it is readily moulded to close dimensional tolerances, being amorphous.

The polyimides (PI) are tough, strong and stiff, with little propensity to creep and a continuous use temperature of the order of 260°C (500°F) and a maximum allowable of ~300°C (570°F) when degradation will commence before softening. Glass reinforcement gives further enhancement of properties. Processing is more difficult than for most other polymeric candidates but applications have included turbo-fan engine backing rings, washers, bearings, bushes, high temperature

insulation and valve seats.[11] Polyphenylene oxide (PPO) is another strong, stiff plastic with good creep resistance for use up to 190°C (375°F). A particular advantage is its low moisture absorption giving good dimensional stability in hot water for applications in domestic appliances, pumps etc. Polyphenylene sulphide (PPS) can operate at up to 250°C (480°F) when reinforced with ~40% glass fibre, with good mechanical properties. It is thus an excellent material for under-bonnet applications in automobiles. Polytetrafluoroethylene (PTFE) remains an attractive material for use to maximum service temperatures of 260°C (500°F) but has the disadvantage that it is not melt processable (Note also the low HDT value in Table 10.3).

New materials of the aromatic polyester class, known as liquid crystal polymers (LCP) are more easily processed and may be of particular interest for elevated temperature applications since they are chemically and thermally stable and flame resistant. The long term service temperature in air without mechanical loading is about 180–250°C (360–480°F) for LCPs. During processing in the molten state a highly ordered rod-like molecular structure is produced giving strength with a high degree of anisotropy.

Magnesium alloys can be used up to 200°C (400°F) and aluminium alloys can top this by a few tens of degrees. Aluminium alloy pistons in diesel and petrol internal combustion engines run at temperatures of 200–250°C (400–480°F). The alloy LM13 in BS 1490 (Lo-Ex) is commonly used for this application – it contains 12% silicon with additions of copper and magnesium: piston alloys and the use of aluminium matrix composites are discussed in Chapter 17. The aluminium casting alloy 4L35, containing 4Cu–1.5Mg–2Ni and commonly known as 'Y' alloy, was developed for stressed parts operating at elevated temperatures. It was the progenitor of the wrought alloy RR58, commonly used for compressor components in gas turbine engines, forgings and the skin of Concorde, and had a nominal temperature limitation of 150°C.

Of the copper-base materials, the addition of arsenic to copper slightly improves its creep performance at moderately elevated temperatures and phosphorus-deoxidized arsenical copper is used in chemical engineering applications where high electrical conductivity is not required. If high electrical conductivity is also required then the Cu–1Cr, Cu–0.1Zr, and Cu–1Cr–0.1Zr alloys are good up to 350°C (660°F). Prior to the Second World War, tin bronzes and phosphor bronzes with tin and phosphorus contents up to 8% and 0.3% respectively were widely used in drawn sections for steam turbine blading (British Standard 369). Monel metal (Ni–30 Cu–2.5 Fe–2 Mn) was also used. More intensive steam conditions have caused these materials to be replaced by stainless irons. Aluminium bronzes maintain their mechanical properties well up to 300°C (570°F), or even 400°C (750°F) in the higher alloys, and are also highly resistant to oxidation and scaling.

For low-pressure turbine casings where the temperature is lower than 250°C (480°F) it is possible to use the higher grades of cast iron in BS1452,[12] preferably with spherulitic graphite, but the use of steam reheat generally causes this to be replaced with plain carbon steel. The upper limit of temperature at which plain carbon or carbon–manganese steels are generally considered serviceable is 425°C (800°F) but for very long service ($\geq$ 20 years) in situations requiring exceptional dimensional fidelity, such as fuel-withdrawal mechanisms in nuclear reactors, the temperature limit is much lower and it is generally necessary to use molybdenum or chromium–molybdenum steels.[9]

For compressor assemblies (i.e. discs and rotor blades) in gas turbines, steels (e.g. low alloy and 12% chromium martensitic steels) have been progressively replaced by titanium alloys, but in order to reduce the risk of titanium fires the titanium rotor blades are alternated with steel or nickel-based alloy stator vanes. Compressor turbine discs in the UK are now generally 12% chromium martensitic steels, but there is continued development of austenitic steel use in the USA.

400–600°C (750–1110°F)

The principal materials employed within this range are titanium alloys and low-alloy ferritic steels. Hanson[13] has reviewed the uses of titanium alloys. It is the alpha close-packed hexagonal phase in titanium alloys that is the most resistant to creep deformation – in contrast, the body-centred-cubic beta phase exhibits poor creep resistance. The most widely used alloy for general purposes, Ti–6Al–4V (IMI 318), is an alpha–beta alloy and the presence of the beta phase limits its maximum operating temperature to 300–450°C (570–840°F). This is true also of the high-strength alloys Ti–4Al–4Mo–2Sn–0.5Si (IMI 550) and Ti–4Al–4Mo–4Sn–0.5Si (IMI551). Superior creep resistance is exhibited by the near-alpha alloys which may be used up to 500°C (930°F) or even 600°C (1110°F), depending on heat treatment. Examples are IMI 679, IMI 685 and IMI 829. IMI 679 (Ti–11Sn–2.25Al–5Zr–1Mo–0.25Si) has given well-established service for use as compressor discs and blades operating at temperatures up to 450°C (840°F). This alloy is heat-treated in the alpha–beta range and requires heavy working to produce a sufficient refinement of the two-phase structure.[14] A further disadvantage is that the additions of tin and molybdenum increase the density. These disadvantages have been overcome in IMI 685 (Ti–6Al–5Zr–0.5Mo–0.25Si) which is heat-treated in the beta field, and possesses good tensile properties and creep resistance up to 530°C[14], combined with high fracture toughness and low fatigue crack growth rates and good weldability.

Titanium alloys are widely used for compressor blades (including the front fan blade), discs and the casings of modern aero-engines. IMI 829 (Ti–5.5Al–3.5Sn–3Zr–1Nb–0.25Mo–0.25Si) extended the range of use up to 600°C (1110°F) and an even more recent alloy, IMI 834 (Ti–5.8Al–4.0Sn–3.5Zr–0.7Nb–0.5Mo–0.35Si–0.06C), has a temperature capability of around 630°C. This development is reflected in the creep curves shown in Figure 10.13 which may also be compared with the titanium alloy values indicated in Figure 10.9. As temperature of use at these levels increases, however, the titanium alloys become more sensitive to embrittlement and to surface oxidation. Coatings will become necessary to protect conventional titanium alloys above 580°C, and for the rear end of high

Figure 10.13 Creep Properties of Titanium alloys. (Courtesy of Timet UK Ltd.)

pressure compressors practice has been to switch at around 520°C to nickel alloys and incur the weight penalty.[15]

Future developments in the use of titanium for aero-engines are likely to centre on titanium alumide intermetallics and titanium base metal matrix composites.

Where cost is a major factor and the higher density is acceptable, low-alloy ferritic steels are preferred. For service at temperatures higher than 400°C (750°F), alloying is necessary and the element which appears to be essential in all creep-resisting steels is molybdenum. For steam piping working at temperatures up to, but not exceeding, 500°C (930°F) it is possible to use the

carbon–0.5 molybdenum steel but owing to its low rupture ductility it has been largely superseded by the chromium–molybdenum and chromium–molybdenum–vanadium types. For intermediate- and high-pressure turbine rotors the 0.2C–1Cr–1Mo–0.25V steel is used universally.[16] Steam piping which needs to be welded requires the carbon content to be reduced to 0.12% and the chromium content to 0.4%.[9] Turbine castings for use with steam temperatures up to 525°C (980°F) can be made from the 1Cr–0.5Mo steel[17] but improved rupture ductility is obtainable from the 0.5Cr–Mo–V steel which can be used for steam temperatures up to 565°C (1050°F). For high-temperature steam piping, and also for chemical plant requiring high resistance to hydrogen attack, the 2.25Cr–1Mo steel has been widely used but for temperatures higher than 575°C (1067°F) scaling and oxidation resistance becomes a major factor and the low-alloy steels are only acceptable on the basis of short-term replacement.

In the gas turbine field[18] heat-treatable martensite stainless steels with ~13% chromium have been used in the range 400–500°C (750–930°F) for discs and blades. These steels have a family relationship with modifications to the composition of BS 970 410 (UNS S41000) to maximize specific properties as required, e.g. Cr and Ni may be increased for better corrosion resistance, increased C for improved strength and hardness, increased S and P for machinability, C decreased for improved toughness, Mo, V, W added for strength and toughness at elevated temperature. Within this family come a large number of proprietary steels as instanced by the historic Firth Vickers grades (FV) (see Table 10.4).

TABLE 10.4.

	C	Cr	Ni	Mo	Nb	V	
FV448	0.10	11.00	0.75	0.70	0.40	0.15	0.05N
FV535	0.07	10.50	0.30	0.70	0.45	0.20	6.00Co
S62	0.25	13.50	0.40	–	–	–	–

575–650°C (1070–1200°F)

Within this temperature range the provision of oxidation resistance becomes as much, perhaps more, of a problem as that of providing creep strength. The usual way of increasing the scaling resistance of iron is to add chromium, and at least 8% is needed to withstand temperatures as high as 650°C (1200°F). Various steels are available with chromium contents varying from 5 to 12% or more. These generally also contain molybdenum but their creep resistance is not particularly good so they are mostly used in chemical engineering applications where their corrosion resistance is of special value. However, the 13% chromium stainless steels with 0.5 Mo can be used for blading in marine steam turbines at temperatures up to 565°C (1050°F) provided measures are taken to deal with attack from chlorides originating from the feed water.

The ferritic chromium steels are much cheaper than the high-alloy austenitic steels but the latter are able to deliver a much better combination of creep resistance and scaling resistance. A goodly number of austenitic stainless steels are well established for use as superheater tubes and reactor heat exchangers. Of the standard compositions, types 304, 321, 347 and 316 (Table 10.5) are used quite generally for piping in power generation and chemical plant, as is also the proprietary FV548 for temperatures up to 650°C (1200°F) or, in some cases, higher. Selection for any given application must take into account special environmental hazards such as fuel ash corrosion.

650–1000°C (1200–1830°F)

Three main groups of alloys are available for use in this temperature range: (1) the austenitic stainless steels; (2) the alloys based on the nickel–chromium and nickel–chromium–iron systems; and (3) the cobalt-based alloys.

Austenitic steels

The upper limit of use for the standard austenitic stainless steels is about 750°C (1380°F). A very popular alloy for high-temperature steam piping is Type 316 and the molybdenum content of this alloy appears to be highly beneficial at this sort of temperature. However, still better results are obtainable with even more highly alloyed compositions containing additions of molybdenum, cobalt, tungsten, vanadium and niobium. These

TABLE 10.5.

	C	Si	Mn	Ni	Cr	Mo	Others
304	0.06	0.20–1.00	0.50–2.00	9.0–11.0	17.5–19.0	–	–
304(ELC)	0.03	0.20–1.00	0.50–2.00	9.0–12.0	17.5–19.0	–	–
321	0.08	0.20–1.00	0.50–2.00	9.0–12.0	17.0–19.0	–	–
316	0.07	0.20–1.00	0.50–2.00	10.0–13.0	16.5–18.5	2.25–3.0	Ti 5X C–0.70
347	0.08	0.20–1.00	0.50–2.00	9.0–12.0	17.0–19.0	–	Nb 10X C–1.00

TABLE 10.6.

	C	Cr	Ni	Mo	Co	W	V	Nb	Mn	B
Esshete 1250	0.1	15	10.5	1.0	–	–	0.25	1.0	6.0	0.005

additions do not necessarily extend the temperature of service but allow higher stresses to be employed. Such alloys as these are mostly proprietary products, as for example, British Steel Esshete 1250 (see Table 10.6).

Nickel-base materials

The high-temperature alloys based on the nickel-chromium system had their origin in the electrical resistance heating alloy Brightray C: Betteridge[19] states that Nimonic 75 was developed by selection from routine production batches of that material and abnormally high titanium contents were established as the important strengthening factor: the development of the precipitation-hardened series followed on from this. The Nimonic alloys are well established for service at temperatures of 700°C (1300°F) and above: the incentive for their development was provided by the invention of the gas turbine, a machine which still provides the most stringent and aggressive conditions of service. Although Nimonic 75 was first used for rotor blades it was rapidly superseded and flame tubes and nozzle guide vanes became its main applications for inlet temperatures around 800°C (1475°F) Later members of the Nimonic series have been used principally for the moving blades in aircraft turbines.

As a simple means of appraising the temperature resistance of these materials it is instructive to make use of the fact that most materials engineers have a clear mental picture of the strength at room temperature of ordinary mild steel. This, measured as the short-term yield stress, can be taken as 200 MPa (29 ksi). The temperature at which Nimonic 75 exhibits this yield stress is about 720°C (1330°F) and the temperature for rupture after 100 hours at the same stress is around 600°C (1110°F), depending on the form of the product. The corresponding figures for Nimonic 120 are 1030°C (1885°F) and 930°C (1700°F). For longer lives the temperatures are, of course, lower.

The compositions shown in Table 10.7 demonstrate the increasing metallurgical complexity as the series extended. Eventually it was found that the alloys could not be mechanically worked and casting then became the only method of production. The high-strength cast nickel-base alloys, such as MAR-M200, extend the 100 hour, 200 MPa (29 ksi) rupture temperature by about 30°C (54°F) over the best wrought alloys and the unidirectionally solidified nickel alloys listed in Table 10.8 provide about a further 30°C.[18] Specially developed single crystal alloys can give further improvement in creep strength over directionally solidified polycrystalline material, possibly by as much as 50°C.

TABLE 10.7.

	C	Cr	Co	Mo	Ti	Al	B	Zr
Nimonic 75	0.12	20.0	–	–	0.40	–	–	–
Nimonic 120	0.04	12.5	10.0	5.70	2.50	4.50	0.03 max	0.05 max

TABLE 10.8.

	C	Cr	Co	Al	Ti	W	Nb	B	Zr
MAR-M200	0.15	9.0	10.0	5.0	2.0	12.5	1.0	0.015	0.05

Many of these alloys might be used at temperatures higher than 1000°C (1830°F) but the 100 hours rupture stress then decreases rapidly below 100 MPa (15 ksi).

Glenny, Northwood and Burwood-Smith[18] give the figures shown in Table 10.9 for stress-rupture tests at 982°C (1800°F) and 206 MPa (30 ksi).

As well as for blades in turbines, nickel superalloys are used as sheet components for combustion chamber liners and for reheat and exhaust components. Oxide dispersion strengthened nickel-base materials such as Inconel MA 754 (Ni–20Cr–1Fe–0.3Al–0.5Ti–0.6Y$_2$O$_3$–0.05C) have also been employed for sheet components, for example, in the high pressure guide vane for the GE F404-400 engine. Combustor liners are heavily cooled and they are also protected by thermal barrier coatings. However, metal temperatures in the primary zone are unlikely to be in excess of 900°C (1650°F) unless there are local hot spots. Nozzle guide vanes are also heavily cooled, but metal temperatures on the aerofoil section may exceed 1000°C (1830°F).

Nickel–Iron-base alloys

One way of increasing the high-temperature stability of the highly alloyed austenitic stainless steels is to increase the nickel content so that it becomes comparable to, or greater than, the iron content. This has given rise to the Inconel and Incoloy series of alloys developed by International Nickel. Incoloy 901 (with molybdenum for solid solution strength and titanium and aluminium for precipitation-hardening) and Inconel 718 (strengthened with the γ' phase Ni$_3$ (Ti, A1, Nb) have been used for compressor and turbine discs (Table 10.10)).

Other alloys bearing the Inconel and Incoloy trade names are not hardened by heat treatment and are selected for their ability to resist corrosion and oxidation in applications such as furnace components and reactor vessels. They are not strictly creep-resisting materials but are often used at temperatures ranging from 500°C (930°F) to 1200°C (2200°F) depending upon the level of stress. For example, Inconel 600 has been used for a reactor vessel operating at 500°C and also for a copper-brazing retort at 1200°C.

Cobalt-base alloys

Although the melting point of cobalt is 1492°C (2700°F), compared with 1455°C (2650°F) for nickel, the cobalt-base superalloys do not now compare well with nickel-base alloys. They can

TABLE 10.9.

	Life (hours)	Elongation (%)	Min. creep rate per hour
Conventionally-cast, equiaxed	35.6	2.6	23.8×10^{-5}
Directionally solidified, polycrystalline	67.0	23.6	25.6×10^{-5}
Directionally solidified, single crystal	107.0	23.6	16.1×10^{-5}

TABLE 10.10.

	C	Cr	Al	Ti	Mo	Fe	B
Incoloy 901	0.05	12.5	0.25	2.8	6.0	42.5	0.015
Inconel 718	0.04	18.6	0.40	0.9	3.1	18.5	–

be used for nozzle guide vanes which, being stationary, are stressed less highly than the moving blades.

Coatings

In the main, the discussion here of materials usage in the context of service temperature has been concerned with bulk properties. As temperature rises, and surface attack through mixtures of erosion and corrosion/oxidation becomes significant, there is an increasing requirement for the provision of protective surface coatings such as zirconia. These not only provide specific surface properties in relation to attack, but also introduce a thermal barrier. Such coatings are routinely employed on first stage and some second stage turbine aerofoil components in aero, marine and industrial gas turbines. Nickel-based materials are frequently used as a bond coat onto superalloy components, allowing an outer thermal barrier coating of ceramic. There is also much interest in the use of the intermetallic compounds, NiAl, Ni_3Al and TiAl, for use as coatings at temperatures greater than 1000°C.

1000°C (1830°F) and above

For stressed applications at temperatures above 1000°C it is necessary to look towards (1) the refractory metals, (2) ceramics and, possibly, (3) *in-situ* composites.

The refractory metals

All of the refractory metals, tungsten, tantalum, niobium and molybdenum are available as commercial materials. Tungsten is used universally for electric lamp filaments, and molybdenum is used for radiation shielding of high-temperature furnaces. These, however, are environmentally protected, unstressed applications. For more advanced uses the refractory metals present problems. Tantalum and tungsten have not been

investigated for gas turbine use because of their high densities. Commercial alloys of molybdenum and niobium have been developed but their low resistance to oxidation necessitates protective coatings which have proved inadequate for molybdenum and severely limiting in the case of niobium. In protected environments these materials can be used at temperatures in excess of 1500°C (2730°F).

Ceramics

Although the term ceramic must cover materials used for furnace linings, pottery, tiles, etc., engineering interest centres on the more specialized materials known as engineering ceramics. These consist of sintered oxides of aluminium, magnesium, beryllium, zirconium, thorium and certain borides, carbides, nitrides and silicides. They are generally polycrystalline, containing little or no glass phase, and their creep behaviour can be described in terms similar to those that apply to metals. Many ceramics possess high strength at temperatures higher than can be sustained by metals (they can be used under stress up to, or above, 1400°C (2550°F) but, like metals, they are not trouble-free. Carbides and borides tend to oxidize rapidly at temperatures above 1000°C (1830°F). All ceramics are hard and brittle and vulnerable to thermal shock.

The most promising ceramics for advanced engineering use are probably silicon carbide, sialon, and silicon nitride. The latter material, having a density of only $3\,Mg/m^3$ and being resistant to oxidation, is a candidate for service temperatures of 1200°C in gas turbines.[20]

10.5 The selection of materials for creep resistance

The designer must specify the desirable levels of temperature, stress and life, but these parameters are negotiable. It has been shown in previous sections that the best creep-resisting materials are complex and therefore expensive. As always, cost will be the final arbiter of selection and it may be worthwhile relaxing one or more of the three design parameters to allow a cheaper material to be used. Reducing the upper limit of temperature will generally be associated with a reduction in the thermal efficiency of engineering operation – the question is whether the cost of this can be offset by the reduced materials costs and reduced failure hazards. Reduction in stress implies weightier structures – this may be acceptable in stationary land-based erections but is likely to be resisted for transport systems, especially in aircraft. Reducing permissible life introduces replacement and down-time costs, and possibly also the prospect of in-service failure. The failure of a single turbine blade in service is not necessarily catastrophic but an exploding turbine disc can wreck a whole engine.

Over the whole subject of selection for creep lies the shadow cast by the problem of inadequate creep data. Since extrapolation is unreliable, it is wise in all cases where required service lives are in excess of times for which data are available to select cautiously from established materials which have been well proved in time-tried applications. Figure 10.9 gives some guidance in this respect, but detailed design data must be obtained from producers.

It must be remembered that environmental hazards are often exacerbated under creep conditions and final selection is often influenced by factors such as fuel ash corrosion in aircraft gas turbines and chloride attack in marine steam turbines.

10.6 Deformation mechanism diagrams

The deformation of a material under stress can be the result of one or more separate mechanisms and, as previously described, the dominant mechanism in any material will be determined by the stress level and temperature. The ranges of stress and temperature, or strain-rate and stress,

within which the different mechanisms occur can be shown in deformation mechanism diagrams.[21,22] Such diagrams have been determined for many metals and ceramics and a generalized example is shown in Figure 10.14. This shows that elastic behaviour can only occur below some critical value of temperature, and that the boundary dividing elastic from creep behaviour is also stress-dependent, the critical temperature for thermally activated deformation processes decreasing as the stress increases. As temperature increases at low stresses, the boundary between the elastic and diffusional creep regimes is more practical than theoretical, since even at low temperatures there is a small creep strain. However, the strain rate is then so small that the material is considered to behave elastically.

Within the range of creep behaviour the area is divided according to the dominance of either dislocation (power-law) creep or diffusional processes. The divisions between the areas are established from the rate equations for the processes concerned.

Materials selection for high-temperature applications is aided by deformation mechanism diagrams since they provide creep information plainly and succinctly.

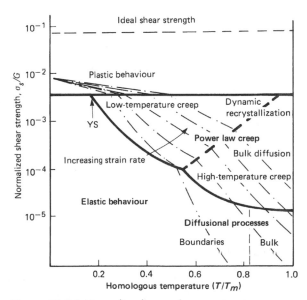

Figure 10.14 Normalized stress/temperature map indicating deformation mechanisms.

References

1. S. S. MANSON and A. M. HAFERD: N.A.C.A., TN 2890, March 1953.
2. F. R. LARSON and J. MILLER: *Trans ASME*, July 1952; **74**, 765.
3. J. E. DORN and L. A. SHEPHERD: A.S.T.M., *S.T.P.* 165, 1954.
4. *THE NIMONIC ALLOYS – DESIGN DATA*. WIGGIN ALLOYS LTD.
5. R. LAGNEBORG: *Int. Met. Rev.*, June 1972; **17**, 130.
6. W. J. PLUMBRIDGE and D. A. RYDER: Metallurgical Review 136, *Met. Mater.*, August 1969.
7. J. E. BOWERS and R. D. S. LUSHEY: *Metallurg. Mater. Technol.*, July 1978.
8. S. G. GLOVER: *Modern Theory in the Design of Alloys*. Institution of Metallurgists/Iliffe, London 1967, p. 85.
9. J. GLEN: *The Problem of the Creep of Metals*. Murex Welding Processes Ltd, Waltham Cross, Herts., 1968.
10. M. C. HOUGH and R. DOLBEY: *The Plastics Compendium Vol 1*. RAPRA Technology Ltd, 1995.
11. A. G. FOLEY and G. S. BOYCE: PROMAT – *Profit through Materials Technology*. Papers and Case Studies, DTI (NEL) 1987, 32.
12. H. T. LEWIS: in *Materials for Marine Machinery* (ed. S. H. Frederick and H. Capper). Institution of Marine Engineers. Marine Media Management Ltd., 1976.
13. B. H. HANSON: *The Selection and Use of Titanium*. Engineering Design Guide No. 39, Oxford University Press, Oxford 1980.
14. P. H. MORTON: *Rosenhain Centenary Conf.*, The Royal Society, 1975.
15. S. MILLER: *Interdisciplinary Science Reviews*. 1996, **21**, 2.
16. M. G. GEMMILL: *Met. Mater.*, July 1968, p. 194.
17. R. CROMBIE: *Metallurg. Mater. Technol.*, July 1978, p. 370.
18. R. J. E. GLENNY, J. E. NORTHWOOD and A. BURWOOD-SMITH: *Int. Met. Rev.*, 1975; **20**, 1.
19. W. BETTERIDGE: *Metallurg. Mater. Technol.*, April 1974.
20. B. WILSHIRE: in *Creep of Engineering Materials* (ed. C. D. Pomeroy), Mechanical Engineering Publications, 1978.
21. M. F. ASHBY: *Acta Met.*, 1972; **20**, 887.
22. H. J. FROST and M. F. ASHBY: *Deformation Mechanism Maps*. Pergamon, Oxford, 1982.

Selection for surface durability

11
Selection for corrosion resistance

11.1 The nature of the corrosion process

Corrosive attack is the result of chemical reaction at the interface between the material and the associated environment. At its simplest it can be regarded in terms of a normal bulk reaction, with the free energy for the reaction, and the thermodynamic activity (i.e. effective concentration) of the reactants providing the driving force for the process, i.e. determining the stability of the system. The actual rate at which the corrosion process occurs, i.e. the reaction kinetics, is controlled by the rates at which transport mechanisms operate within the reactants at a common interface and within the corrosion product developing between them.

The corrosion reaction is dictated by the chemical nature of the environment and the effective concentration of reactive species, whether major or minor. In some cases of corrosion by acids the presence of oxygen is required and the degree of aeration of the system, i.e. the oxygen concentration, can be controlling in determining whether corrosion will occur or not.

As regards the material being corroded, the overall composition alone does not necessarily indicate the activity of individual elements in solution. The structure of an alloy, for example, may be very heterogeneous, with several different phases present, each of differing composition and distributed in different forms. Individual phases within metals are themselves non-uniform, the more reactive sites being associated with disorder, such as grain boundaries and structural defects produced by mechanical deformation (dislocations).

Whilst corrosion is sometimes considered only in the context of metallic materials, in the more general sense of deterioration of materials through reaction with an environment it also includes the behaviour of glasses, ionic solids, polymers, concrete, etc. in a range of environments, including electrolytes and non-electrolytes, molten metals and gases.

It is difficult to classify the various types of corrosive attack. Traditionally, a broad division into 'wet' and 'dry' corrosion reactions has been employed, determined by the presence or absence of water or an aqueous solution. A more rational classification for metals has been given by Shreir[1] as follows:

(1) Film-free chemical interaction in which there is direct chemical reaction of a metal with its environment. The metal remains film-free and there is no transport of charge.
(2) Electrolytic systems:
  (a) Inseparable anode/cathode (insep. A/C) type. The anodes and cathodes cannot be distinguished by experimental methods although their presence is postulated by theory, i.e. the *uniform* dissolution of metals in acid, alkaline or neutral aqueous solutions, in non-aqueous solution or in fused salts.
  (b) Separable anode/cathode type (sep. A/C). Certain areas of the metal can be distinguished experimentally as predominantly anodic or cathodic, although the distances of separation of these areas may be as small as fractions of a millimetre. In these reactions there will be a macroscopic flow of charge through the metal.
  (c) Interfacial anode/cathode type (interfacial A/C). One entire interface will be the anode and the other will be the cathode. Thus a metal/metal oxide interface might be regarded as the anode and the metal oxide/oxygen interface as the cathode.

In general, 2(a) and 2(b) include corrosion reactions which are normally classified as 'wet' while 2(c) includes those which are normally designated 'dry'.

In the case of 'dry' corrosion by oxidation, category 2(c), the metal oxides which are formed are ionic in character, with the metal and oxygen ions regularly arranged on a specific crystal lattice. Such lattices are normally defective with either a deficiency of cations or anions such that vacant sites exist and ions are able to diffuse through oxides by way of these vacancies in the lattice. Thus metal cations produced by reaction at the interface can enter the oxide and migrate via vacant cationic sites towards the oxide/gas interface, the rate of movement being a function of temperature. A compensating flux of vacancies moves towards the metal surface, and may even produce voids there. At the same time interstitial electrons enter a positive hole in the oxide lattice at the metal surface. These electrons pass through the metal oxide by electron switches on the cations. Oxygen molecules absorbed on the oxide layer then become ions by capturing electrons that have been conducted through the oxide from the metal. Oxygen anions and arriving metal cations then together produce the growing oxide film. It is clear here that overall electrical neutrality is maintained and that the process can be considered to be electrochemical, with the metal/oxide film interface as the anode and the oxide/gas interface as the cathode. While such mechanisms based on the formation of cation defective lattices are most general in dry attack producing a film, cases of cation excess lattices (e.g. zinc oxide) and anion defective lattices do exist (e.g. Fe_2O_3).

Clearly, the mechanisms will differ in these cases. In Fe_2O_3 formed on iron, for example, the anionic deficiency causes oxygen anions to migrate towards the metal, rather than the metal moving out through the oxide as previously explained.

Since the growth of the oxide film depends on the movement of ions and electrons it can be likened to the passage of current (i) under the control of a potential (E) for the cell metal/metal oxide/oxygen based on the driving force (free energy, ΔG) for the oxidation reaction. Taking the electrical analogy further, the electrical resistance to the total conduction by ions and electrons (R), will determine this current and thus the rate of growth in thickness (x) of the film and

$$\frac{dx}{dt} = \frac{AE}{R} = \frac{BE}{x}$$

where A and B are constants, assuming the electrical resistance to be directly proportional to its thickness. This parabolic law, where the rate of growth varies inversely as the thickness, is also found experimentally for a large number of systems. It also indicates that if the oxide film has a high resistance the rate of oxidation will be low, also confirmed experimentally.

Such an assessment assumes, of course, a continuous adherent oxide film. If the oxide is volatile (e.g. MoO_3) or if stresses developed within the film due to differences of specific volume give rise to rupturing stresses, exposing fresh metal surface, a rectilinear law may be followed with the rate maintained at that at which the oxidation is initiated.

There are, of course, other non-aqueous systems where no interposed product film is generated at all (category 1) where direct chemical reaction also continues at the initial rate. Examples include the reaction of solid metals with their fused halides or with liquid metals, slags with refractories and even with some metals and organic liquids.

'Wet' corrosion occurs by electrochemical mechanisms. There may be overall attack (2(a) Insep. A/C) or corrosion may be concentrated at specific regions due to heterogeneity of metal structure, or of a required species in the electrolyte (e.g. oxygen), which result in some areas being specifically anodic to the rest, (2(b) sep. A/C). Current will flow from such anodes, resulting in dissolution of the metal at the anodes.

$$M \rightarrow M^{n+} + ne^- \text{ (an oxidation reaction)}$$
$$\text{e.g. Fe} \rightarrow Fe^{2+} + 2e^-$$

There are several possible cathodic (reduction) reactions by which the cell is completed, depending on the acidity or alkalinity of the electrolyte (i.e. pH) and the degree of oxygenation of the electrolyte. The reaction

$$H^+ + e^- \rightarrow H$$

followed by

$$H + H \rightarrow H_2$$

takes place in strong non-oxidizing acid solutions containing high concentrations of hydrogen ions, or where oxygen is not available for the usual cathodic reaction

$$O_2 + 4H^+ + 4e^- = 2H_2O$$

In neutral solutions, the usual cathodic reaction is

$$O_2 + 2H_2O + 4e^- = 4OH^-$$

and this again will give way to the hydrogen evolution reaction if the level of oxygenation is low. In alkaline solutions corrosion again occurs with the evolution of hydrogen, in spite of the low concentration of H^+ ions. In this case it is possible that the cathodic reaction is

$$H_2O + e^- \rightarrow OH^- + H$$

The cathodic reactions yielding hydrogen are of great importance in relation to the effect that absorbed hydrogen may have on the properties of the metal involved, and of particular significance in relation to high strength steels. The subject of hydrogen embrittlement is considered in some detail on p. 155.

As corrosion takes place the processes occurring at anode and cathode involving changes of state, charge transfer and the availability (i.e. activity) and movement of ions in the electrolyte will themselves affect the potential of an anodic or cathodic area. Depending on its effect in relation to either or both of the electrode processes, as a function of the activation or deactivation of ions going into or coming out of solution and their movement to or away from the electrodes, the passage of current in the corrosion cell will 'polarize' either the anodic or cathodic reactions, changing their potentials such that there is a smaller potential difference between them. The maximum corrosion current (*i*) occurs when the anode and cathode are both polarized to the same potential (Figure 11.1)

Figure 11.1 Polarization diagram for a corrosion cell.

assuming no ohmic resistance in the circuit. The more strongly the electrode reactions are polarized, and the steeper the polarization curves, the smaller will be the maximum corrosion currents. Any resistance (R) in the electrolyte, metallic paths and through any interface films would reduce the potential between cathode and anode by IR, and the corrosion current from i to I. The function of many corrosion protection coatings is to increase this resistance in the system. Clearly, also, the closer the anodic and cathodic areas are to each other the lower will be the resistance and the greater the corrosion current.

Polarization of electrodes arises from two main mechanisms. Activation polarization arises simply from the energy consumed as charged ions enter solution from a solid metal surface, or are discharged from solution. This consumption of energy is subject to kinetic control arising from the rate of transfer of charges as the change of state proceeds. The other mechanism is concerned with the mass transfer of species to and from electrodes in the electrolyte, and particularly through the stagnant boundary layer associated with the surface. This affects the overall rate of an electrode reaction in all but the most vigorously stirred or impinging systems. These requirements of mass transfer frequently lead to what is called concentration polarization.

One of the most important examples of concentration polarization is cathodic polarization in neutral electrolytes (e.g. salt solution) where the rate at which oxygen can be reduced at the cathode will depend on the degree of oxygenation of the solution and on the degree of stirring, natural or forced, in relation to the electrode surface. Such cathodic polarization produces a potential/current diagram (Evans diagram) of the form shown in Figure 11.2 where the cathodic curve becomes very steep and the cathode current assumes an almost limiting value.

Figure 11.2 Effect of oxygen concentration on cathodic polarization.

Some electrode reactions are sensitive to the presence of specific impurities; for example, the cathodic reaction on steel, yielding hydrogen, is facilitated by the presence of arsenic, as is the anodic dissolution of iron in the presence of traces of sulphide in an acid solution.

If oxygen is excluded from solution the oxygen absorption reaction becomes impossible and the cathodic reaction has to be that of the reduction of hydrogen ions to hydrogen. Only if the solution is substantially acid, however, will the concentration of H^+ ions be significant, and thus the rate of corrosion will be low. This is why it is important to avoid air ingress into central heating systems, boilers and car radiators. It is important to note that since the cathodic reaction in neutral solutions is normally the reduction of oxygen ions, regions which have a high level of oxygen access will be cathodes and be unattacked. So whilst a high oxygen content increases the amount of corrosion, the attack takes place in areas which are of lowest oxygen content, i.e. these become anodes. The concept of differential aeration in leading to the setting up of substantial potential differences and the corrosion of non-aerated regions is important, for example, in riveted or bolted joints where 'crevice corrosion' occurs if, say, salt water can penetrate the joint. Attack will take place inside the joint, usually around the stem of the rivet or bolt. Similar effects are produced where deposits settle or adhere to the walls in tanks, preventing the movement of liquid carrying oxygen to the surface beneath, but where oxygen is elsewhere available.

Passivation

Just as in dry oxidation, where the continued attack on the metal depends very much upon the continuity and structure of the oxide film produced, so the product of attack on a metal in an aqueous solution may form a more or less non-porous, sparingly soluble film on the surface which reduces the rate of attack to negligible proportions. This is termed passivity.

Figure 11.3 shows the anodic polarization curve of iron in dilute sulphuric acid. If the potential of the specimen is raised from the reversible equilibrium value an anodic current flows and corrosion occurs. This continues with increasing potential to a point where a limiting current density of the order of $20\,A/dm^2$ is reached at about $+600\,mV$. On further increase in potential the rate of corrosion drops markedly, and the iron is said to have become 'passive'. The limiting current density, corresponding to the maximum corrosive attack, is related to a steady state in which the corrosion product $FeSO_4$ is removed by convective diffusion from the surface of the anode at the same rate as it is formed.

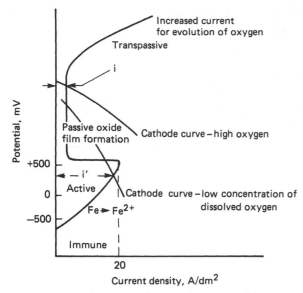

Figure 11.3 Polarization diagram for iron in dilute H_2SO_4.

At higher potentials the nature of the film changes and an oxide is formed as a thin, non-porous, passivating layer. The porous solid $FeSO_4$ previously present is dissolved into the solution very quickly, and further corrosion is now controlled by the transport of ions through the oxide film. The current remains at a constant low value, i, until the potential is raised to a new value at which further oxidation occurs; e.g. producing oxygen gas from the dissociation of water or chromate on chromium-rich materials if formed.

Clearly, since the formation of an anodic film is the basis of passivity for metals in aqueous solutions, with the thickness of film required for passivation depending on the metal involved and the environment, there are effects associated with the supply of oxygen to the system. Stagnant conditions with poor oxygen supply are less likely to lead to the passivation of surfaces. Similarly, the presence of oxidizing agents at the anode surface will help promote and maintain passivity. The cathodic reaction will also be affected by the oxygenation of the electrolyte. At low oxygen concentrations the cathodic and anodic polarization curves can intersect in the active rather than the passive regions giving a substantial corrosion (Figure 11.3), corresponding to current i'.

Factors which produce partial or complete removal of the passivating film will restart corrosive attack. There may be electrochemical dissolution of an oxidative or reductive nature, leading either to the formation of a more soluble higher oxide (e.g. chromate) or the metal itself. In fact, the most useful higher valency passivating oxide films are often theoretically chemically soluble, even in the solutions in which they are formed, and their protective property is mainly due to the extreme slowness of their dissolution under many conditions, in turn a function of the ionic structure of the film. Passivating films can also, of course, be undermined from breaks or pores, particularly when the attack on the metal is rapid. This emphasizes the value of self-healing films which can survive the inevitable mechanical damage that will occur in plant, whether or not the attack at breaks in the film is particularly rapid.

Pitting

Sometimes with the breakdown of passivation, pitting occurs. For example, a clean surface of 18/8 stainless steel will pit in a stagnant solution of NaCl containing the depassivating Cl^- anion. The pits will be randomly distributed unless there is a variation in oxygen access in relation to the rehealing process. Where there are crevices, for example, the pitting attack will be concentrated in these crevices. Pitting is of particular significance in the use of stainless steels, which rely on the passive nature of the oxide film for their corrosion resistance. The likely response of a material can be judged by a measure of the breakdown or pitting potential of an oxide under any given environment conditions. Pitting will occur in due time if the redox potential of the solution is more positive than the pitting potential for the same conditions. Similarly, at a given potential there will be a given environmental condition, e.g. chloride ion concentration, when pitting will occur.

Stainless steels offer good resistance to pitting, but it is essential that the correct grade be selected for the particular environment of use. Higher chromium, nickel and the addition of molybdenum increase resistance to pitting. If it is likely that BS1449 304 will pit then BS1449 316 will be preferred. The effect of molybdenum may be due to the absorption of MoO_4^{2-} ions on the surface after the dissolution of Mo, rather than by a change in the composition and characteristics of the surface oxide film.

Heterogeneous metal systems

Whilst the corrosion of a completely heterogeneous material could be caused by an electrolyte condition such as differential aeration, the attack is normally due to the heterogeneity of the metal system itself. In the case of two dissimilar metals joined together we speak of a bimetallic couple and the polarities of the two members of the cell are usefully indicated by their position in the galvanic series (Table 11.1). For metals that are far apart in the series it may be expected that the base reactive member of the couple will corrode. Strongly polarized electrode reactions can, in fact, shift the electrode potential so far with the onset of corrosion that polarity can be reversed from the theoretical equilibrium relationship. The formation of a thin, stable high-resistance oxide film on an electrode of aluminium, titanium or stainless steel can, in some cases, make it cathodic when intrinsically it is the more reactive member of a couple and should theoretically act as the anode.

The existence of bimetal couples in a composite component or structure is a very common cause of electrochemical corrosion attack. Whilst the undesirability of linking iron and copper might appear obvious, some seemingly safe combinations may not be so. For example, aluminium and magnesium are both strongly electro-negative, but the cell potential between them is, in fact 0.71 volts, quite sufficient to cause appreciable intensification of the attack on the magnesium, a material which in any case is not corrosion-resistant in a non-surface-treated

TABLE 11.1. Galvanic series for metals and alloys in sea water[2]

Noble
Titanium
Monel (67% Ni, 30% Cu, 1.2% Mn, 1.2% Fe) plus C, Si.
Passive stainless steel (18% Cr, 8% Ni) – covered with oxide film
Silver
Inconel (80% Ni, 13% Cr, 6.5% Fe)
Nickel
Copper
α-brass (70% Cu, 30% Zn)
α/β brass (Muntz metal 60% Cu, 40% Zn)
Tin
Lead
Active stainless steel (18% Cr, 8% Ni) – oxide film destroyed
Cast iron
Mild steel
Aluminium
Zinc
Magnesium
Base

Note: This table does not show the metals in quite the same order as one for the standard electrode potential against a reference electrode. This is because of the nature of the oxide film, as shown by the two positions given for stainless steel.

condition. This is important since there is a common tendency to use these two light alloys together, perhaps as wrought aluminium alloy sections fixed to magnesium castings.

The linking of two metals together occurs in the context of surface coatings, e.g. electroplates. A gold plated finish, for example, is popular for bathroom fittings, but can lead to unsightly and rapid blistering if it is insufficiently thick to completely exclude any corrosive environment from the substrate beneath, the noble gold being the cathode of any cell set up with the substrate at an imperfection in the plating. Gold is, of course, an extreme case, but similar principles apply to tin or nickel on steel, where coatings are only protective where they exclude the environment from the substrate. Apart from inadequate thickness and compactness, damage to such coatings which reveals the substrate will lead to enhanced corrosion pitting as in the localized rusting of tin cans at points of abrasion, etc.

On the other hand, coatings of metals which are anodic to the metal beneath, for example zinc or cadmium on steel, will protect the underlying surface, even to the point of preventing attack where damage to the coating has exposed small areas of the steel to the environment, the zinc or cadmium being the anodes of any cells set up. Corrosive attack on the zinc or cadmium coatings themselves can be reduced by passivating treatments.

Zinc coatings may be applied by hot dipping (galvanizing), by electroplating, or by solid state diffusion (sheradizing). Whilst galvanizing is widely applied to improve the corrosion resistance of steel structures and components, it is not without its own dangers, particularly when applied to higher strength steels. Where it is applied to a cold-deformed part, the heating involved in immersion in a zinc bath can produce some reduction in ductility, known as strain ageing, where there is diffusion of interstitial solutes, such as carbon and hydrogen, to imperfections in the lattice (dislocations) requiring a greater force for the further movement of the dislocations, which is the basis of deformation. Further, the generation of the zinc layer is achieved by attack on the iron, particularly at locations such as grain boundaries, which can then act as incipient surface grain boundary cracks. In steels of high ductility, which are not notch-sensitive, then embrittlement should not result, but with higher strength steels imperfections in steel surface caused by zinc attack, leading to stress concentrations, could be harmful. Also, although the zinc is anodic to the steel and this protects it against corrosion at an imperfection in the coating, the fact that the steel is then the cathode of a cell may lead to the generation of hydrogen, mainly a function of the degree of oxygenation, as discussed on p. 147. Whilst exposed steel, e.g. at a deformed thread root may not corrode, it may be made susceptible to hydrogen embrittlement if under stress.

Also, the pickling process, which usually precedes hot dip galvanizing, can lead to hydrogen absorbtion by the steel and consequent embrittlement (see p. 157).

The effect of metal microstructure

Corrosive attack at a metal may be initiated by the variations in microstructure of a metal. Generally, overall corrosion tends to be more rapid at sites of high energy, such as grain boundaries and structural defects (dislocations), even in a pure metal. In alloys, the segregation of alloying elements within phases, or the proximity of separate phases of varying chemical composition, can initiate or enhance and localize corrosion by creating an anode/cathode system which leads to attack at the anode sites. In an alloy, segregation of solutes to grain boundaries can lead to the boundary itself being either anodic or cathodic to the adjacent grain, depending on the solute concerned. Grain boundaries are often sites for the precipitation of second-phase particles, leading to a form of localized corrosive attack that can be very damaging. In the sensitization of stainless steel, for example, the precipitation of chromium carbide at grain boundaries during heating and cooling in the range of 425–600°C depletes the immediately surrounding matrix in chromium, such that a complete and protective passive film is not formed. This phenomenon of sensitization is most widely seen associated with welding, where part of the heat-affected zone of the weld is within the critical temperature range long enough for precipitation to occur. The resulting intergranular attack is known as 'weld decay'. There are various methods of avoiding the precipitation of chromium carbides. Clearly, the lower the carbon content the less likely attack will be, and heat treatment with quench could be employed for small components, although welding as a means of fabrication would then be unlikely. The full solution is found in adding an element to the steel which preferentially forms a more stable carbide, e.g. Ti or Nb. Thus for welding fabrication BS1449 304 S12 of lower carbon (0.03% maximum) would be preferred to BS1449 304 S15 (0.06% maximum carbon content). For weldments in very aggressive environments the steel employed would be 321 S12 containing titanium equal to five times the carbon content but not more than 0.7%.

In aluminium alloys, electrolytic attack is caused by the precipitation of inter-metallic compounds of electrode potentials different from that of the matrix in which they are set. With a finely dispersed intragranular precipitate the effect is not so serious, but coarse precipitation of the intermetallics at grain boundaries can lead to severe intergranular corrosion. Alloys, for example, which feature the compounds $CuAl_2$, Mg_2Al_3 and $MgZn_2$ may be susceptible, the first because it is more noble than aluminium and the others because they are more base. The sensitivity of aluminium alloys to corrosive attack has much, therefore, to do with the composition, mode of fabrication and heat treatment of a particular alloy, since these dictate the phases present and their distribution.

In the case of both carbon and stainless steels the initiation of attack and pitting is attributed by Wranglen[3] to the absorption of activating ions, particularly chloride ions, on certain defect sites in the oxide film, the effect being similar to slag inclusions or precipitates of secondary phases. When the pitting potential is reached, the electrical field strength above the thinnest parts of the oxide film (at the defect sites) will be so high that the chloride ions can penetrate. In steels generally, localized pitting may be associated with the presence of sulphides in the structure. In contrast to silicates and most oxide inclusions, sulphide inclusions in steel are conductors, with a low hydrogen overvoltage. Accordingly they act as local cathodes and initiate anodic dissolution of the nearby matrix. Some sulphide inclusions appear to be much more effective in initiating localized pitting attack than others. They contain different amounts of Fe, Mn, Cr, Cu, S and O content, not only within different parts of the same but even within one microstructural area and may, of course, be modified to CaS, CeS by purposeful additions for shape control in rolling. Furthermore, the chemical action of sulphides may locally alter the composition of the associated matrix giving, for example, manganese depletion in some cases.

In aluminium alloys the significance of intermetallic compounds in promoting corrosive attack has already been mentioned. Wranglen[3] has proposed detailed mechanisms, for example, for the creation of a pit associated with a more noble intermetallic precipitate such as Al_3Fe, which acts as the cathode for the attack. Pitting attack in aluminium is enhanced if the water contains traces of copper ions, and he indicates a mechanism whereby the deposition of copper on to an aluminium surface and which subsequently enters the oxide lattice, produces an effective cathode for the anodic dissolution at the pit.

Stress corrosion cracking, corrosion fatigue and fretting

The conjoint action of corrosive attack and stress can frequently lead to the failure of materials that would not corrode in the unstressed state which, at the level of stress applied, would not fail mechanically in the absence of corrosion. There are several possible mechanisms by which stress corrosion cracking could start. There could be a breakdown of passivity at the point of stress concentration in the oxide film through the preferential adsorption or absorption of specific ionic species on to or into the film at these higher-energy points. Alternatively, the stress concentration, as in a stainless steel, could be related to the presence of second-phase particles or regions such as non-metallic inclusions, carbides, sigma phase, martensite, leading to local electrochemical cells being set up, the breakdown of passivity and anodic attack being coincident with the point of stress concentration. Again, the passivating film may merely rupture mechanically as the stress concentration develops. This last explanation fits best with the fact that the stress intensity factor threshold (K_{ISCC}), corresponding to the strain to be developed before the film will crack, exposing bare metal, is found to vary with environment, composition of alloy and microstructure. Subsequent growth of the crack may well be rate-controlled by electrochemical factors until the onset of fracture instability. Calculations for the electrochemical growth rate of the crack after initiation are in good agreement with observed velocities.

Whether the stress corrosion crack follows an inter- or transgranular path will depend on the characteristics of the matrix in terms of the dislocation population and the dispersal of segregates (i.e. segregation) and second-phase particles.

Austenitic stainless steels usually resist a wide range of chloride environments and remain free from this form of corrosion. It may occur, however, in transgranular form in highly stressed austenitic steels, because of unique metallurgical and chemical circumstances. In particular, stress corrosion can occur in austenitic stainless steels when they are operated under tensional stress in chloride environments at above 60°C (140°F). The stress may arise from conditions of service, as in a pressure vessel, or from internal stress left by the fabrication method, e.g. cold working. Alloys susceptible to stress corrosion cracking often have low stacking fault energies. In such systems the process of recovery, whereby the stored energy in the system can be reduced by the movement of dislocations, is more difficult. This applies to austenitic steel and some brass and magnesium alloys. The 'season cracking' of brass pressings, spinnings, condenser tubes, bolts, etc. is another form of stress corrosion cracking, caused by the association of residual stress from cold work with ammoniacal environments. Failure here is normally intercrystalline.

Aluminium–magnesium alloys can be subject to intercrystalline corrosion, accentuated by applied stress or cold work. The rapidly falling solubility of magnesium in aluminium (~1.5% at room temperature) means that in useful alloys (e.g. 5–7% Mg) the β phase (Mg_2Al_3) can be precipitated as a film at grain boundaries with heating in the region of 100°C (212°F), as for example in heat-affected zones in welding, leading to attack and failure, particularly under stress. Heat treatment and controlled cooling is employed to remove the precipitate.

In the context of the stress corrosion cracking of high strength steels it is important to take into account the possibility of hydrogen having been introduced by the corrosion process or by preceding treatments of the surface, since very small amounts of absorbed hydrogen can lead to serious embrittlement (see Section 11.2).

Caustic embrittlement is a related form of failure to stress corrosion and describes the attack on carbon steel boiler plates around punched rivet holes, associated with residual internal stress from the punching and the presence of alkali in the boiler water. This was originally the most common example quoted, but with oxygen cutting and welding replacing so much punching and riveting it is now much less evident.

With cyclic stresses, the conjoint action of corrosion and fatigue can lead to fracture at an appreciably lower number of reversals than is normally associated with pure fatigue at a given load, and the normal fatigue limit does not apply (Figure 11.4). Corrosion fatigue occurs in most corrosive media and even air itself can be considered corrosive as compared with vacuum. For practical reasons, however, behaviour in air is taken as the standard for comparison. In several respects the mechanisms of enhanced crack growth in corrosion fatigue can be related to those discussed for stress corrosion cracking. As the crack opens during a cycle, the oxide film ruptures exposing reactive (i.e. plastically worked) bare metal to the corrosive action. On closing and re-opening the corroding agent will be expelled and sucked back, ensuring a steady supply of non-stagnant (e.g. oxygenated) fluid.

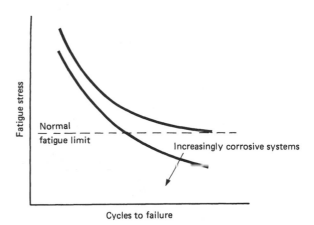

Figure 11.4 Effect of corrosion on fatigue behaviour.

Fretting corrosion

There is a variety and confusion of terms which describe the deterioration of mating surfaces where there is no intentional relative movement, as in shrink fits, bolted assemblies, components keyed to shafts, etc. 'fretting' is perhaps best restricted to an actual wear process, with 'fretting corrosion' used to describe situations where either or both mating surfaces, or the wear particles produced from them, react with their environment.

Dry fretting corrosion is a phenomenon in which the protective oxide film is being continually broken off by the surface asperities of a mating surface under conditions of very slight relative slip. Oxide debris will tend to build up between the surfaces, and in the case of steel, will be mainly α-Fe_2O_3 of red, brown to black colour. Fretting still occurs in an inert atmosphere, with the formation of debris, and in a high vacuum can lead to seizure. Contrary to normal expectation the presence of moisture generally decreases rather than increases the surface attack, dispersing the debris and producing softer hydrated oxides, which may act as lubricants. If a protective oxide glaze is produced by, say, increased temperature of operation, this can minimize intermetallic contact and the early adhesion stage fundamental to the wear process. In line with this it is clear that the better the surface finish the lower the incidence of fretting damage.

The avoidance of fretting resides in preventing slip in joints that are not intended to have relative movement, by increasing the friction through an improved machined fit or by interposing a soft electrodeposit between the surfaces. Alternatively an elastic material may be interposed which will accommodate some relative movement, but without slip at the surfaces. Clearly a reduction of vibration in the system is always important. In the context of purposefully moving interfaces the problem then becomes one of bearing surfaces where control of friction and wear becomes largely associated with surface lubrication.

Fretting fatigue

Fretting corrosion can result in surface microcracks which can act as notches, with the virtual elimination of Stage I of fatigue crack growth, so that fast growth during Stage II is predominant and the life of the component is greatly reduced (see page 102). Very little surface damage is necessary before there is a very considerable reduction in fatigue strength as compared to a smooth finish. Clearly, every attempt should be made to execute the design so that the interface does not enter into a region of stress concentration.

The worst conditions in relation to the effect of fretting or fatigue behaviour seem unfortunately to be when there is a low slip amplitude, which is most likely in the context of 'fixed' joints. In controlling fatigue life, as well as controlling the slip, surface hardening may be useful.

Impingement and erosion

Stress corrosion cracking and corrosion fatigue have been described above as being substantially controlled by the effect of the mechanical stress system within the metal on the oxide film covering it, at the point where the crack is initiating and then propagating. It must be noted, however, that forces can also be exerted by the large-scale movement of the corrosive media and these can be harmful to the protective film or the conditions leading to its formation. The more stagnant the conditions the worse will be the localized attack, associated for example, with differential aeration. If fluid velocities are raised, however, cathodic and anodic polarization can be decreased through dispersion of cathodic and anodic electrode reaction products, with increased general corrosion. On the other hand, if the fluid movement can increase the supply of oxygen, passivation of such materials as stainless steel can be more readily obtained and corrosion resistance is improved.

At high flow rates, under turbulent flow conditions, erosion corrosion may occur. This takes the form of impingement attack where, in addition to the rapid transport of solution reactants and products, the action of the turbulent flow removes the product of corrosion from the metal surface mechanically. This mechanical action is made worse by air bubbles or solid particles entrained in the impinging liquid. At very high superficial flow rates cavitation in the liquid may occur at the surface. The subsequent collapse and re-forming of vapour bubbles again produces a mechanical force destroying the passivating surface oxide film and maintaining corrosion attack.

11.2 The problem of hydrogen embrittlement of steel

Where steel embrittlement is encountered as a result of absorbed hydrogen, the most common source of the hydrogen is electrochemical reaction, as in corrosion. Since, also, all mechanical properties are adversely affected, it is convenient to consider the problem in the context of this corrosion section, rather than elsewhere.

The influence of hydrogen on the properties of steel

This is large subject and can only be very briefly dealt with here, but an appreciation of the general principles governing the behaviour and influence of hydrogen in steel may be helpful in understanding the way in which failures under investigation may have occurred. Hydrogen has only a very limited solubility in the ferrite matrix in steel, but is accepted by interfaces within the steel where the atomic mismatch between phases provides interatomic space at the boundary. The degree of mismatch (cf. coherency) between different phases or 'grains' of the same phase (e.g. ferrite) and the length of the boundaries is the main factor determining the hydrogen solubility. Similarly, the diffusion of hydrogen is significantly affected by the type and extent of

interphase boundaries. Where the interface is truly incoherent, e.g. between a non-metallic inclusion and the ferrite matrix, then the hydrogen can the more readily re-associate into molecular form and force the interface apart, creating voids and internal cracks. This is known as 'hydrogen-induced cracking' and may occur even in the absence of applied or internal stress, although the generation of such internal flaws will, of course, influence the subsequent mechanical behaviour of the material detrimentally.

More usually, however, hydrogen reduces the toughness of the steel without producing internal damage prior to the application of stress. This is known as 'hydrogen embrittlement'. There are several well-known experimental observations associated with the phenomenon; it is time dependent, loading mode dependent, strain-rate dependent and temperature dependent. The extent of the embrittlement increases with increased strength (hardness) of the steel, decreased strain rate, increased hydrogen content and increased triaxiality of stressing as, for example, when notches are present.

There is clearly interaction between the deformation processes in progress resulting from applied stress and the movement of the hydrogen within the steel. Apart from this interaction relating to the movement of dislocations in the metal crystal structure, by which deformation occurs, and of hydrogen atoms, there is also a dependence on the microstructure of the steel, as affecting both the deformation process and the hydrogen mobility. Whether there are a larger number of coherent interfaces providing small individual spaces for hydrogen or, at the other extreme, a small number of incoherent interfaces ('strong' traps) for a given overall hydrogen content, will influence the activity of hydrogen in a given region (i.e. the driving force for diffusion) and the average distance of a hydrogen source from an approaching crack or a stress concentration. The general order of resistance to hydrogen embrittlement is considered to be martensite, upper bainite, pearlite, lower bainite and the best, fully tempered martensite. This latter structure gives an intrinsically tough structure and provides a very large number of weak traps. It is therefore capable of accepting the greatest amount of hydrogen without serious embrittlement. Lightly tempered martensite of high hardness, or untempered martensite or bainite have poor resistance to hydrogen-induced crack propagation.

The resistance of steels to hydrogen uptake and thus cracking is affected by the directionality of the microstructure in the material. In relation to the exposed surface, diffusion of hydrogen occurs most rapidly into the steel when grain boundaries and phase interfaces are predominantly set at right angles to that surface.[4] This can affect the performance of components and is a factor which should be considered in the selection of material and processing to the finished shape and size. A pearlitic microstructure elongated along the axis of a bar has a good resistance to hydrogen-induced crack propagation across the bar, as does a quenched and tempered structure, providing the hardnesses are not too high.

There is extensive literature on hydrogen cracking in steels and an aspect which has received considerable attention is delayed cracking. This occurs when a steel containing hydrogen is subjected to static loading in the presence of a stress concentrator such as a notch. It is found that the steel may withstand the initial application of the stress without failure, but then after a certain period of time, failure occurs. The time of failure decreases as the stress level increases. In detail what happens in such cases is that, after a certain time, a crack initiates at the notch root, and grows very slowly at first, and then at an increasingly rapid rate to final fracture. Typical behaviour for a number of steels is shown in Figure 11.5. Several features can be noted.

(1) The notched tensile strengths of the various steels are lowered by the presence of hydrogen to different extents and in some steels it is a very marked effect.
(2) The time to failure increases with decrease in the level of applied stress.

Figure 11.5 Constant load rupture strength of different steels in the quenched condition to simulate HAZ microstructures, and charged with hydrogen.[5]

(3) Below a certain stress, which depends on the steel, the time to failure increases markedly so that a plateau stress is obtained below which the probability of failure is very low and tending to zero.
(4) The plateaus are reached for different steels after different times but in Figure 11.5 the times are all less than a day, i.e. if in these tests stressed specimens survived a day, they would endure for at least 6 days, which is the time limit of the tests shown.
(5) Specimens stressed to levels below the plateau stress would not suffer any damage, but stressing above the plateau for even a short time could induce damage which might then propagate at lower stresses, including stresses below the plateau.

Thus hydrogen embrittlement depends on the transportation of hydrogen to initial microstructural locations where it modifies the local deformation behaviour by contributing to crack nucleation and propagation. How does it get into the steel in the first place?

The source of hydrogen in steel

(a) Hydrogen will, of course, be introduced during the steelmaking process, to an extent depending on the process being employed. In particular there will be reaction between molten steel and atmospheric humidity, which will mean that unless a steelmaking process is excluding air (e.g. oxygen/fuel gas blowing), or a degassing process is carried out after steelmaking immediately prior to casting, the steel is likely to contain 4–5 cc H_2/100 gms on solidification. Where the steel is to be subsequently hot-rolled to small section this is unlikely to cause any further problem, since the amount of hydrogen in the solid steel, in equilibrium with the atmosphere at lower temperatures, will continually reduce and hydrogen will diffuse out as the steel is worked. There are, however, problems if these high solidification hydrogen values are associated with heavy forging ingots, where thick section, and the thermal stresses associated with reheating, can cause cracking. It is unlikely that bar sections will carry significant hydrogen contents from the original steelmaking operation.

The pick-up of hydrogen during welding is also usually associated with liquid steel/moisture interaction.

(b) Although not frequently encountered, hydrogen can be introduced into steel by pressure charging in a gaseous hydrogen environment. The degree of dissociation of the molecular hydrogen and the concentration of absorbed hydrogen atoms on the steel surface and subsequent absorption by the steel will be controlled by temperature and pressure. Even at room temperature there is a finite take-up of hydrogen from a pressurized atmosphere (cf. hydrogen gas cylinders). At high temperatures and pressures, as could be encountered in oil refinery equipment, cracking may result from the formation of CH_4 in internal voids. The most resistant steels contain chromium and molybdenum[6].

(c) Very commonly electrochemical charging is involved, in a number of differing circumstances: corrosion, pickling, plating. In all cases the hydrogen is being generated at the steel surface, or a part of the steel surface which is the cathode of an electrolytic cell. These cathodic reactions have been discussed earlier on p. 147. In laboratory investigations it is common to charge steel with hydrogen either by driving a cell with the steel specimen as the cathode using an external current, or by an intensive corrosion process, such as with NACE solution (NaCl/dilute acetic acid solution saturated with hydrogen sulphide).

Pickling

Prior to the application of a surface coating to steel, as for example zinc in the galvanizing process, it is necessary to remove all surface scale and oxide corrosion products. This is achieved by acid pickling. Although H_2SO_4 is used in the sheet industry, HCl is used prior to the galvanizing of fabricated articles and wire.

It can be seen that the conditions of electrolytic cathodic hydrogen charging are very likely to obtain. During the dissolution of the scale the steel is anodic and the oxide cathodic, but on exposed steel surfaces free of scale, and once scale has been removed, there will be overall attack with ferrous ions being produced in anodic areas of the steel and hydrogen in the cathodic areas. The differing polarity of different surface regions is produced by the underlying microstructure. Even if the steel is single phase, for example, ferrite only, at very low carbon content, there will be a difference between grain boundaries and grains and between regions containing otherwise differing concentration of defects.

Although the majority of the hydrogen produced may escape as gas bubbles, some of it diffuses into the steel in atomic form. The current in the cell can be substantial and result in significant hydrogen ingress. With hardened or high carbon steels the effect may be so pronounced that internal voids and cracks may be produced, some evident as surface 'blistering'.

The electrode reactions are sensitive to the presence of impurities in the acids. Arsenic, for example, raises the overvoltage for the hydrogen evolution reaction and increases hydrogen absorption and diffusion into the steel. Sulphide in the acid promotes the anode reaction and the formation of ferrous ions.

Although hydrochloric acid is more expensive than sulphuric acid, it is becoming increasingly used instead of sulphuric acid, because the waste liquor can be recovered more economically. It is more active than sulphuric acid at an equivalent concentration and temperature, probably because the rates of diffusion of acid to, and ferrous ions from, the steel surface are greater. It is also more suitable for pickling prior to applying a coating, since it gives less 'smut' on the steel and any residual iron chloride can be rinsed off more readily than residual iron sulphate deposits. However, HCl dissolves detached magnetite and haematite more rapidly and hence the ferric ion produced increases the rate of attack on the steel, with greater hydrogen evolution and acid consumption.

In order to reduce the acid attack on exposed steel to a minimum, organic inhibitors are used; it would be expected that since the hydrogen evolution is reduced the amount of hydrogen absorption and embrittlement would also be reduced. This is not always the case; thiocyanate inhibitors, for example, actually increase the absorption of hydrogen.

A prime consideration must be that if pickling is employed then the pickling time should not be longer than is required for oxide removal. Protracted immersion will lead to hydrogen charging.

If the pickled steel is then immersed in a zinc bath and a coating produced, hydrogen is essentially 'locked in', since the diffusivity of hydrogen through the zinc is slow. Coatings of cadmium and copper have a similar effect and are used to retain hydrogen in steel after charging in laboratory experiments investigating the effect of high hydrogen concentrations which would otherwise diffuse out.

Thus absorbed hydrogen in a galvanized steel will tend to concentrate below the zinc coating and may lead to hydrogen embrittlement in steel of high strength.

Electroplating

Since the current efficiency of many electroplating systems is significantly below 100%, hydrogen is generated as part of the cathodic reaction, with absorption into the substrate. The efficiency can be maximized by high current density and careful control of the bath composition. The application of an initial high current density 'flash' accelerates the deposition of, for example, cadmium, and reduces the ingress of any hydrogen produced at any stage of the operation, since the formation of a fine microstructure at the start provides a barrier, and this microstructure is maintained during lower current density operations subsequently. Baking such an initial flash coat, or complete plating of zinc, cadmium, nickel or chromium, at 200°C will facilitate the escape of hydrogen, which may otherwise be trapped behind the plate.

11.3 The selection of materials for resistance to atmospheric corrosion

The most significant factor controlling the probability of corrosive attack is whether or not an aqueous electrolyte is likely to be provided by condensation of moisture under prevailing climatic conditions. Clearly, hot, dry or cold, icy conditions give less attack than wet, as does a clean atmosphere as compared to the industrial or marine atmospheres containing sulphur dioxide and salt respectively. Even within given areas, differing microclimates can exist as a function of direction of exposure to sun, wind and polluting sources. In the case of sulphurous acid attack the effect is more noticeable in the winter, when more fuel is burned and conditions are generally wet.

Untreated steel is very prone to rust in damp environments, at a rate depending mainly on the level of atmospheric pollutants. The initial rate of attack tends to be the same, but diminishes in cleaner, less aggressive environments. In the UK the attack, once established, will generally be between 0.05 and 0.18 mm/year inland. In less temperate regions the range may fall to 0.005–0.1 mm/year. Coastal situations may markedly raise the attack rate, particularly in hot climates.

Such attack rates may be acceptable for plant which is designed to have a finite working life, particularly bearing in mind that the concentration of corrosive atmospheric pollutants in industrialized countries has tended to drop in recent years. Whilst painting may be desirable for appearance, it may not be justified in terms of economics. The rate of rusting will, in general, be of the same order for all mild steels and low-alloy steels. The presence of up to about 0.2% copper in such steels has a marked improving effect. Amounts between 0.1% and 0.2% are commonly present in steels anyway, arising from recycled scrap steel, with the amount tending to build up, since it is not removed in the process of steelmaking. As the rusting rate decreases, so the rust film is more dense and protective. The superior corrosion resistance of wrought-iron was attributed by Chilton and Evans[7] to a build–up of copper and nickel at the surface and segregation within the matrix, produced by the particular method of iron manufacture in the solid state. Phosphorus, silicon and chromium present in carbon and low-alloy steels are also considered to improve corrosion resistance, particularly in combination with copper. More highly alloyed steels, such as the maraging steels containing 18% Ni, rust uniformly and somewhat more slowly.

Stainless steels of all types are used satisfactorily indoors, but if outdoor service requires the maintenance of a bright finish without any rust staining or fine pitting corrosion, then the martensitic stainless steels (e.g. BS970, 403, 405, 420) should be avoided. The general corrosion resistance of the stainless steels is much affected by the uniformity of chromium content, and in the martensitic steels the retention of δ-ferrite leads to segregation of the chromium to higher levels in this phase, with reduced corrosion resistance overall.

Ferritic stainless steels such as BS1449, 434 S19 or 442 S19, generally retain a satisfactory finish without staining and are being used for motor car trim. 430 S15 of lower chromium than 442 and without Mo (as is present in 434) may not be suitable.

The austenitic stainless steels are generally superior, although, of course, pitting may occur if second phases such as δ-ferrite or chromium carbide generate lower chromium regions in the associated matrix, thus lowering the chromium oxide in the surface film.

Cast iron, usually of heavy section, is attacked only slowly and rusting is not a problem. This resistance is due to the inertness of the main microstructural constituents, particularly the graphite and iron phosphide eutectic. Aluminium alloys generally behave well, but with some superficial pitting in early years in alloys containing precipitates.

Copper gives excellent service and is widely used for roofing and flashings. Such attack as does occur produces the attractive green patina, which with time moves towards the composition of the mineral brochantite, $CuSO_4.3Cu(OH)_2$. Lead is also popular for roofing and for facing panels and the electrically insulating and protective corrosion products that are produced ensure good service. Zinc also gives excellent service, but being more reactive, and the products of corrosion somewhat less protective, it is more sensitive to wet and polluted conditions.

Plastic materials are used extensively in the building industry. For example, traditional materials for waste systems such as copper, steel and cast iron are typically replaced with plastic pipes and fittings. There are several advantages to using plastics; corrosion resistance is one. uPVC has been in use for the manufacture of double-glazed window frames for several years.

11.4 The selection of materials for resistance to oxidation at elevated temperatures

Whilst traditionally corrosion has been treated on a 'wet' and 'dry' basis, 'dry' oxidation corrosion can be considered as an electrolytic process of an interfacial anode/cathode type (see p. 146). Since the corrosion rate is governed by the transport of ions and electrons through the produced film, equatable to a current, it is clear that the oxidation rate will be low where the oxide film has a high electrical resistance and where it is not prone to mechanical rupture.

The oxidation characteristics of metals and alloys have been summarized by Kubaschewski and Hopkins[8] and are also dealt with very thoroughly by Shreir.[1]

Iron and low-alloy steels behave in a similar way with a layered film $Fe/FeO/Fe_3O_4/Fe_2O_3$ film developing parabolically. Outward cationic diffusion of Fe^{2+} and Fe^{3+} predominates in the growth of the FeO and Fe_3O_4 layers with O^{2-} anions diffusing inwards in the Fe_2O_3 layer. The effect of alloying elements and impurities relates to the influence that they have on these diffusion rates and on the mechanical integrity of the film. Briefly, carbon has no major effect, unless CO produced by any decarburization at the metal surface results in the scale breaking away.

Silicon, producing fayalite (silicate) in the scale, decreases the oxidation rate as does aluminium and chromium (Figure 11.5). Of these, chromium is most commonly employed, modifying the oxide film to iron chromite. It is not very effective at low concentrations, but at substantial levels gives protection to a wide range of steels.

The composition of the gas in which the dry oxidation is occurring will also have an effect.

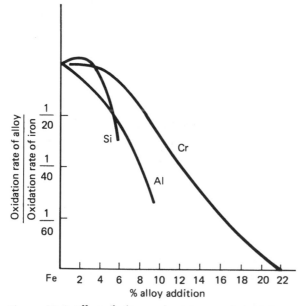

Figure 11.6 Effect of silicon, aluminium and chromium on the oxidation rate of iron at 900–1100°C. (Shreir[1], after Kubaschewski and Hopkins[8])

The oxidative power of the atmosphere will vary, steam and carbon dioxide or sulphur dioxide may be present and there may be deposits of ash or volatile oxides from a combustion source. The interaction of these constituents is complex, but their influence must be considered in any assessment of a material for high-temperature operation, since in some cases their effects are quite specific.

In the case of graphitic cast irons, the high-temperature oxidation process producing surface scale is accompanied by progressive internal oxidation of the iron along the graphite flakes, with swelling and distortion of the casting, this effect being additional to any increase in volume due to breakdown of carbide in the microstructure to graphite, as a result of the high-temperature service. The morphology of the graphite will clearly be important, a finely dispersed graphite giving improved resistance to internal oxidation and swelling. Increased carbide stability, as achieved by alloying additions, will also be important in maintaining dimensional stability. The characteristics of the scale will, of course, be similarly affected by alloying additions as for steel, with silicon, chromium and aluminium additions producing the most advantage. The most commonly employed heat-resisting cast irons include Silal (2.5% C, 6.0% Si), which has a fine graphite structure in a silico-ferrite matrix. The material can be produced with nodular graphite by inoculation with magnesium, and the stability is thus further increased. Nicrosilal and Niresist irons are both of high nickel content, and thus have austenitic matrices. Silicon, chromium and copper are variously added to improve resistance. High chromium cast irons of a ferritic type, containing up to 30% Cr, offer useful service.

11.5 The selection of materials for resistance to corrosion in the soil

The aggressiveness of soils can vary substantially. First, the texture of the soil governs the access of the air necessary for the corrosive process and, second, the presence of water is required for the ionization of the mineral in the soil and the oxidation product at the metal surface. The amount of water present can, of course, also affect the availability of oxygen. Below the water table the rate of oxygen diffusion will be substantially less than in the porous air-filled layers above, leading to differential aeration and attack below the water line in the presence of electrolytes. Recognizing that increased soil porosity and water accumulation can be important in encouraging attack, it is clear that the soil disturbance in trenching for pipe-laying and back-filling, after some inevitable consolidation at the base of the trench, can actually increase the risk of attack. The aeration normally associated with loose backfill also encourages the activity of aerobic bacteria, the activity of which can lead to local variations in aeration and even the consumption of some organic protection systems on pipes, e.g. asphalts. Corrosion by wet H_2S can also be produced in steel by an aerobic bacterium which reduce sulphates (e.g. gypsum).

The chemical nature of the soil minerals, and the presence of biological products, both organic and inorganic, will lead to a wide range of basicity and acidity.

In general, dry, sandy or chalky soils of high electrical resistance are the least corrosive, with heavy clays and saline soils the worst.

Most of the material underground is steel, cast iron or concrete, but an increasing role is being played by plastics. Normally, the rate of attack on buried bare steel and cast iron is considerably lower than in the atmosphere and is approximately the same in each case.

Many underground pipes carry water. Most of Britain's water mains are old and made of unlined cast iron. Reporting a survey in the Midlands[9] in 1982 it was stated that more than 20% of the region's 20,000 km of iron mains were so corroded that half their capacity was being lost. However, it was not only the old mains that were cracking up. Mains laid before 1930 were thick-walled, as necessary when made by vertically casting the grey iron. In later years thinner-walled centrifugally cast pipes became standard and even more recently authorities have used even thinner-walled pipes from ductile, nodular graphite, iron. All the pipes were strong enough when laid, but since the corrosive attack rate is similar in all the ferrous materials used there is likely to be a serious bunching of failures when the older thicker-walled and the more modern thin-walled pipes are penetrated by attack or fail under stress or pressure at much the same time. It may be that this stage has now been reached in the UK, with the serious level of leakage in water supplies emphasized by recent drought conditions and water shortage.

Underground steel pipelines for oil, natural gas or trunk mains for water can be given cathodic protection, which is capable of controlling corrosive attack totally. The pipeline is made the cathode of a galvanic cell. There are two methods by which this can be achieved.

In one method the steel pipe is associated with a more electronegative and reactive material as a sacrificial anode. In the case of galvanizing or sherardizing with zinc the steel is directly coated with the anode metal. With a permanent structure such as a pipeline it is more convenient to connect the pipe electrically to separate buried anodes, which can be renewed as necessary. These anodes may be alloys of magnesium, aluminium or zinc.

In the other method the steel pipe is made the cathode of a cell, which is powered with direct current from the surface. The anode is of carbon or some other suitably inert material. By this means the potential of the steel is controlled at a level which ensures that, instead of the anodic reactions, oxygen reduction occurs.

Usually, cathodic protection is combined with protective wrappings and insulating coatings. By this means the protective current can be concentrated at any defects in the coating, pre-existing or developing, and the current required for a satisfactory system will be small and easily applied. Such an arrangement is particularly satisfactory since it allows tolerance in the coating system where expense would increase greatly if it were necessary to provide total insulation and impermeability. A minimum-cost protection system combining good, but not necessarily total, metal coverage and light cathodic protection can be developed in this way.

A similar method of corrosion protection is applied to ships' hulls (see p. 261).

Plastics

Whilst the use of plastics has been limited for the distribution of potable water, gas distribution lines are increasingly constructed from medium-density polyethylene (MDPE) or similar material, rather than iron or steel. The properties in plastics are optimized through control of chain length (i.e. molecular weight) and the degree of chain branching (density) and crystallinity, the aim being to provide a material which gives maximum flow rate in a fusion-welded system together with good impact strength, resistance to environmental stress cracking and low temperature toughness.

The advantage of MDPE is that it is flexible enough for jointing, etc., to be carried out on flexed pipe above ground, and can be temporarily sealed by squeezing between clamps. Further, the low coefficient of friction between gas and the smooth bore means that smaller pipes can be employed. The change from town gas to natural gas of higher calorific value implies a lower flow requirement and this has meant that, in some cases, smaller plastic pipe can be threaded through existing iron mains (e.g. at busy road junctions) without loss of pressure.

Side connections can be made under pressure through saddles heat-welded on externally, which are provided with a built-in cutter to give entry through the wall of the main once the connection has been made. (Figure 11.7). Interference gas-tight fits are not possible and all joints have to be fused. This is readily accomplished using a heating tool operating between cooling collars.

Plastic gas mains are generally cheaper to install and the prediction is that the material will be stable for typically 50 years minimum service and require little maintenance.

Figure 11.7 Polyethylene Aldyl 'A' pipe with fittings (By courtesy of Du Pont (UK) Ltd.)

Cement and concrete

Cement and concrete are generally regarded as being very stable, particularly when buried in soil and not subject to extreme changes of temperature. Calcium hydroxide and hydrated calcium aluminate in cement are, however, chemically attacked by sulphates, for example, by calcium sulphate often present in ground water. The products of reaction are of higher specific volume and cracking results. Resistance to sulphate attack is maximized by minimizing the amount of tricalcium aluminates (C_3A) and maximizing the tetracalcium aluminoferrite (C_4AF). Sulphate-resisting cement is designated Type V in the ASTM classification.

Concrete is often used underground as a protection for steel structures or pipes, and steel is used within concrete, as reinforcement. The alkali generated during the setting of the concrete initially produces a film of ferric oxide on the steel surface, which is protective. Concretes are sometimes porous and may be locally markedly so, and under suitable conditions of moisture the alkalinity can be replaced by acidity, giving rapid localized attack on the steel under the current concentrating conditions of a small anode/large cathode. As rust develops porosity will generally give way to cracking with further ingress of water and accelerated attack. If the steel is under stress, as in prestressed concrete, the attack is further accelerated (stress corrosion).

A particular danger results when calcium chloride is added to accelerate cement hardening, since this donates chloride ions to the corrosion system. Galvanizing reinforcing bars is useful, but ensuring an adequate covering of dense, sound, concrete should give satisfactory performance. In the case of high strength steel bar reinforcement, pickling prior to galvanizing could lead to hydrogen embrittlement (see p. 157)

Another specific danger is in the use of concrete-covered steel structures where the steel that they contain is directly coupled to steel which is not covered. This can generate differential aeration, with intensified attack at any points of weakness in the concrete covering.

Similar comments apply to the use of concrete-covered or reinforced-concrete structures in atmospheric or water environments, exacerbated, of course, in salt water by the high level of Cl^- anion and the electrolytic conductivity with hydrogen generation, which it confers. Again hydrogen absorbtion by the steel can result in embrittlement.

Other materials

There are, inevitably, occasions when other engineering materials are buried during use. Aluminium and its alloys are liable to attack in some soils and it is usual to provide protective coatings such as bitumen.

As instanced by the widespread use of copper water pipes, unprotected copper gives satisfactory service in a wide range of soil conditions. The most aggressive conditions are highly acid peaty soils or made-up ground containing cinders. Under such conditions it may be wise to provide organic coatings or wrappings.

Brasses give variable performance underground and are generally not to be recommended. All are subject to dezincification by a wide range of soil conditions, particularly those high enough in zinc to give duplex structures, which promote galvanic attack.

Stray currents are always a danger when metallic components are immersed in an electrolyte, be this water, damp soil or damp screed. Cases have been known where severe attack on copper-sheathed power cables has occurred in damp conditions due to stray currents in the system producing anodic conditions at some surfaces.

Lead pipes and cable sheaths are particularly prone to attack for this reason, possibly because they often exist in long runs and also in some cases because they are used as earthing points for alternating currents. It is generally not a good idea to use uncoated and uninsulated lead systems where there are likely to be stray currents associated with them.

11.6 The selection of materials for resistance to corrosion in water

The corrosion of materials in water depends, of course, on the substances that are dissolved, or suspended, in it and also upon its temperature. Dissolved oxygen is most important since in neutral solutions it must be reduced at the cathode for the corrosion reaction to proceed, and it also accounts for the development of passivating oxide films, where these can be produced. Since oxygen enters the system by dissolution from the air, its concentration in large masses of water can vary appreciably both in terms of flow and depth.

Carbon dioxide dissolved in natural water is usually associated with calcium carbonate or bicarbonate. Where the dissolved carbon dioxide is not high enough to maintain the bicarbonate state in solution, the change in pH at cathodic areas will cause the carbonate to precipitate on to the metal surface and if the 'fur' so produced is adherent, further corrosion will be restricted. 'Soft' waters, usually derived from upland open reservoirs of low carbonate content, are therefore more aggressive, particularly since they often contain organic acids deriving from moss and peat which give a low pH.

Other dissolved salts can, of course, have very important effects, and this is particularly significant in the case of sea water, where the chloride ions present decrease the electrical resistivity of the water, so that corrosion currents will be larger. As discussed in relation to passivation and stress-corrosion cracking, the presence of such conductive ions can also affect the properties of a normally protective surface film. Sulphates are also an important constituent of both inland and sea water in relation to corrosion, generally producing the same disadvantage as chloride, although sulphate attack on concrete comes into a separate category. On the positive side, the presence of the SO_4^{2-} ions in feed water may be advantageous in countering alkali-induced stress-corrosion cracking (caustic cracking) in boilers.

Organic matter, both living and dead, will be present in natural water. These materials may deposit on surfaces and if the covering developed is continuous, the blanketing effect may reduce corrosion. More usually, however, the effect of organic films and bacteria or algae coatings is to produce strong regions of local deaeration and thus accelerate attack.

Increasing the temperature of the water will markedly increase the rate of corrosive attack, unless the increase in reaction rate is offset by some opposing effect which also increases with temperature, such as the more rapid and complete coverage of the metal surface by deposits which act to reduce the availability of reactants and increase the corrosion cell resistance.

As discussed earlier in this chapter, aqueous corrosion will be affected by the flow characteristics of the electrolyte, since these control the transport of reactants and products to and from the metal surface, and in extreme cases produce erosion and impingement effects.

All ferrous structural materials of low alloy content corrode in natural waters at about the same rate, although wrought iron shows some superiority.[1] The average rate of corrosion in sea water falls during the first year from an initial value of ~0.3 mm/y to ~0.15 mm/y after 6 months and to ~0.1 mm/y over many years. It is suggested that a figure of 0.13 mm/y can be taken as a reasonable estimate of the expected rate for continuous immersion in sea water. These values do not, of course, predict the depths of any pitting

which might occur. For comparison, total immersion in fresh water gives average values of between ~0.01 and ~0.07 mm/y, depending particularly on the carbonate content. More acid waters are the most aggressive.

Only chromium has been found to reduce the rate of rusting occurring in sea water, with 3% Cr halving the attack rate. Stainless steels, particularly the molybdenum-containing austenitic grades, have good corrosion resistance in sea water. The martensitic 13% Cr types are, however, generally considered to be unsuitable. When corrosion occurs it normally is by pitting or by crevice attack, exacerbated by low velocity or stagnant water conditions.

Cast irons

Cast irons corrode a little more slowly in water than steel, with the rate depending again on the pH, the level of carbon dioxide, carbonate and chloride ion present and the degree of aeration (particularly as a function of water velocity), but commonly of the order of 0.05–0.1 mm/y in sea water.

The corrosion of cast iron in sea water, albeit a slow process, produces remarkable effects when an object such as a cannon ball is recovered after hundreds of years immersion. The ball is covered with a calcareous and ferruginous shell of corrosion products and encrustation; when this is removed and the 'ball' is exposed to air, it becomes hot owing to the oxidation of some of the particulate iron in an external graphite skeletal layer, retaining the original dimensions of the ball, on a residual cast iron core. (Figure 11.8).

Austenitic cast irons, with substantial nickel content, have superior corrosion resistance; they would be chosen for sea water use or where the water to be handled contained high levels of carbon dioxide or pollutants. In sea water the rate of attack is similar to that on gunmetal (88% Cu, 10% Sn, 2% Zn) and may be between one-third and one-tenth of that for low alloy ferritic cast iron. The austenitic irons containing higher chromium and silicon contents are particularly good since they passivate more readily. This

Figure 11.8 Section of cast iron cannonball recovered from the sea, showing outer ferruginous accretion and graphitic shell retaining original dimensions from which iron has been largely removed.

leads to a consideration of the high-chromium cast irons *per se*. If the iron is to passivate and not rust, enough chromium must be left in the matrix, after carbide formation, to produce a chromium oxide film at the surface. It is commonly held that this requires a minimum chromium content given by $\% \, Cr = (\% \, C \times 10) + 12$. This means that a 2.0% C iron should contain 32% Cr. Such irons are difficult and thus expensive to cast owing to the rapid rate of oxidation of the chromium on melting and pouring, which can lead to dross incorporation; in addition, high shrinkage makes the castings difficult to feed to avoid porosity. These alloys are thus used only in critical applications. Silicon improves the castability and, like molybdenum, assists in producing a more continuous passivating chromium oxide film by refining the form of the carbides present. Up to 2.5% Si may be incorporated. Molybdenum may also replace some of the chromium in the carbide, making more of the latter available in the matrix, and thus aiding passivation.

Aluminium

Aluminium and its alloys do not corrode in pure water, although they may stain somewhat in natural fresh water, and only those containing copper as a major alloying element will corrode significantly in normal sea water. For freedom from localized pitting associated with the most aggressive sea water conditions (pollutants, etc.) the Al–Mg alloys offer the best choice.

Copper and copper alloys

These have a traditional and extensive use in the handling of natural waters. Copper is widely used for distributing cold and hot water both in domestic installations and industrial plant, and a wide range of copper-based alloys are employed for such items as tubes for condensers in power stations and desalination plant, and for propellers, valves, pumps, etc.

The mechanism of the protective film formation associated with the use of copper alloys in water is not fully understood, but it is clear that it is sensitive to water movement; impingement attack due to turbulent flow, particularly if carrying air bubbles, is a common cause of failure of, for example, Admiralty brass (70% Cu, 29% Zn, 1% Sn) condenser tubes and high tensile brass (manganese 'bronze') marine propellers. The inclusion of aluminium in brasses for sea water service (e.g. 76% Cu, 22% Zn, 2% Al, 0.04% As) greatly improves resistance to impingement attack, presumably through modification of the oxide film, making it more tenacious and impervious by the incorporation of alumina. Cupronickels and tin bronzes both have generally good resistance to impingement attack and the former are widely used in aggressive conditions at high water velocities.

Another common form of attack in water concerns the dezincification of brasses. Selective attack on the zinc content of the brass leaves behind a porous plug of copper, which is very weak and which may completely penetrate a component. In the single-phase α brasses the attack takes place uniformly, but in the α/β alloy the zinc-rich β phase is attacked preferentially until the zinc level is much reduced. The α phase may be attacked subsequently. The zinc salts produced may be removed by water flow or produce bulky deposits on the surface. This can be a particular nuisance in brass valve components where clearances are important or where small apertures may become blocked. An example from practice is the slow ignition valve in water-operated gas water heaters, where the water flow actuates the gas valve through the build-up of pressure behind a diaphragm. The valve contains a ball, the movement of which ensures a slow build-up of water pressure and thus the slow introduction of gas and a rapid turn-off. Dezincification in some domestic waters containing relatively high concentrations of chloride ion and little carbonate hardness, causes the ball to stick in the valve, making the heater unusable. The component is now usually produced from a resin, such as polyacetal.

The chloride ion content of the water is therefore significant in this form of attack, as is water temperature and velocity. It is clear, therefore, that dezincification attack can be considerable in sea water, especially where there are surface deposits or encrustations, such as barnacles, which lead to differential aeration. Arsenic addition (~0.04%) inhibits dezincification in the α but not α/β brasses. In the latter, tin addition reduces the rate of attack (naval brass 61% Cu, 38% Zn, 1% Sn). The complex high-tensile brasses containing aluminium, manganese, iron, tin, and nickel are similarly resistant but not immune to dezincification attack.

Pitting corrosion in copper and copper alloys can be caused by surface deposits which lead to differential aeration. The dezincification of brasses may be associated with the presence of barnacles, but these and other encrustations and deposits particularly encountered at low water velocity give rise to attack on copper and most other copper-base alloys also. Pitting is also encountered in domestic copper water pipes supplying deep, cold well water of high SO_4^{2-}/Cl^- ratio which is virtually free of any naturally occurring organic inhibitor, which means that Cu_2O forms as a loose, rather than a protective, deposit. Attack is concentrated beneath carbon residues left in the tubes, which has its origin in drawing lubricant, carbonized during heat treatment. The surface below the cathodic residue, depleted in oxygen, is rapidly attacked, to give pinhole perforations. The British Standard for copper water pipes now specifies that the internal surfaces should be cleaned of any deposits. This is usually achieved by scouring with sand entrained in high-velocity water.

Nickel

Nickel and nickel alloys are generally resistant to corrosion in fresh water, except under conditions of high acidity and stagnant conditions where the passive oxide film cannot be maintained. The same principle applies in sea water, attack being low (~0.01 mm/y) in neutral chloride-containing environments where the oxygen supply is adequate through active flow conditions in relation to the metal surface. Widely used alloys for marine service are the nickel-copper series (Monel) and nickel-chromium. Both have the particular advantage for pumps, valves, etc. in sea water that the passive film is tough and resists turbulent, high-velocity flow conditions, i.e. they are resistant to impingement and erosion attack.

Zinc

Zinc is of interest, particularly in relation to the behaviour of zinc coatings and sacrificial anodes (cathodic protection). Under active flow conditions uniform attack produces a protective film of zinc hydroxide, which will be reinforced by scale where calcium and magnesium salts are in solution, as in hard land water and in sea water. Where corrosion does occur it is normally by pitting, but zinc is generally attacked only slowly. This natural resistance to attack in sea water, and the potential of about −0.25 V with respect to steel, means that it is a natural choice for sacrificial anodes in the application of cathodic protection where there is the good electrolyte present. In soil, on the other hand, resistances are generally higher and an anode with a higher driving potential such as magnesium may be necessary (see p. 167). The zinc used for anodes has to be of high purity, particularly with respect to iron, to avoid the formation of a dense adherent film of high electrical resistance. The zinc may, however, be purposefully alloyed with cadmium, aluminium and silicon to modify the effect of any iron on the corrosion product and to ensure an active zinc surface.

11.7 The selection of materials for chemical plant

There are few generalizations to be made about the resistance of materials to chemicals, such as are found in processing plants. The number of chemicals that might be involved is very large and the conditions of use, temperature, concentration, fluid velocity, degree of aeration, purity, stress state, etc. can differ from one application of the same material to another. Unfortunately, the science of corrosion is not yet able to predict the behaviour of a system solely on the basis of fundamental relationships, and although a vast amount of empirical data is available, it is not complete and may not be readily accessible.

A most useful data source is the *Corrosion Guide*.[10] There are also standard specifications dealing with, for example, the linings of vessels and equipment for chemical processing. Further information can be obtained from materials suppliers (e.g. International Nickel) and from monographs published by materials development associations such as the Zinc and Lead Development Association, and direct assistance can usually be obtained from such organizations. Where resistance to corrosion is a prime requirement for a materials application, it is important to recognize that the use of data from such compilations requires understanding of the process involved, so that the effect of somewhat different conditions can be assessed.

In general, attack follows where a protective oxide film, or other corrosion product on the metal, is dissolved or becomes locally unstable (to give pitting). The halides are particularly bad in causing pitting.

A remarkable corrosion test has been established by Burstein[11] in which the oxide film on a metal sample is scratched by a diamond under the given environmental conditions. The reformation of the oxide film at the scratch is monitored by sensitive electrical instrumentation. Where repassivation is rapid and complete, good corrosion resistance can be predicted.

An interesting approach to the problem of data handling was pursued by Edeleanu and co-workers.[12,13] They asserted that appropriate experience and knowledge is not easily located and retrieved just when it is required. This means that when decisions have to be made it is difficult to predict with certainty what will happen in service. In an attempt to deal with these difficulties in the case of stainless steel they used basic equations and data to develop a model for the corrosion process, leading to a computer program which predicts polarization curves under a wide range of conditions and from which the circumstances under which loss of passivity is most likely to occur can be established. Estimates can thus be made for the corrosion behaviour of a given stainless steel in a particular situation, by people who are familiar with the interpretation of polarization curves, but with future extension it is foreseen that it can play a more directly predictive role.

Metals

Where service conditions are not especially aggressive, cast iron is widely used, as in mixers, digesters, pump bodies, etc., and has some advantage over stainless steel for salt or caustic soda evaporating pans as it is not subject to stress corrosion cracking.

It should also be remembered that there are special irons with improved corrosion resistance to meet specific duties. The high nickel austenitic cast irons containing up to 35% Ni (BS3468: 1962) are substantially more resistant to attack by both dilute sulphuric acid and caustic soda than standard ferritic irons, and this is typical of a wide range of aggressive situations. The high-chromium cast irons (and similarly the high-chromium steels), developing an impervious and tenacious protective oxide film, are most useful in environments containing plenty of oxygen or oxidizing agents. They are not so good in solutions containing those anions which can penetrate this film, such as halides, and do not offer resistance to hydrochloric or sulphuric acids under most conditions. They have some advantage with dilute nitric acid. Their main attraction, however, lies in the resistance to high-temperature corrosion, which enables use for furnace parts, heat exchangers, etc.

High-silicon cast irons, similarly, rely on the impervious and tenacious silica film on the metal for their improved corrosion resistance. Again, attack will follow if this film is damaged or destroyed by the environment. As would be expected, the film is destroyed by hydrofluoric acid, but attack also occurs in hydrochloric, hydrobromic and sulphurous acids. In the case of the halogen acids the relevant anion seems to be able to penetrate and modify the character of the silica film so that it is no longer passivating. Other conditions which produce halogenic anions may also be corrosive. Good service is given in nitric and sulphuric acids and in mixtures of the two. As might be expected the silicon irons are not recommended for alkalis.

Stainless steels find a wide range of uses in the chemical engineering industry and a great deal of data concerning response to specific chemicals and conditions is available. In general, performance is improved with higher chromium and nickel contents, lower carbon content and by the presence of molybdenum. In sulphuric acid useful service can be obtained, particularly under conditions of plentiful oxygen supply or small additions of an oxidant (acting as a cathodic reactant) e.g. $CuSO_4$ or HNO_3. Hydrochloric acid is more aggressive to stainless steels and use is restricted to dilute systems. Reasonable service is given in nitric acid with low attack rates, particularly in dilute systems and austenitic stainless steels are used extensively in nitric acid plant.

In practice, despite its relatively poor resistance to corrosion, low-carbon steel is the most widely used material for chemical plant operating up to temperatures of 400°C.[18] Where the corrosive attack is unacceptably severe, either in terms of the life of the plant or because of adulteration of the product by the iron salts produced, then the move will be made to more highly alloyed systems as described, or to low-carbon structures with specifically selected linings. This solution is, of course, at its most advantageous when strength requirements dictate a heavy section. Titanium-lined equipment has found favour for some applications in the chemical industry, the titanium often being explosively bonded to thick steel sheet. It is, for example, very resistant to corrosion in nitric acid, even at the boiling point. Like stainless steel, its performance in sulphuric and hydrochloric acids which produce hydrogen on the metal is considerably enhanced by small amounts of oxidizing agents capable of providing an oxidizing reaction (e.g. ferric and cupric salts).

Corrosion and hydrogen problems in the oil and gas industry

Environment-sensitive fracture (ESF) in general, and that associated with hydrogen embrittlement in particular, constitutes over 25% of failures experienced by the oil and gas industry. Kermani[14] provides an assessment of failure in the petroleum related industries which indicates the dominance of ESF, see Tables 11.2 and 11.3.

TABLE 11.2. Analysis of selected number of failures in petroleum related industries (Kermani[14])

Type of failure	Frequency (%)
Corrosion (all types)	33
Fatigue	18
Mechanical damage/overload	14
Brittle fracture	9
Fabrication defects (excluding welding defects)	9
Welding defects	7
Others	10

TABLE 11.3 Cause of corrosion related failure in petroleum related industries (Kermani[14])

Type of failure	Frequency (%)
CO_2 related	28
H_2S related	18
Preferential weld	18
Pitting	12
Erosion corrosion	9
Galvanic	6
Crevice	3
Impingement	3
Stress corrosion	3

Much of the material employed in, for example, drilling is high strength low alloy steel and in the failure of drill string components crack formation is caused by the susceptibility of these steels, in the presence of the drilling fluid, to crack formation from hydrogen embrittlement, and the growth of the crack through corrosion fatigue and finally a pure fatigue process. As Kermani points out, the stages of crack growth are mechanically, physically and environmentally dependent, where the cathodic generation of hydrogen under certain fluid conditions leads to brittle fracture and reduced fatigue life.

'Sour gas'

The problems of stress corrosion cracking in the oil and gas industry, associated with hydrogen embrittlement, are exacerbated by the presence of hydrogen disulphide in the environment, an increasingly common feature, the wells becoming increasingly sour as they are drilled deeper. The H_2S has two effects, it lowers the pH of the system (i.e. more acid), facilitating the cathodic hydrogen evolution reaction, and also catalyzes the entry of atomic hydrogen into the metal, increasing absorption and thus creating embrittlement. Bacterial action, resulting in the formation of sulphides at the metal surface, will enhance the hydrogen diffusion into steels, particularly when the sulphide films are loose and bulky and not masking the surface themselves.[15].

Kermani et al.[16,17] have developed graphical domain type relationships correlating pH with H_2S partial pressure relationships for a range of low alloy and carbon steels which should be useful in relating the selection of materials and conditions of use. There is also a NACE standard relating to this field – 'Sulphide stress cracking resistant metallic materials for oil field equipment', MRO 175–94, 1994.

Composite solutions to corrosion

Composite solutions to the problem of maintaining a stable surface are, of course, very widespread, and are often most easily applied to specialized fittings.

Interesting examples arise in the case of valves for the control of fluids. In some circumstances unlined materials represent the best solution in terms of initial cost and maintenance. For water there is cast 60/40 brass (where attack by dezincification can be a problem) but cast bronze is better. In many cases where superior resistance is requested choice could move to a cast austenitic stainless-steel valve, as for example in the brewery and dairy industries, where any contamination through corrosion would taste, even if not toxic.

The Saunders 'A' FD valve for food and pharmaceuticals is such a cast austenitic stainless steel valve (Figure 11.9a). In the Saunders diaphragm valve system it is important to notice that as the fluid concerned is separated from the operating mechanism by the diaphragm, the bonnet assembly can be of cheaper materials, in this instance epoxy-coated aluminium silicon alloy. The diaphragm is white butyl rubber which has good resistance not only to the product, but also to steam sterilization. At one time the body was a shell mould casting but the internal surfaces then had a slightly rough surface which had to be brought to the finished, polished condition essential for handling foodstuffs by expensive hand polishing. Much of this

Figure 11.9 Diaphragm valves for varying service: (a) stainless steel; (b) glass-lined; (c) plastics-lined; (d) rubber-lined; (e) butterfly type. (By courtesy of Saunders Valve Co. Ltd.)

polishing requirement has been removed by using investment casting to produce a much better initial finish.

Glass-lined valves

It may be that we cannot find a single material for the valve which is economically adequate in strength, rigidity and corrosion resistance as compared to composite solutions with linings to provide the necessary surface stability. A borosilicate glass-lined cast iron valve is shown in Figure 11.9b. Such valves are widely used because of their good chemical resistance, particularly to acids. Glass linings may also be

specified from a cleanliness requirement as, for instance, in the manufacture of pharmaceuticals, antibiotics, etc. Their main disadvantage is their limited resistance to rapidly changing temperatures. Also, from the outside they appear to be wholly metallic valves and it is not always easy for operating staff to remember that they are, in fact, glass-lined, and they may be hammered to dislodge solids. Some colour coding with a distinguishing mark on glass-lined components is clearly indicated. As in all the Saunders valves, the operating mechanism in the bonnet does not have to be specially treated since it is separated from the fluid by the diaphragm which, in the case illustrated, is of a black butyl rubber.

Plastic-lined valves

One of the most chemically inert polymers is PTFE (polytetrafluoroethylene), and valves are available lined with this material. It is, however, rather difficult to process as a lining. For this reason there has been in the last few years a substantial growth in injection-moulded fluoropolymers such as PVDF (polyvinylidenefluoride), ETFE (ethylene/tetrafluoroethylene copolymer) and PFA (perfluoroalkoxy polymer). The chemical resistance of these materials approaches that of PTFE, and they are much tougher; furthermore, since they are injection mouldable the finished parts are made more quickly than PTFE and are cheaper. (Figure 11.9c)

All of these injection-mouldable fluoropolymers are too rigid to use as a diaphragm. For that, PTFE is still the best but does not seal to atmosphere well unless there is a resilient rubber cushion behind it. Thus in ETFE-lined valves a PTFE/rubber composite diaphragm is employed.

Rubber-lined valves

Rubber withstands abrasion well. In the illustration, Figure 11.9d, a valve is shown which is intended for handling abrasive materials such as cement, fertilizers, coal and ore slurries, etc. The lining and the diaphragm here are made of polybutadiene rubber.

Clearly the diaphragm valve is ideally suited to lining with a wide range of materials because of its relatively simple shape. In the butterfly valve shown as Figure 11.8e a high styrene/butadiene synthetic rubber has been chosen for an application involving dilute acids since it is easy to mould to the rather complex shapes required.

Non-metallic materials

This discussion of the use of composite solutions to corrosion problems in chemical plant has introduced the concept of using non-metallic materials as linings. They can, of course, be considered in their own right. Concrete construction is generally limited to water handling (cooling towers, etc.) or as back-up for plastic or tiled linings. The corrosion resistance of aluminous cement makes it valuable for flooring, etc. in cases where Portland cement would deteriorate. Wooden vats are much less used, although giving excellent service for many applications. Certain coniferous woods are still employed for cooling towers.

Glass linings have been mentioned, but to an increasing extent wholly glass equipment is being used in highly corrosive situations where the alternatives are very expensive refractory metals. Fragility and brittleness are problems, and for large-scale production plant glass-lined or vitreous enamelled vessels, pipes and valves are likely to be more satisfactory. An account of vitreous enamelling follows in the next section.

Engineering ceramics have been mentioned in the context of creep and temperature resistance (p. 140), but it is worth noting that improved ceramics are of increasing importance in chemical and process plant, where they provide unique combinations of temperature resistance, wear resistance and chemical inertness. Sintered silicon nitride is used in the handling of molten metal in casting machines, nozzles, etc. and is resistant to most chemicals, the main exceptions being hydrofluoric acid, phosphoric acid and molten caustic alkalis. Partially stabilized zirconia at fine grain size also has extremely good wear resistance and is in use for such items as wire guides at high speeds. With a thermal expansion similar to that for steel, it can be used as a bearer surface in steel equipment when wear resistance and fluctuating temperatures are involved. Zirconia is also a good thermal insulator. Certain grades of the material have been developed for the production of small components, where toughness has been greatly improved ($\sim 20\,MPa\,m^{1/2}$), but it is, of course, less hard and is not capable of withstanding the same high temperatures as the yttria partially stabilized zirconia (1000°C), or the cubic stabilized zirconia (2000°), where the toughness is low (~ 9 and $\sim 1\,MPa\,m^{1/2}$, respectively). Zirconia is attacked by hydrofluoric acid but is otherwise generally resistant to chemicals, although oxide ceramics may be susceptible to corrosion in non-oxidizing atmospheres or in contact with reducing species, as a result of partial reduction of the oxide which can result in low melting point phases being formed.

Plastics find a wide range of uses in chemical plant. Unplasticized polyvinylchloride (PVC) is resistant to attack in hydrochloric, sulphuric and chromic acid; it is not resistant to aromatic hydrocarbons, chlorinated hydrocarbons, ketones and esters. It is, unfortunately, hard and brittle, but plasticized PVC is less useful for resistance to chemical attack, although quite satisfactory for normal atmospheric and water applications such as guttering, downpipes, buckets, etc. Flexible PVC is widely used instead of rubber for water hose.

An attack-resistant plastic is polyurethane, obtainable in a range of forms from high-density hard thermoset solids to rubbery foams, being resistant to weak acids and alkalis, greases and aliphatic hydrocarbons; they are not resistant to strong acids and alkalis, alcohols and hot water. Polyurethane coatings can be employed on metals. Polyphenylene sulphide (PPS) is also resistant to chemical attack and hence finds application in under-bonnet parts in cars exposed to oil, petrol and brake fluid, as well as in pumps and valves for the chemical industry. PPS is usually supplied reinforced with glass fibres or a mix of glass fibres and mineral fillers. PTFE is the most inert polymer of all, and can be used up to temperatures of about 250°C. It is, however, very expensive; of the order of 10 times the cost of polyethylene or PVC and three times the cost of nylon. For this reason it is almost always used as a lining. As indicated in relation to the Saunders valve, PTFE does not stick readily to metals and mechanical keying is usually necessary.

The outstanding chemical resistance of the new liquid crystal polymers (p. 134), which arises from their tight molecular structure, is enabling them to challenge ceramics in chemical process plant for the packings in gas/liquid contact columns. The 'saddles' employed can be injection moulded and do not readily damage to give chips or debris as does ceramic, with resultant loss of flow efficiency. The material is tough and strong and operating temperatures can be as high as 190°C. Fields of application have been identified in the handling of mineral acids, bases, alcohols, aromatics, hydrocarbons, esters, aldehydes, and organic acids. The high chemical resistance of LCPs also applies at elevated temperatures.

Rubber, as covered in its various forms by BS 5176 and ASTM D2000, which assist specification for purpose, has many engineering uses and can be useful in chemical plant. The resistance to hot oil attack, allowable service temperature and overall resistance to hostile chemical environments increases through the range from natural rubber, chloroprene, neoprene, ethylene acrylic to acrylic and fluorosilicone, although in terms of chemical attack much will depend on the solvent involved. The silicone rubbers have high service temperatures (~200°C) but are considerably less oil resistant than the fluorosilicone. Nitrile rubbers, on the other hand, have limited service temperatures (~100°C) but moderately good hot oil resistance. Bonding of rubbers to both metals and to polymers capable of withstanding short periods at 200°C is readily achieved.

Vitreous enamels

These coatings are useful in protecting steel or iron in a wide range of chemical plant situations. It is also the most satisfactory finish for a range of domestic appliances, satisfying appearance and withstanding abrasive and caustic cleaners. In the latter respects they are much more durable than the acrylic or other paint finishes which are used in some competitive situations.

Whilst fused silica itself can be used in extremely corrosive situations, enamels are silicate and borosilicate glass, usually containing some fluorine. The basic ingredients are quartz (sand), kaolin ($Al_2O_3.2SiO_2.2H_2O$) and felspar ($K_2O. Al_2O_3. 6SiO_2$). Additions of fluxes are made from borax ($Na_2B_4O_7$), soda ash (Na_2CO_3), cryolite (Na_3AlF_6), fluorspar (CaF_2), litharge (Pb_3O_4) and whiting ($Ca(OH)_2$). They are thus of similar basic compositions to the glazes applied to ceramic ware. Fusion temperatures must be low, as firing temperatures usually range from 800 to 850°C for mild steels and 650–750°C for cast irons, and firing times are short (1–5 min.). Generally opaque enamels are required, and as in ordinary glasses this can be achieved by adding constituents insoluble in the glass, or by producing enamels which become opaline on cooling due to the rejection of a constituent from solution. The most common opacifiers are SnO_2, Sb_2O_5 and, in special grades, ZrO_2. For opalescent enamels, cryolite and fluorspar are added in considerably greater quantity.

For acid-resisting enamels the flux additions are low and for 'glass-lined' chemical equipment borosilicate compositions are preferred.

As in the case of glazes the composition of the enamel or glass bonded to the metal must be adjusted to give as nearly as possible the same coefficient of expansion as the metal to prevent cracking during temperature changes, bearing in mind that it is the otherwise excellent performance of the vitreous finish at elevated temperatures (~250°C), as in ovens, which is particularly attractive. The first coat, or groundcoat, is usually a dull blackish-grey, as it contains no colouring agents, but may contain up to 0.5% cobalt or nickel oxides (for coating steel) and lead oxide (for coating cast iron), since it has been found that these promote adhesion. A thin electroplated nickel 'flash' is also said to improve adhesion in some vitreous enamelling systems where a separate ground coat is not applied. Usually the ground coat contains a higher proportion of fluxes than the surface coats, giving a coefficient of expansion intermediate between that of the metal and that of the surface coat. In common with glazes the main ingredients are 'fritted' together first and then ground with the remainder, usually opacifiers, kaolin and colouring agents.

Steel for use in enamel ware usually requires a very low carbon content, pickled and 'process' or 'close' annealed. This means that the surface is cleaned of scale and the structure is of recrystallized ferrite. Although normal process annealing, after cold rolling to required dimensions, is carried out in a controlled atmosphere to prevent surface decarburization, for enamelling use decarburization is purposefully allowed. The structure thus consists of polyhedral ferrite and a little elongated pearlite, which shows a tendency to be spheroidized, and with the surface layers almost wholly ferritic. The aim is, of course, to prevent the evolution of gas and blistering of the enamel by reaction between the metallic solution of carbon and the molten oxide.

In recent years, however, it has become a common practice to use specially alloyed steel compositions for enamelling, for example with the addition of titanium. This combines with all the carbon, nitrogen and oxygen to produce inert compounds in the metallic base and prevent both the evolution of gases from metallic solution and the reaction of carbon with oxygen supplied via the molten enamel. Further, the addition eliminates the yield point and strain ageing, so that bending prior to enamelling can be effected smoothly without the cold-rolling of the sheet otherwise necessary.

A secondary effect of vitreous enamelling is that it imparts stiffness and under suitable conditions a high damping capacity to the sheet steel component, a factor which can be recognized in the design synthesis. Sharp edges should be avoided in design for enamelling since

otherwise there can be difficulty in coating uniformly up to and around the edge. The fluid frit may thicken at various points on an edge and thus cause easy failure by flaking – sometimes only the ground coat, higher in fluxes, is applied right around an edge, giving the characteristic blue/black thinner edge covering to the otherwise coloured (usually white) main finish.

A major field for enamel ware is, of course, on cast iron. Quite obviously, since the carbon content is high there is always the risk of interaction between the carbon and the oxide enamel to give bubbles in the coating. This applies particularly to the combined carbon, and the problem is greater with pearlitic irons than with ferritic matrices, and with low rather than high silicon.

Graphitic carbon does not seem to be particularly reactive in this context, although there have been suggestions that the coarseness of the graphite has some effect. The most likely reason for the reduced sensitivity to carbon content as compared to steels is that at the operative temperatures, silicon, manganese and phosphorus in the matrix react with available oxygen preferentially, and in so doing improve the adhesion of acid-resisting glazes (low flux) by giving a transition interface.

A major problem with cast iron is, however, hydrogen evolution. Whereas during steel solidification the hydrogen will generally escape if present above the maximum solid solubility, in the solidification of the iron/graphite eutectic the hydrogen evolved may be retained by the carbon, to be evolved slowly during subsequent reheating. Alternatively, hydrogen pick-up through surface reaction with water from slurry application, with the graphite acting as a host site until the firing operation, is advanced as a cause. Avoidance of difficulty due to hydrogen is said to be achieved through surface oxidation by heating for a few minutes at 760°C before applying the enamel slurry.

The articles to be enamelled are first cleaned. Sheet is frequently pickled with phosphoric acid and iron castings are shot-blasted. Where the shot is white iron of high combined carbon there is a risk of blistering of the enamel if any of the shot becomes embedded, and this applies to both steel sheet and to castings. A thin film of rust after cleaning, or the phosphate film left behind after phosphoric acid pickling, are usually considered advantageous in promoting adhesion.

The enamel is usually applied as a slurry by spraying or dipping. After drying the enamel is then fused on to the surface. Alternatively, finely powdered frit is sieved on to the preheated surface, so that it sticks, and is then glazed by insertion into a furnace for a short period.

Developments in vitreous enamelling have been concerned with both new enamels and enamelling techniques.

Small changes in chemical composition are always being tested by the industry to improve particular properties or to reduce costs. More significantly, however, there have been developments in control of the microstructure of the enamels[19]. The introduction of a second phase such as silica or alumina to improve the refractory properties of enamels has been employed for some time, but more recently the further introduction of a metallic phase, to produce a form of 'cermet', has been shown to improve the resistance of the coatings to flaking and chipping, common forms of failure[20]. The temperature resistance is also improved, permitting them to be used at temperatures approaching their firing temperature. The composition of such cermet coatings is given by Faust[21] as 15–16 wt.% glassy enamel frit, 2–20 wt.% refractory oxide, 20–80 wt.% aluminium or aluminium alloy powder. The lower metal content compositions have a higher re-fire temperature of 870°C (1600°F) and are more acid-resisting. The higher metal content compositions have lower firing temperatures, 680°C (1250°F), and are more ductile. They have been used for a wide variety of lining applications such as silencers, exhaust and heat exchanger parts.

It has been shown[22] that the introduction of the metallic oxides CoO, V_2O_3, Al_2O_3 and TiO_2 into the surface of enamels by deposition as organometallic derivatives and heat treatment increased surface hardness by 30–40%. Such surface treatment is less costly than a major change in the overall composition. As Maskall

and White[19] point out it appears possible that any of the normal methods used for the modification of glass surfaces can be used, i.e. ion-exchange, thermal toughening or crystallization.

The moulding and subsequent crystallization of glass for hard-wearing and shock-resistant ovenware is well established, but only recently has this principle been applied to one area of the vitreous enamelling field. To produce special chemically-resistant enamels the frit is applied normally but is then heat-treated to produce nuclei and then crystallize the glass. This improves both the abrasion and thermal shock behaviour of the coatings.

In terms of enamelling technique the objectives are always to reduce the number of operations, particularly the metal pretreatment stages which involve the use of acids etc., generating effluent problems, high capital and handling costs. Obviating the need for preliminary pickling or nickel application is a major objective in compositional changes. The application of the dark ground coat and the true colour cover coat as powders with simultaneous firing of the two, clearly achieves economy. Coating with the frit by either dry electrostatic application, or by electrophoretic deposition from suspension, models developments in the application of paints elsewhere, giving good control of coating and economy in materials.

11.8 The degradation of polymeric materials

In so far as the failure of polymers may occur partly or wholly as the result of chemical changes, often associated with environmental conditions, degradation processes taking place may be likened to corrosion in metallic systems.

In the structure of a polymer, monomeric units are joined by chemical bonds, which are established during the polymerization process. The degradation of the polymer, where it involves the removal of monomer from chain ends, is thus equivalent to depolymerization, although it more commonly results from scission of main chains or cross-links. The removal of side groups from the structure is also an important form of degradation, as in the thermal degradation of PVC where dehydrochlorination can occur at temperatures in the region of only 150°C. Susceptibility to degradation can be affected by the polymerization process itself, in that weak links can be introduced where subsequent breakdown of bonds is more likely, as, for example, where tertiary chlorines exist in PVC through branching, or where impurities or even additives enter the structure.

The degradation of polymers is influenced by external heat, mechanical stress, ozone and radiation. The first two are, of course, normally experienced in processing, and the stability of the polymer under specific conditions of thermomechanical treatment is an important consideration. The introduction of stabilizers and antioxidants can be important in ensuring that degradation does not occur under processing conditions where the temperature is raised sufficiently to allow rapid production (e.g. in bottle forming).

Whilst raising temperature may cause structural changes within the material, producing degradation, it also increases the likelihood of chemical reactions with the atmosphere, particularly oxygen, but also in some circumstances with such gases as sulphur dioxide. In the context of oxidative degradation it is also known that the morphology of the material is important, i.e. how far crystallinity has developed, since oxidation is initiated and is at a higher rate in the amorphous phase of a dual amorphous/crystalline system,[23] the diffusion of oxygen being more rapid in the less dense structure. Antioxidants, typically hindered phenolic compounds are used to retard thermal and oxidative degradation of polyolefins during polymerization, processing and end-use.

As far as mechanical stresses are concerned the effect on degradation can be substantial. Whether the stress is internal residual stress remaining after processing, or externally applied (particularly over long-term service), it can initiate and exacerbate chemical degradation processes (cf. stress corrosion).

As far as radiation is concerned the most common effect is that of UV in sunlight, where the energy input is capable of dissociating polymer bonds. This is particularly true, for example, where oxygen introduced has formed carbonyl groups which absorb UV light and which render the material sensitive to what is termed photo-degradation. Certain monomers do not absorb UV light (e.g. monomer units of PVC, PMMA and PS), but these polymers almost invariably contain traces of UV-absorbing impurities. Hence for prolonged outdoor use, most polymers contain light stabilizers; hindered amine light stabilizers (HALS) and carbon black are frequently used.

Weather is, of course, so variable. Seasonal variations such as higher temperature and increased UV radiation due to a higher sun angle can cause summer exposure to be 2–7 times as severe as winter exposure in the same place. Real-time data from natural environment exposure is most useful and for plastics two of the commonly used ageing sites are Arizona and Florida. Arizona is very hot and dry; Florida is very hot and humid. They are considered to represent the worst case for applications in the Northern Hemisphere. The Plastics Design Library[24] has collated a good deal of weathering data for plastics.

Following the initial degradation steps of chain scission, etc., bonds may be reformed in cross-linkage giving hardening and embrittlement, with changes in colour and transparency, or the smaller units may remain stable with a general weakening of the structure.

In selecting polymers for long life in the context of chemical stability, high purity in the initial monomer is obviously important and the use of additives has to be carefully controlled. Stability against degradation can be improved, with specific stabilizers employed to counter the differing modes of degradation. Particularly where the polymeric material is in domestic use the stabilizer must be non-toxic, tasteless and odourless. Recyclability is another issue. As an example, the most efficient stabilizers for PVC against degradation by heat (and thus important in relation to rapid fabrication as well as stability in service, as PVC is extremely sensitive to oxidation) are certain metal carboxylate compounds and tetravalent derivatives of tin.

There is, of course, another side to the coin of polymer stability. With its wide-spread dispersion in service, particularly as packaging, a controlled but rapid degradation of some classes of polymer on weathering (i.e. exposure to sun and rain and bacterial action) could be an advantage.

In a condensed treatment of the subject such as this only general comments can be made about the suitability of various materials for differing types of service. Major texts such as Shreir,[1] and more detailed data services, should be consulted in considering the suitability of a material for specific corrosive service.

References

1. L. L. SHREIR, R. A. JARMAN and G. T. BURSTEIN eds: *Corrosion* 3rd Edition, Butterworth-Heinemann, Oxford, 1994.

2. J. P. CHILTON: *Principles of Metallic Corrosion*. Royal Institute of Chemistry. Monographs for Teachers No. 4, p. 27.

3. G. WRANGLEN: *An Introduction to Corrosion and Protection of Metals*. Institut für Metallskyd. Stockholm, 1972, p. 105.

4. H-L. LEE and S. L-I. CHAN: *Mat. Sci. and Eng.* 1991 **A 142**, 193–201

5. R. G. BAKER: *Rosenhain Centenary Conference, Phil. Trans. Series A.* **282**, 1307, 215.

6. A. TURNBULL and M. SAENZ DE SANTA MARIA: *Metall. Trans*, 1988, **19A** , 1795.

7. J. P. CHILTON and U. R. EVANS: *J. Iron Steel Inst.*, 1955; **181**, 113–122.

8. O. KUBASCHEWSKI and B. E. HOPKINS: *Oxidation of Metals and Alloys*. Butterworths, 1962, p. 233.

9. *New Scientist*, 11 February 1982, p. 376.

10. C. RABALD: *Corrosion Guide*, 2nd edn, Elsevier, 1968.

11. G. T. BURSTEIN and G. W. ASHLEY: *Corrosion*, 1983; **39** (6), 241–247.

12. C. EDELEANU and J. G. HINES: *Br. Corr. J.*, 1983; **18**, 6.

13. J. CLELAND and C. EDELEANU: *Br. Corr. J.*, 1983; **18**, 15.

14. M. B. KERMANI: in *Hydrogen Transport and Cracking in Metals*, ed. A. Turnbull. Inst. Materials Pub. No. 605, 1995, pp. 1–8.

15. J. BENSON and R. G. J. EDYVEAN, ibid. pp. 289–298.

16. M. B. KERMANI, D. HARROP, M. L. R. TRUCHON and J. L. CROLET: *NACE Annual Conf., Corrosion 91, Paper 21*. Cincinatti, 1991.

17. M. B. KERMANI, J. W. MARTIN and D. F. WAITE: *Proc SPE Tech. Conf. Bahrain SPE 21364*, 1991.

18. B. HOOPER: *Metallurg. Mater. Technol.* 1993, **5**, 355.

19. K. A. MASKALL and D. WHITE: *Vitreous Enamelling*, Pergamon, 1986, p. 102.

20. J. WRATIL: *The Vitreous Enameller*, 1981, **32** (2), 42–47.

21. W. D. FAUST: Proc. 47th Porcelain Enamel Institute Technical Forum, American Ceramic Society, 1985.

22. R. STEVENSON: British Patent 53567/72, British Gas.

23. T. KELEN: *Polymer Degradation*, Van Nostrand, 1983, p. 114.

24. PLASTICS DESIGN LIBRARY: *The Effect of UV Light and Weather on Plastics and Elastomers*, 1994.

12

Selection of materials for resistance to wear

Several models have been proposed to describe the processes occurring at moving surfaces in contact. As a result of the interfacial forces there may either be displacement of material at the surfaces, with a change in shape and dimension, or else there will be removal of material from surfaces to produce debris, or a mixture of both. Where debris is generated the wear rate may be assessed as the amount of material removed per unit time or sliding distance.

The normal engineering finish provided on surfaces cannot be regarded as truly flat. Microscopically it consists of asperities and depressions, which may be arranged randomly or in ridges, depending on the finishing techniques employed. The better the finish or polish the less will be this surface roughness. In bringing two surfaces together the asperities will touch at only a fraction of the total nominal contact area and subsequent behaviour at the asperities will be controlled by the characteristics of the material and the load applied. Friction results where the sliding forces have to act against the bonds developed between contacting points. Thus in lubrication we seek to interpose a film of lubricant between the two surfaces, to minimize the number of points of contact, and to replace them with a system where the bonds to be broken are of much lower strength.

12.1 The mechanisms of wear

Some clues as to mechanisms of wear are to be found in the shape of debris particles produced in a wear process.[1] These can often be of plate form. In this case it is proposed that there is plastic deformation of a surface asperity, smoothing it somewhat. As strain accumulates on the surface, cracks are nucleated below the surface which eventually shear on the surface, at weak points,[2] i.e. the deformed material delaminates. The subsurface cracking may be nucleated by second-phase particles such as inclusions, where in general the work of adhesion to the matrix is already low and the inclusion itself may be directionally deformed parallel to the surface. If a surface layer becomes embrittled by work-hardening prior to delamination, fracture to the surface may be more widespread, resulting in flat elongated particles.

Other particle shapes found include round-end 'wedges' (rather than flakes) and spheres. The former are explained as the result of low-cycle, high-stress fatigue cracks initiated at the surface in rolling which is associated with sliding in the opposite direction, the cracks propagating at an angle of 35°.[3] Spherical particles are variously ascribed to a polishing action on previously irregular particles trapped in cavities or cracks, to the spheroidization of irregular particles by the action of heat developed, or to the agglomeration of finer particles by constituents in oil.

Ribbon-like particles, of similar form to swarf from a machining operation, are attributed to an abrasion mechanism where an embedded particle or an asperity, harder and stiffer than the opposing surface, acts as a cutting tool and removes material in the same fashion. Such abrasive action on metals is particularly associated with non-metallic contaminants which may be non-metallic inclusions from the microstructure itself, dirt particles, or adhesive wear particles which have themselves oxidized on detachment from the surface.

A hard second phase present in a microstructure, such as a carbide, becoming an asperity, can promote abrasive wear in the opposing surface, if this presents areas of softer phase which can be gouged. When machining strong steels the wear on the tool is affected by the relative carbide size in the two materials. Where the carbide size is small and uniformly distributed in the tool, with small intercarbide spacing, and the carbide is coarsened in the workpiece, as by a spheroidizing treatment, the wear on the tool is minimized. Where the carbide spacing is smaller in the workpiece than in the tool, increased wear of the tool results.

Adhesive wear is widely accepted as being an important concept, giving rise to irregular-shaped wear particles. Where the asperities on mating surfaces come into contact under load, they will deform to an area of contact as a function of the elastic and plastic flow stresses of the two matrices and the load applied. Bonding will occur across the area of contact, to a degree dictated by the nature of the materials and the degree of oxidation. If the materials in contact are the same, bonding across the interface may be facilitated and asperities will deform equally. With continued traction the bonded asperities will shear at a point away from the junction on one side and a detached fragment will be carried away on the other asperity. This may detach fairly rapidly on further sliding, or it may even grow for a while, picking up more material from the opposing surface, until it becomes unstable.

In order to minimize adhesive wear the area of contact developed at asperities has to be reduced. Clearly, reducing the load on the junction for a given material will reduce the amount of deformation of the asperity and thus the contact area. Increasing the yield stress, i.e. the hardness, will, similarly, reduce adhesive wear.

There are some difficulties with the simple adhesive wear theory. It is not clear, for example, how the shear crack develops across an asperity, and to what extent fatigue processes are involved. Some have even claimed that mechanical interlocking between asperities, and the resultant flow and shearing which occurs in either or both sides, provides a satisfactory explanation for the observed phenomena, without actual adhesion. In some cases the formation and repeated removal of an oxide film by an abrasive wear process could be a significant factor. This is the case in fretting, where corro-sion occurs between two unlubricated surfaces subjected to a small relative oscillatory motion.

Having outlined the basic mechanisms of wear it is possible to make general statements about the selection of systems in which wear will be minimized. Minimized normal load and good lubrication have already been mentioned. Plastic deformation and fracture processes occur during wear, and the starting yield strength or hardness of the surface, or the hardness it will attain by deformation without fracture, are important characteristics. Materials which do not deform or fracture readily, i.e. strong, tough materials, are resistant to wear and the microstructural features associated with increase in yield strength, *viz.* fine grain size, fine strengthening precipitates, etc., are similarly associated with good wear resistance. In abrasive wear, if the surface hardness is, or becomes, higher than the contaminating particle, then the latter is deformed or fractured, and wear will be prevented. Thus very hard mating surfaces are more tolerant of grit, etc. As an example, valves for handling sewage have been produced with silicon carbide parts, primarily, however, to resist erosive rather than sliding wear.

The development of the subsurface crack which produces the delamination form of wear mentioned earlier may well be by a fatigue mechanism. Similarly, the shear crack which separates an asperity from the surface may be the result of several cycles of traction through contact with an opposing surface and may be of a low-cycle, high-strain-fatigue nature. So far, however, a quantitative relationship between wear resistance and fatigue resistance has not been proved.

12.2 The effect of environment on wear

There is a very marked effect of gaseous environment on wear in 'dry', unlubricated, systems. Oxidation at the surface, whilst representing a degree of degradation, may provide a protective film which gives a lower coefficient of friction and less wear, and atmospheres which limit or exclude oxidation may result in increased wear. Just as the presence of oxide can reduce the degree of metal contact, so an increase in temperature can increase wear by increasing asperity deformation and thus the true area of contact.

In aqueous systems there will be a combination of corrosion and mechanical mechanisms operating at the surface, with the mechanically worked asperity material being preferentially attacked. The continuation of attack will depend particularly on the nature of the corrosion product, but there will often be similarity to the conjoint action of stress and corrosion, as in stress corrosion cracking and corrosion fatigue.

12.3 Surface treatment to reduce wear

As in surface degradation by corrosion, the technical or economic solution to failure of the surface by wear may well be one of localized treatment, rather than by manufacturing the whole item from a wear-resistant material, which might not give the overall properties required, or which would be more expensive. Such local treatment may take the form of surface alloying and/or surface heat treatment, or the application of a surface coating.

There are two distinct approaches to the problem of wear. One is to produce hard surfaces, which resist wear by resisting the deformation and fracture processes which characterize the process; the second is to apply soft lubricating films which interpose between the asperities and reduce the opportunity for adhesion.

The production of hard surfaces

Surface alloying and heat treatment

This approach is particularly familiar in the case of steel, where the surfaces of low- or medium-carbon steels may be carburized or nitrided (see p. 316). Modern developments in carburizing include the use of plasma surface heating with a low pressure of gaseous carburizing medium, which is said to accelerate the rate of surface acceptance of carbon (see p. 317).

Specialized treatments may also be applied in which a range of elements, including carbon, can be introduced into the surface of a non-heated component by forming ions and accelerating them towards the surface (ion implantation). More traditional treatments introduce alloying elements into a surface by diffusion at elevated temperatures. Not only does this apply to carbon and nitrogen, but boron (boronizing), vanadium (vanadizing) and chromium (chromizing) may be similarly introduced, enabling a higher level of hardness to be achieved.

Provided a steel with a sufficiently high carbon content and hardenability is employed, surface hardening at its simplest may take the form of a surface heat treatment where rapid surface heating by oxy-fuel gas (flame hardening) or by an HF-induced current (induction hardening) is followed by a water spray quench. An important aspect of such treatment in obtaining a uniform surface hardness is that the Fe_3C should be uniformly distributed in the workpiece on a fine scale for uniform solution prior to the quench. Ideally, therefore, through-hardening (quenching and tempering) to the required core properties precedes the flame or induction hardening process, although in less critical applications adequate results may be obtainable using properly normalized material of fine grain size and fine pearlite. Extremely rapid local surface heating by laser has also been employed to develop hardened surfaces, through normal transformation hardening, and also even by local surface fusion. In the latter the fused layer may be alloyed by the incorporation of applied films such as graphite or tungsten carbide.

Applied surface films

If the base material is not steel, treatable by altering the characteristics of the surface layers, or if there is need for even harder surfaces on steel, the solution may be to apply an external coating of a hard material. A whole range of possibilities then exist, both as regards the material to be applied, and the method by which the coating is produced. Special alloys, such as Stellite, may be applied by a surface 'brazing' technique (useful for building up previously worn surfaces); chromium can be applied by electroplating; hard nickel by electro- or electroless plating; alumina can be plasma sprayed. Vapour deposition, ion plating, sputtering, surface reactions (e.g. pyrolitic decomposition of organic compounds) are other examples of the techniques by which required surface films can be developed, and the compounds for use may be carbides, nitrides, oxides or borides.

Such applied films or coatings need to be well bonded to the base material, and it is here that particular claims are made between competitive systems. It is generally considered that thick hard coatings are more likely to fail by spalling. Thin coats limit possible temperature gradients and should have better mechanical properties, acting as a compliant layer. With a very hard material it should not be necessary to have coatings in excess of ~10 μm. This has the advantage, also, that allowance is not usually necessary for the coating within the design dimensions, since tolerances are usually considerably greater than this.[1]

Surface hardening, whether by alloying and/ or heat treatment or by an applied coating, is not particularly useful in resisting surface damage where the surface loading is highly localized and of an impact nature. Deformation damage may then occur behind the surface layer, with a change in the local surface profile, which appears as wear, or in the extreme, failure of the surface film through sub-surface cracking.

Where the component itself is subject to cyclic stresses, intrinsic surface alloying or hardening is to be preferred as the method, since it is known to improve the fatigue performance.

The application of soft, lubricating films

This relates to the application of such materials as PTFE, molybdenum disulphide, graphite, lead and indium in such a way as to produce a surface film on the material which reduces the friction coefficient (i.e. it has a low shear strength so that the force to break bonds at contacting asperities is small) and/or which will embed particles and reduce abrasive wear. The suitability of these materials (e.g. MoS_2 and graphite) is often associated with their having a dominant low-energy slip plane which is inherent or develops at the surface. Once again the integrity of the coating is important; it must adhere to the base material well, and has to be of sufficient thickness under conditions of abrasive wear to embed particles so that they are not available as asperities to contact the opposing surface. In relation to adherence and general performance as a solid lubricant, molybdenum disulphide, for example, is not equally good on all surfaces and will behave differently in vacuum as opposed to in air. Graphite fails completely in vacuum.

Graphite and lead are frequently generated as solid surface lubricants by incorporation in the bulk material as a dispersed insoluble phase, using a powder metallurgy technique for production of the component, typically a leaded or graphitic bronze bush or bearing. Aluminium–tin bearings may be surface-coated with indium or lead to reduce friction and to allow some accommodation of misalignment of the shaft.

12.4 Wear-resistant polymers

PTFE has a low surface energy and the lowest coefficient of friction of all solid materials. Hence its wide use in anti-wear coatings, as above, and sometimes, suitably filled (e.g. with graphite, glass fibre or molybdenum disulphide), as a low-load bearing component. The PTFE tends to fill any asperities present in the mating material, resulting in PTFE–PTFE contact and therefore very low friction. If, however, the bearing is used in situations where the PTFE cannot embed into the mating material, e.g. in the presence of fluids, the wear rate of the PTFE will be high.

Other polymers that can be used for low-load bearing materials and in wear-resisting applications include polyacetals and polyamides (nylons). For example, both polyacetals and polyamides have been used for low-load sprockets and gears, sometimes in combination with metal gears (see Section 21.1: Investigative case studies, Black and Decker chainsaw). The materials with the best wear characteristics tend to be filled, for example, with silicone oil, PTFE, MoS_2 or carbon fibres. The ability to injection mould small components accurately in complex shapes, together with the favourable wear properties, has resulted in the widespread use of polyacetals, for example, in moving parts in office equipment, clocks and speedometers. It has been commented that polyamides have better wear characteristics than aluminium when subjected to impact from liquid drops or cavitation-erosion, experienced, for example in pumps[4]. The tough, elastic response of the polyamides is also an advantage for ventilators or car spoilers subjected to mild blasting by particles.

12.5 Erosive wear

The wear produced on materials by the impact of solid particles is an important factor in selection for gas turbine components, chemical plant, coal combustion gasification and liquefaction, and in many other systems involving the movement of solid particles in fluids. Even in the generation of shape in exposed rock by the impingement of wind-borne sand, erosion has been long recognized as a specific form of surface degradation.

Where grit particles impinge on a surface the amount of material removed has been shown to depend on the velocity of impact and the angle at which the particle strikes, with a different angular dependence for brittle as opposed to ductile materials. In the equation $E = kV^n f(\theta)$ (where E is mass eroded per unit mass of impinging particles, V is the particle velocity, and θ is the impact angle between the plane of the surface and the incident motion of the grit), the value of n for ductile materials is usually in the range 2.3–2.5. Brittle materials show a much larger variation of velocity dependence with reported values of between 2 and 4 (Hutchings[5]).

The difference in the influence of angle of impact as between ductile and brittle materials is particularly striking (Figure 12.1) and, as Hutchings indicates, this is related to the mechanisms of wear occurring. In ductile materials there is extensive plastic flow of the surface with detachment of metal fragments, the shape of which suggest that they were associated with the eventual fracture of lips raised around the impact sites. With brittle materials and rounded impacting particles, giving elastic contact, an axially symmetric conical crack is produced, also typical of the static loading of a hard sphere on a brittle surface. With pointed particles, however, some plastic flow occurs at the contact points and cracks are generated from the plastic zone both radially ahead, and laterally as a result of residual stresses produced by the plastic zone as the particle rebounds. Both forms of crack can lead to surface removal as cone cracks intersect and lateral cracks propagate.

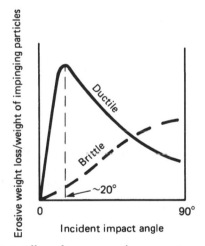

Figure 12.1 Effect of impact angle on erosive wear in ductile and brittle materials.

As in other wear systems, erosion may be coupled with corrosive attack, with erosion removing the passivating or protective films as they form. Only where the oxide film is thickly developed, strong and adherent, protecting the underlying surface from damage, will the erosive contribution to overall degradation be

181

reduced. Such a system could exist where scaling in a hot system might give some protection against erosion.

Erosive wear due to the impingement of liquid droplets has similar mechanistic features to the cavitation phenomenon discussed on p. 155.

12.6 Selection of materials for resistance to erosive wear

Although wear by particle impact involves processes of surface deformation and fracture, it is notable that with hard erosive particles the mass of material removed per mass of impinging particles is very similar for pure metals and metallic alloys of widely differing microstructure and hardness,[5] although the angular dependence characteristics may change as between ductile and brittle systems. If, however, the surface is appreciably harder than the abrasive particle then resistance to erosive wear results. In practical terms this means that there is no scope for

development in metallic systems, but that coatings or whole components in non-metals, i.e. ceramics, carbides, nitrides and borides would be expected to provide marked improvement, particularly at high density and fine grain size. As in coatings for normal wear resistance, good adhesion between coating and substrate is essential, and methods such as chemical vapour deposition may be expected to give better results than, for example, plasma-sprayed layers.

References

1. L. E. SAMUELS, E. D. DOYLE and D. M. TURLEY: *Fundamentals of Friction and Wear of Materials.* ASM. Ohio, USA, 1981, 15.
2. N. P. SUH: *Wear,* 1977; **44**, 1.
3. M. L. ATKIN, R. A. CUMMINS E. D. DOYLE and G. R. SHARP: *Trans. Inst. Eng. Aust.,* 1979; **ME4**, 40.
4. H. DOMININGHAUS: *Plastics for Engineers,* Hauser, 1993.
5. I. M. HUTCHINGS: *Tribology, Friction and Wear of Engineering Materials.* Arnold, London, 1992.

13
The relationship between materials selection and materials processing

There is no profit in selecting a material which offers ideal properties for the job in hand only to find that it cannot be manufactured economically into the required form. Processing (value-added) costs are often many times the basic material costs of a part and since there exists a great number and diversity of manufacturing processes from which to choose, each of which will give better results with some materials than with others, it is essential to match materials to processes very carefully. Materials selection and process selection go hand in hand.

For technical reasons, selecting a manufacturing process is frequently not an entirely free choice. Many metallic alloys – for example, permanent magnet materials and advanced creep-resisting nickel-base alloys – are too hard and strong to be mechanically worked and must, therefore, be formed by casting or by powder metallurgy; timber can sometimes be shaped by steaming and bending but more normally only by cutting and adhesive joining; concrete can only be cast; natural stone can only be cut. Processing also influences material properties. For example, short fibre reinforced plastics will tend to display regions of anisotropy when injection moulded; rolling of metals will alter the grain structure; casting conditions will influence the grain structure, and so on. But these are not disqualifications, merely limitations within which the materials engineer must work.

There are other limiting factors. The reasons for preferring one process to another should ideally be based on considerations that are purely technical and economic: unfortunately, expediency often supervenes. The reasons for this are manifold, sometimes resulting from crisis situations such as supply failures or trade disputes: a constant factor is the influence of the size and nature of the manufacturer. This reflects a conflict between the flexibility and control associated with in-house production, as opposed to buying in components from specialist suppliers, and the capital equipment necessary to manufacture with a wide range of production methods. The medium-sized business, maximizing in-house production, will favour simple materials and processes because of the high cost of providing narrowly-specialized and less widely useful capital equipment, an approach which lacks technical edge. In contrast, large organizations can develop and use more advanced materials and methods, by virtue of the greater turnover. In a special category are the small specialist firms, working over a narrow range of activities. High technology is often involved, requiring advanced equipment, and their products may be sold widely direct to the market, or they may service the needs of large companies, in the latter case with less freedom of action.

The materials engineer, therefore, although he should always consider all the possible processing options, frequently finds that his final decisions are less than ideal for reasons that are outside his control.

13.1 The purpose of materials processing

Materials processing has three principal aims. These are the achievement of (1) shape and dimensions, (2) properties, and (3) finish (this last in the sense of ready-to-use quality).

Shape and dimensions

These are obtained by three basic methods: (1) rheological (flow) processes, (2) fabrication (the assembly of ready-made constituent parts by joining), and (3) machining.

Flow processing

This method can be used to shape liquids, fluids and solids: it includes the liquid casting of metals, injection moulding of plastics, slip-casting of ceramics, mechanical working of metals and the densification of powder-metallurgy compacts. The technologies of these various processes are very different and it is a little academic to classify them together – they are therefore discussed separately in later sections.

Fabrication

This is accomplished by mechanical, metallurgical or chemical methods of joining. Mechanical methods, include riveting and bolting and other diverse methods of clipping and fixing. These methods are widely used, sometimes because of the need for a demountable joint, or because of the simplicity and convenience of assembly (e.g. self-tapping and hammer-driven screws); some alloy systems are, in any case, unweldable.

Metallurgical techniques embrace welding, brazing and soldering and each has a range of applications in which it is the preferred method of permanent assembly: thus, welding for heavy engineering, soldering for electrical circuits. Chemical methods involve the use of adhesives, glues or cements (the terms have the same meaning; adhesive is preferred as it reflects the physical principle involved). For timber and metallic materials, adhesion is a well-established method of fabrication and for most joint pairs within these groups of materials it is possible to specify a suitable adhesive. However, some materials, notably plastics, are difficult to join by adhesives; special surface pretreatments are invariably required to raise the surface energy and improve wetting.

The fabrication of large-scale metallic structures, for which welding is the usual method of joining, presents problems in relation to dimensional tolerances. Overall limits may vary by several millimetres. This must be considered in relation to the fact that when steels susceptible to heat-affected-zone (HAZ) cracking are being welded, a fit-up accuracy worse than 0.4 mm is considered unsatisfactory.

There are obstacles to the use of adhesives for the assembly of large structures: if joints of good integrity are to be assured the precision of dimensional fit and surface preparation must be of very high quality. Fabrication methods are also discussed later in this chapter.

Machining

Machining represents the failure of the processes that have preceded it. Expensive in terms of energy and labour, wasteful of basic resources and requiring a good deal of costly capital equipment, it retains its major position within production engineering only because of its flexibility and convenience, and for its ability to make up for the shortcomings of other processes. Naturally enough, reduction in machining by other means of near net-shape forming, with improved surface finishes, is constantly sought.

In normal manufacturing, machining has the ability to combine high quality with large throughput. Its technical flexibility is such that almost any shape can be produced from a solid block of material provided the price can be paid (although hollow shapes are limited), and machining is frequently adopted for the manufacture of prototypes and one-off items. Sometimes, machining is used for the bulk manufacture of a part which has a shape inappropriate for any other forming process: in this case redesign should be sought if at all possible.

Reynolds[1] has given an analysis for the costs of machining a bought-in blank or semi-finished product. Suppose the unit cost and weight of the blank are C_B and W_B, respectively, whilst the corresponding quantities for the finished part are C_F and W_F. Let the cost of producing unit weight of swarf be C_M. This will be made up of the total machining cost less the re-sale value of the swarf produced. Then

$$C_F W_F = C_B W_B + C_M (W_B - W_F)$$

If the yield of the process, W_F / W_B, is denoted by y then

$$C_F = \frac{C_B}{y} + C_M \frac{(1 - y)}{y}$$

This is the equation of a straight line, of slope $(1-y)/y$ and intercept C_B/y. Three such curves are shown in Figure 13.1 We are interested in making a choice between, on the one hand, achieving a given shape by machining it from a simple, largely unformed blank and, on the other hand, carrying out a mainly finish-machining operation on a blank which has already received much of its shape from some other process. In the first case the cost of the blank is low but the machining yield is also low. In the second case the reverse applies, with the unit cost of the preformed blank generally being lower, the larger the scale of production. If there is to be a real choice between two such processes then the two curves must intersect, and this requires that the intercept of curve 1 be less, and the slope greater, than the corresponding values for curve 2, i.e.

$$\frac{C_{B_1}}{y_1} < \frac{C_{B_2}}{y_2} \quad \text{and} \quad \frac{1 - y_1}{y_1} > \frac{1 - y_2}{y_2}$$

This means that $C_{B_2} > C_{B_1}$ and $y_2 > y_1$. If these conditions are not met, as between curve 3 and either of the others, there is no intersection and therefore no choice, since process 3 will always be the more expensive. Which of processes 1 and 2 is to be preferred depends upon a number of factors. Consider, for example, a steel part which may be produced with equally satisfactory properties by automatic machining from plain bar stock (process 1), or finish machining of steel forgings (process 2). If we assume initially that C_{B_1}, C_{B_2}, y_1 and y_2 are fixed, then process 1 would be preferred to process 2 if it happens that the real machining costs are smaller than the value given by the intersection of curves 1 and 2. One factor which greatly influences machining

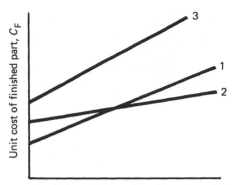

Figure 13.1 Cost relationships for machining processes.

costs is the machinability of the material. This can be influenced by the metallurgist, since if it is desired to favour process 1, the purchase of free-machining steel bar stock containing sulphides greatly reduces machining costs (although at the expense of some degradation of mechanical properties as compared with the forgings).

Another way of influencing the situation is to alter the position of the intersection, either by altering C_{B_1} or C_{B_2}, or one or other of the slopes. With similar materials it is difficult to influence the relative values of C_{B_1} and C_{B_2}. However, the values of the yield in each process can be influenced by suitable design. The smaller the yield of process 1 and the higher it is in process 2, the less likely it is that process 1 will be favoured.

Considering competition between different materials it may be noted that a high scrap value of the swarf reduces net machining costs. Titanium is expensive to buy, so that C_B for this material must be high, but the scrap value of titanium swarf is negligible, and it is therefore not economic to shape titanium extensively by machining methods. This is not true of aluminium alloys, which are often competitive with titanium.

Properties

The properties of an engineering part derive mainly from the basic nature of the material of which it is made, but where metallic materials are concerned, properties can generally be greatly modified during the successive stages of a manufacturing process. This is impossible with unprocessed natural materials such as timber and stone, but the approach of modifying structure by processing can be applied to products where the basic ingredient is wood or mineral (e.g. chipboard, plywood, reconstituted stone and cement products). It is also an approach which is increasingly applied to non-metallic manufactured materials, i.e. ceramics, glasses and plastics.

The ability to control properties of a part during manufacture often allows these to be better matched to application than might otherwise be the case, especially in respect of the magnitude and directionality of mechanical properties. A shaped metallic casting is a primary product where only the solidification process is available to modify the potential properties of the basic material (control of solidification, however, must not be underrated since the use of chills and denseners to control feeding of shrinkage, directional solidification, and grain refiners, etc. can profoundly modify the properties of castings). On the other hand, the separate processes that culminate in a metal sheet will have included solidification in an ingot, reheating, hot-rolling, cold-rolling and annealing in a complex sequence of operations, at every stage of which its properties will have been manipulated to suit its final use, whether this be for a transformer lamination, an aeroplane wing, a deep-drawn can or a simple machine cover. In contrast, once the melt for making an injection-moulded plastics component has been prepared there is less in the subsequent manufacturing procedure that can significantly modify its properties.

Such limitation can be accepted because, as Beeley[2] has pointed out, there are two aspects to the quality of a manufactured artefact, one concerned with the quality of the material of which it is made, the other with the quality of the artefact as an engineering component, determined by the integrity of its shape, dimensions and finish. It is the second of these two aspects which is often the more important and which must take precedence if there is any conflict between them.

Finish

This includes engineering tolerances, surface quality, surface protection and appearance.

In so far as it is essential to the proper mechanical functioning of a component, finish is a property that can be precisely specified, for example, in terms of standard limits and fits and parameters of surface topography (Talysurf – centre-line average). Desirable levels of surface protection and appearance are a little more difficult to quantify and the choice between, say, galvanized or cadmium-plated steel and anodized aluminium or chromium-plated plastics may present problems, notwithstanding the easily-determined variations in cost.

Surface processing purely for appearance is entirely a subjective matter, and decisions can hardly be taken without the benefit of market research. However, for light reflectors, and other applications requiring highly finished surfaces, quality can be assessed quantitatively in terms of the relative proportions of specular and diffused reflection, using standardized methods of measurement.

13.2 The background to process selection

Before choosing a process for the manufacture of a given article, apart from taking into account the material selection, it is necessary to know (1) how many are required, (2) the size and weight per piece, (3) the geometrical complexity, (4) the required dimensional tolerances and (5) the desired surface finish.

Effect of numbers required

Except for a prototype, it is rare to manufacture only one of a given part: usually larger numbers are required, varying from, say, ten to hundreds of thousands. A large production reduces the unit cost, i.e. the cost of each individual piece, since larger numbers permit the use of more complex machinery and more advanced manufacturing methods. However, coping with a large production demands more highly developed techniques of inspection and quality control, not only of manufacturing methods throughout the factory but also of incoming material. To achieve overall benefit, the additional costs thereby incurred must be smaller than the savings accomplished by high-volume production.

The effect of production numbers on costs can be analysed as follows: The total cost, P, of a batch of N pieces can be expressed as

$$P = T + xN \qquad (13.1)$$

in which T is the cost of tools and equipment and x represents the costs associated with each individual piece.

T varies very greatly from one process to another. At one extreme the cost of a wooden pattern and moulding box for a metal casting might be less than £100. At the other extreme the cost of tools for producing an injection moulded thermoplastic article could exceed £20,000. x is made up of several components. These are M, the unit material cost: F, the unit cost of finishing and rectification: L, the proportion of labour and factory overhead costs borne by each piece, expressed as cost per unit time; and R, the rate at which the pieces are produced. Expanding equation (13.1):

$$P = T + N\left(M + F + \frac{L}{R}\right) \qquad (13.2)$$

Comparing one process with another, M can be taken as constant. L is not constant since more advanced processing machinery and methods will require more elaborate factory support systems, including staff for inspection, quality control and maintenance. R is a measure of productivity and should increase with more advanced machinery.

The effect of installing more efficient processing machinery should be to reduce the value of x in equation (13.1). However, T will simultaneously increase and this means, referring to Figure 13.2, that x must decrease by an amount sufficient to make the curves intersect at a useful

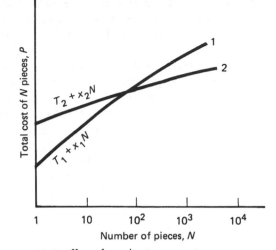

Figure 13.2 Effect of production quantity on manufacturing costs.

value of N. The point of intersection in Figure 13.2 gives the minimum batch size which makes it worthwhile to replace process 1 with process 2. From equation (13.2) this critical number is given by

$$N_c = \frac{T_2 - T_1}{(F_1 - F_2) + L\left[(1/R_1) - (1/R_2)\right]} \qquad (13.3)$$

To allow maximum utilization of the more advanced process, N_c should be as small as possible and for this to be so the latter process should produce pieces that require less finishing and rectification and at a faster rate so that any increase in L is more than offset by larger R. The effect of production numbers on tooling costs is seen more clearly if equation (13.2) is rewritten to give unit cost:

$$\frac{P}{N} = \frac{T}{N} + \left(M + F + \frac{L}{R}\right) \qquad (13.4)$$

If the injection-moulding tools referred to earlier were used to make a batch of 100 pieces, then it is inevitable that the cost of each piece must exceed £200 – such expensive tools could not be justified for such a small number. But if N is large enough there is no limit to the permissible

187

expenditure on tools because however large T is, T/N becomes negligible. Sufficiently large numbers could even permit the building of a new factory. For low volume production tooling, it may be possible to utilize 'soft-tooling' methods, facilitated by the application of rapid prototyping techniques. Among the manufacturing methods for soft-tooling are metal spray tools, cast resin or metal tools and electro discharge machined (EDM) electrodes for tools. For plastics, vacuum forming, blow moulding and resin injection moulding can utilize soft-tooling. For metal sheet forming, fluid cell and rubber pad pressing and superplastic forming can take advantage of low-cost soft-tooling.

Figure 13.3 Effect of production quantity on unit cost for different processes.

Figure 13.3 shows unit costs as a function of production quantity for three processes of increasing tool costs.

Effect of size and weight

Each process and material has its own characteristic limits of size. The upper limit on size is most restrictive in those processes which require closed metal moulds, such as shell moulding, diecasting, and closed die forging. On the other hand, sand castings are limited in size only by the available supply of liquid metal, although very large castings must be skilfully designed, first, to persuade liquid metal to flow for long distances through the mould cavity and, second, to avoid

unsatisfactory mechanical properties in thick sections, especially if these are joined by thinner sections. According to Jackson[3] steel castings can be produced in weights up to 250 tonnes but in most jobbing foundries a 25 tonne casting would generally be described as large. The largest forgings in regular production are steel alternator rotors, the bodies for which may attain 200 tonnes, although the low yield characteristic of these products would necessitate an as-cast ingot weight in excess of 300 tonnes. In airframes, a forging would be regarded as large if it weighed more than 2 tonnes: most aircraft forgings weigh between 25 and 250 kg (50–500 lb).

At the other extreme, it is difficult to produce forgings smaller than 100 g or so, whereas casting processes, e.g. pressure diecasting, can produce pieces three orders of magnitude smaller than this. Technically, this is because metals flow into small channels much more readily when they are liquid than when they are solid, but economically it is facilitated by the fact that all casting processes can employ multi-cavity moulds – it is possible, for example, to obtain a hundred or so castings from a single investment-casting mould (lost wax process).

Stampings and pressings, which involve very little material flow, can be made in a wide range of component sizes.

Complexity of shape

There appears to be no single parameter which can give quantitative assessment of complexity of shape. Factors which contribute to complexity are: (1) minimum section thickness, (2) presence of undercuts, (3) presence of internal hollows.

Dimensional tolerances

Different manufacturing processes vary widely in the dimensional tolerances of which they are capable. Reynolds[4] has presented evidence that in practice designers tend to use certain constant values of tolerances, such as 0.010 in, 0.020 in, etc., regardless of the overall magnitude of the

TABLE 13.1

Process	Dimensional tolerances ISO tolerance System IT	Draft allowance	Machine finish allowance mm	Surface smoothness μm RMS
Conventional Closed Die Forging	15–18 (From +0.5mm −0.25 mm)	5°	From 1 mm	≮3.2
Precise Form Forging (Impact machining)	11–15	Can be zero	None on forged faces	1–1.5
Fine Blanking	6–9	Zero	None	0.3–1.5
Green and Dry Sand Castings	Al, Mg 13–15; Cu 15–16; Grey Iron 14–16; Malleable 13–16; Steel 16–18	1–3°	Non Fe / Iron / Steel — 0–150: 1.5 / 2.5 / 3; 150–300: 1.5 / 3 / 5; 300–500: 2.5 / 4 / 6; 500–1.5m: 3–6 / 5 / 6	2.5–25
Full Mould and Fluid Sand Process	Steel, often 16 but 18 attainable	0–0.5°	0.8–6	2.5–25
CO$_2$ and Furan	Intermediate between green sand and shell mould castings	0–3°	Approximately 50% of green sand processes	2.5–5
Shell Mould (Croning Process)	12–14 parting line error 0.25–0.5mm	0.1° attainable 0.25–3° usual	Often none required	1–4
Gravity Die Casting (Permanent Mould)	Al 12–14; Iron 12–15	0.1° attainable but high die wear expensive; 0.2–3° usual; 5° preferred in recesses	0–100 mm 0.8; over 100 mm 1.5	2.5–6.5
High Pressure Die Casting	Zn 11–13; Al 11–14; Fe 11–14; +0.05 mm parting line error	~2°	0.25–0.80	1–2
Investment Casting (Lost Wax Process)	11–14 usual; 10 attainable	Usually zero; 0.5–1° required for exceptionally long cores	0.80 machining; 0.35 grinding	1–3
Hot Extrusion	12	Straightness approx. 0.3 mm/m	Usually zero	1–1.25
Cold Extrusion	9			0.5–0.075
Impact Extrusion	Length 6; Diameter 5	Zero		0.25–1.7
Sheet Metal Work Cutting	11–12	Zero	Zero	–
Powder Sintering	8–11 (sintered and coined)	Can be zero	Can be zero	≮1

Data from Fulmer Materials Optimizer 1974

dimensions to which they apply. He discusses the International Standards Organization proposal that the tolerance applied to any dimension should be related to the magnitude of that dimension; because the larger a dimension is, the more difficult it is to achieve a given fixed tolerance. ISO propose empirical relationships of the following type:

$$\text{Tolerance (microns)} = 0.45x^{1/3} + 0.001x$$

where x is the nominal dimension in millimetres. This defines Grade 1 of the ISO tolerance grades (designated IT). In successive tolerance grades, the tolerances increase by multiples of $10^{1/5}$ that is, they increase by an order of magnitude for every five tolerance grades. Thus, if the characteristic dimension of a casting is 40 mm then tolerance grade IT 1 corresponds to a tolerance of 1.58 μm. The best casting process in this respect is pressure diecasting which, at its best, meets tolerance grade IT 11: this for the 40 mm casting gives tolerances of 158 μm. The same casting, manufactured in steel from a green sand mould, could, at its worst, require tolerances according to tolerance grade IT 18, i.e. 4 mm. Other processes produce results that are intermediate between these two extremes (see Table 13.1). It is seen that the lower the IT number the better is the quality of tolerances, and that hardly any process is capable of attaining a tolerance grade as good as IT 9.

Surface finish

Surface finish, or the degree of approach to perfect smoothness of a surface, is generally expressed as some sort of average measurement of the surface profile about a central mean, either centre-line-average (CLA) or root-mean-square (RMS). Once again, pressure die casting and sand casting offer the extremes of results, with possible figures of 1 and 25 μm RMS, respectively. Some alloys are much worse than others: sand-cast phosphor-bronze produces especially rough castings. Poor surfaces due to metal-mould reaction are obtained from several combinations of metal and mould.

13.3 The casting of metals and alloys

As a method of shaping, casting has two great advantages: almost any shape can be produced (although cost increases with increasing complexity) and there are many available processes, covering a wide range of high and low capitalization, with continued development into new techniques (e.g. rheo-casting and squeeze casting).

Although castings in small batches can be obtained economically with quite modest equipment, high-volume production calls for a good deal of mechanization: the tooling and associated equipment required for a consumer-durable component could cost tens of thousands of pounds sterling.

The score or more of different casting processes from which it is now possible to choose differ mainly in the material of which the mould is made and the size and number of castings which can economically be produced.

Three factors influence the choice of mould material: (1) cost, (2) fidelity of shape and dimensions, and (3) thermal properties.

Cost

Where many identical castings are required, it is sensible to reduce unit cost by employing a mould which can be used many times. The initial cost will be greater, but provided this can be spread over a sufficient number of castings then, as discussed previously, there will be a reduction in total cost. Sand moulds can be used only once, and this is true also of plaster moulds. Some ceramic moulds can be used a few times if the casting is of simple design but, of course, the longest runs are obtained from dies of steel or heat-resisting nickel alloys. Unfortunately there are clear economic disadvantages in the use of metal dies; the lead time for die manufacture is long and cost is high. The cheapest metal moulds are those intended for manual operation (permanent moulds or gravity dies), but the production rate with such dies is generally low and can be substantially lower than when filling sand moulds. In contrast, the dies used for high-pressure die-casting are extremely expensive since they are pressure-filled from complex machines and contain mechanisms for withdrawing cores, opening the mould and ejecting the castings at high speed. The cost of such a die must be spread over many thousands of castings. The working life of a re-usable mould is the number of times it can be used before deterioration of its working face causes unacceptable surface blemishes on the castings.

Fidelity of reproduction

The different casting processes vary widely in their abilities to reproduce dimensional tolerances and surface qualities. In general it used to be considered that pressure die-casting gives the best results and sand casting the worst but this statement now requires considerable qualification. In the case of pressure die casting the good accuracy, surface finish and thin walls are often obtained at the expense of internal soundness, because of the high velocity of flow of metal which is required to fill the die. The consequential turbulence entrains both air and oxide. Modifications to dies to improve gating design etc., are difficult. In terms of dimensional accuracy, expansion and distortion of the die can cause difficulties, generating non-flat surfaces and, with likely variation in die coat thickness, this means wall thicknesses are usually limited to 5–6 mm minimum.

On the other hand, the image of sand casting is changing for the better and it is becoming increasingly employed for the production of quality castings of good soundness. In the use of green sand there has been considerable progress in moulding methods with properly plate-mounted and accurate patterns including gating systems and good sand control. It is, however, in the use of a core assembly approach, whereby all the mould components are accurately produced by blowing on core machines, with subsequent careful assembly, that major advances have been made. This is the basis of the Cosworth Process

used for major engine castings, where, with bound low expansion zircon sand, wall thickness can be held to 4 mm. The high conductivity of this sand simulates a metal die to an extent and the slower cooling possible with silica sand can enable the feeding of even thinner sections (2 mm – Zeus Process). It is an interesting point that if the wall thickness is lower, the heat input to cores is reduced and thermal distortion normally associated with bound silica is reduced, giving higher accuracy. For these processes the integrity of the castings is ensured by avoiding, as far as possible, breaking of the melt surface and turbulence during filling of the mould. With conventional stream pouring for any type of mould, where the metal surface is broken, the frequently associated incorporation of oxide dross inclusions in the casting can be greatly reduced by employing ceramic filters in the feeder heads.

The disadvantage of core assembly for simpler castings is the cost of the binder chemical which is lost each time the mould is broken up to release the casting. In the Lost Foam process a polystyrene foam pattern, complete with attached running system, is coated with a refractory wash, dried and then immersed in loose dry sand. The pouring cup is attached to the running system protruding above the loose sand. By glueing together separately-produced foam mouldings complex geometries can be assembled which may be difficult to reproduce by any other casting technique. As the molten metal is poured into the mould it vaporizes and replaces the foam progressively. Uniform wall thickness can be obtained, although, clearly, close control has to be exercised over the polystyrene pattern production in this and other tolerance and finish respects.

Another binderless sand moulding method of great potential involves containing the sand within plastic film, with air evacuation, which results in the sand being compacted firmly against the pattern and forming a mould which can then be handled as long as the vacuum is maintained. After casting and release of the vacuum the sand falls away and is not significantly adulterated by the polymer film vapour products.

Thermal properties of the mould

The mechanical properties of a casting are strongly influenced by the rate at which it solidifies: this is powerfully influenced by the thermal properties of the mould. Metal moulds provide the highest chilling power and this can be modified and controlled by refractory coatings sprayed on to the mould face. Non-metallic moulds, e.g. sand or ceramic, are much less effective as heat sinks, and it is frequently necessary to incorporate metal chills into selected areas of the mould face to speed up solidification, should this be required.

The properties of castings

Historically, it has generally been considered that the mechanical properties of castings must be inherently inferior to those of the corresponding wrought materials. This need not be true, but where it is, the main contributory factors are unsoundness, high level of impurities, coarseness of microstructure, and residual stress.

The causes of casting unsoundness are the entrainment of air and oxide, as previously discussed, the rejection of dissolved hydrogen on solidification and the decrease in specific volume that most engineering alloys experience as they cool through the liquid to solid phase change. If the casting is to be sound, there are two main requirements in relation to the shrinkage on solidification. First, it must be supplied, as it solidifies, with compensatory liquid metal from localized reservoirs, called feeder heads, connected to the mould cavity. The second requirement is directional solidification, which is a steady movement of the solidification front towards the feeder head from regions remote from it, and terminating within it. Clearly, if some part of a casting away from a feeder head is isolated from it by premature solidification of a section closer to it then unsoundness must result. To prevent this, it may be necessary to delay the solidification of some parts by the localized application of heat (exothermic mould materials) or localized thickening of sections (padding), whilst in other parts solidification may be speeded up by local chilling. Even so, in 'shrink-prone' alloys (alloys with long solidification temperature ranges) it is impossible to avoid small residual amounts of microporosity. If the additional expense can be justified, this final porosity can be removed by isostatically pressing the hot casting, thereby enhancing its properties (see p. 274).

Stresses which develop during the solidification of a casting as a result of contraction during cooling may result in 'hot tears', occurring at temperatures where interdendritic liquid still exists. Stresses developed in fully solid material at lower temperatures can produce distortion and may remain locked in the final component and seriously affect the response to externally applied stress.

Coarseness of microstructure is caused by slow solidification and is therefore associated with thick cross-sections and non-metallic moulds. In alloys which solidify as dendrites (this includes the majority of engineering alloys) the best measure of microstructural coarseness is the secondary dendrite arm spacing. It is found that finer dendrite arm spacings produce stronger, more ductile, castings, and these can be achieved by incorporating chills into the sand moulds. This procedure, together with strategic location of feeder heads, permits the production of 'high-integrity' or 'premium-quality' castings which exhibit a reliability and level of mechanical properties sufficient to enable them to compete on equal terms with forgings. Flemings[5] has shown that the yield strengths of many materials can be nearly doubled whilst ductility can be improved still more. The expense of premium-quality castings has limited them mainly to applications in aerospace. Campbell[6] has published a scientific examination of the whole casting process, from the introduction of the liquid metal into a mould onwards, which through understanding should lead to best practice.

It is clear from the foregoing that although castings can be produced with properties that are comparable to those of forgings, the level of expertise and process complexity that is required is considerable. In many applications this is not justified, and Wild[7] suggests that the designer would be satisfied with lower mechanical properties in castings provided they are consistent and reliably guaranteed. He further suggests that castings may be used (1) when the strength is adequate, (2) where the design is based on factors other than strength e.g. stiffness, size or weight-saving, (3) where complex shape makes other manufacturing methods uneconomic or even impossible, (4) where the use of a casting saves development time and (5) where casting gives substantial cost saving over other methods of manufacture.

Although it is often said that if a metal can be melted then it can be cast, the ease with which high-quality castings can be obtained varies widely from one alloy to another, and within a given alloy system it often happens that the alloys with the highest potential mechanical properties are the most difficult to cast. Casting difficulty increases with the range of temperature over which the alloy solidifies. Eutectic alloys are easy but a solidification range of 200°C (360°F) or more may make it impossible to guarantee pressure tightness and freedom from unsoundness and hot tearing.

A great deal of effort is currently being put into the development of computer modelling of the casting process by physics-based mathematical analysis, to include fluid flow in the filling of the mould, the temperature distribution at all stages to complete solidification, the form of the micro-structure produced and the level of any residual stresses. The more complex the shape and, again, the longer the freezing range of the alloy, the more difficult the modelling becomes. Considerable progress has been made however, with many instances of better control of casting production and economic advantage.

13.4 Wrought products

Wrought products include forgings, extrusions, rolled plate, strip, bar and sections and drawn wire. Of these, forgings are distinct in having complex three-dimensional shapes – the others are prismatic or axi-symmetrical. Forgings are also distinct in being designed individually for a given job – the others are semi-finished products which are used as starting materials for more complex parts and may be purchased in conformity with manufacturers' standards.

There are two main reasons for using wrought material. One is the generally higher level, and greater uniformity and reliability, of mechanical properties. The other is the convenient form of many wrought products for later manufacturing. The first of these reasons accounts for the use of forgings in preference to castings for critical parts in aerospace: there are, of course, many similar examples in other fields. The second reason explains why domestic aluminium pots and pans are deep-drawn from rolled sheet – the small strength requirement and the need for lightness means that they are too thin to be conveniently cast. In contrast, the increased strength requirement in domestic pressure cookers means that they are thick enough to be cast. Castings would not be envisaged for industrial pressure vessels, however, because of the more stringent conditions and the danger of a crack propagating from internal unsoundness, and such vessels are always wrought.

Wrought material is manufactured by applying heavy deformation to cast ingots with the aim of eliminating, or at least reducing, the undesirable features of the as-cast condition. The character of the wrought product depends largely upon how effectively, and in what manner, this deformation is applied, because there are, of course, practical limits to what can be achieved. If the undesirable characteristics of the as-cast structure are summarized as porosity, coarseness of microstructure and chemical segregation then only the first of these can be totally removed (and then not always). Matrix grain size can be effectively refined but not non-metallic inclusions or stable inter-metallics (except in the rare examples of fragmentation): those are usually merely altered in shape and moved around. Segregation is merely redistributed according to the overall geometry of the applied deformation. The effect of large plastic deformations, especially if they are all in the same sense and direction, may be to develop considerable anisotropy of properties.

There are two distinct types of anisotropy: crystallographic and microstructural – the former being developed by crystalline preferred orientations, the latter by preferential alignment of second-phase particles and segregation. Crystallographic anisotropy is sometimes desirable, as in sheet for transformer laminations and for deep-drawing purposes; microstructural anisotropy is usually undesirable since it causes poor toughness and ductility in directions normal to the alignment of the microstructural features. Rolled or extruded blooms are often used as starting material for forgings and the preliminary stages in forging should be 'kneading' operations designed to reduce the anisotropy pre-existing in the bloom before starting to produce the shape required in the final forging. Isotropy of properties is desirable in critical components because there is always a degree of uncertainty concerning the directions of stresses caused by service conditions. Where aligned microstructures are unavoidable, as where the shape or size of the part does not permit effective 'kneading', then it is essential to ensure that the direction of alignment does not cut surfaces and is not normal to tensile stresses, especially where fatigue loading occurs. This means that screw threads should be rolled, and changes of cross-section in shafts should be forged to size. Total application of this principle would limit the role of machining to the establishment of final dimensions but, of course, for many small machine parts it is not economic so to forgo the advantages of machining as a major shaping method. The criterion of choice as between forging and machining from solid must be the foreseeable penalties of failure.

Mechanical working provides the metallurgist with considerable scope for the control of mechanical properties. The simplest method of strengthening metals is by strain-hardening and the upper limit of strengthening by this method is set by the ductility, which is decreased as the hardness increases. In sheet products there are usually five commercial grades of strength from soft (i.e. fully annealed) to fully hard. Strain-hardened metals are available as flat products (sheet, strip and foil) and rod and wire.

Because strain-hardened sheet is both strong and of low ductility, further forming is extremely difficult and, in addition, as the ratio of strength to elastic modulus increases, greater allowances must be made for spring-back. Sheet for deep-drawing purposes is normally annealed prior to drawing. Some sheet materials must be formed at elevated temperatures, e.g. Ti–6Al–4V (650–700°C, 1200–1300°F).

Structural steelwork, i.e. thick plate and sections, is normally hot-rolled and is not strengthened by cold work. Properties depend upon whether it is finished in the as-rolled, controlled-rolled or normalized conditions. Micro-alloying with niobium and/or vanadium, together with control of finish rolling temperature, allows a wide range of strengths to be obtained in combination with good toughness.

Low-carbon, and some low-alloy steels, can be formed cold by extrusion-forging processes. With reductions of 60% or so it is possible to double the yield strength of parts which are not subsequently heat-treated. A major advantage of these processes is the excellent available surface finish.

Mechanical working cannot be applied to metals which are too strong for available equipment, or too brittle. For the first of these reasons, advanced nickel-base alloys for use in aeroengines cannot be wrought (there are also other reasons which favour castings). Alnico permanent magnet alloys are too hard and brittle to be wrought; cast irons are too brittle.

Generally, face-centred-cubic and body-centred-cubic metals are easily worked but close-packed-hexagonal metals need care because of the lack of suitable slip systems.

13.5 The processing of polymers

Plastics materials are manufactured by methods which are superficially analogous to metal-working processes. Injection moulding is similar to metal die-casting, thermoforming of plastics sheet is similar to metal pressing (although plastics are invariably heated whereas metals are not), rotational moulding is similar to centrifugal casting (although the plastics feed is in powder form whereas metal for casting is liquid), and blow moulding is similar to superplastic forming in metals. The extrusion of plastics is similar to metal extrusion except that plastics are sometimes molten prior to entering the die, are forced through the die by the action of a screw and can take the form of flat and corrugated sheet and film as well as the more compact shapes characteristic of metal extrusions. Compression moulding of thermosets is analogous to powder metallurgy.

For engineering purposes, injection moulding and thermoforming are probably the most important processes, since many plastics products such as curtain rods, buckets and bowls, packaging film and containers, etc., are not relevant to the present work. However, whatever the end-product may be it is important to know (1) whether a particular plastics material is amenable to a particular method of processing, and (2) whether the chosen method of processing can be controlled so as to make the properties of the end-product more suitable for its purpose.

Polymers, characteristically, are materials of high molecular weight and the properties of a given polymer are influenced by its average molecular weight (and molecular weight distribution, since there is usually a range of molecular weights present within a given material). Thermoplastics soften when they are heated and can be made to flow if the temperature is high enough, but the viscosity of the resulting melt varies from one material to another. At constant temperature, viscosity increases generally with molecular weight but the nature of the molecular chain is also important since greater entanglement raises the rate of increase of viscosity.

The glass transition temperature (T_g) is a most useful feature of an amorphous polymer in reflecting the flexibility of the polymer molecule and the rate at which it will respond to deformation forces at a particular temperature and thus its fields of application. T_g is also obviously of significance in relation to the processability of polymers. Forming below T_g is difficult; for example, where fibres are drawn below T_g

uneven deformation or necking occurs, presumably at points where concentrated internal friction raises material above T_g. For semi-crystalline polymers, it is the crystalline melting temperature that is of greatest significance in relation to processing. For several different types of plastic, the melt flow index (MFI), generally measured in grams of material passing through a standardized die under prescribed processing conditions in 10 minutes, can be useful for comparative purposes.

Different manufacturing processes require different viscosities for satisfactory operation and since processing at different viscosities can produce different properties it follows that products of the same basic polymer manufactured by different processes can exhibit different properties. There is sometimes a conflict with economics, for in injection moulding the strongest products are obtained when the melt temperature and mould temperature are high and the injection speed is slow,[8] but processing costs are then high. The toughness of products can be aided by polymerization to high molecular weights but not to the point at which the melt becomes unmanageable. Some control of melt viscosity can be achieved by chemical means, an example of the fact that many features of melt synthesis are aimed at improving processability: some of the resulting effects on product properties are not necessarily beneficial.

Many polymers exhibit a degree of crystallinity. Although it is common to refer to some polymers as semi-crystalline (nylons, polypropylene, polyacetal, polyethylene) and others as amorphous (polystyrene, polycarbonate, ABS, SAN, acrylics, polysulphones, PVC) the degree of crystallinity from one to another may vary from perhaps 30 to 80%. The extent of crystallinity and the size and structure of the crystals are influenced by processing conditions with consequent strong effects on the resulting mechanical properties of the products.[9] In moulding processes, crystallinity is affected by the temperature of the tools. Production rates can be increased if nucleation rates in crystalline polymers such as nylons and polypropylene are increased by the addition of heterogeneous

195

nucleating agents. However, crystalline polymers tend to shrink more during solidification than amorphous polymers and mouldings of, e.g. polypropylene, often exhibit unsightly sinks and draws in places of locally increased thickness. It is also more difficult to achieve close dimensional tolerances in materials of high shrinkage.

The long-chain nature of polymers makes them very suitable for the production of highly directional properties. Parallel alignment of polymer molecules is achieved in synthetic fibres by axial stretching during production, and by this means greatly enhanced strength can be developed, as exemplified by the reinforcing fibre Kevlar. Although fibres can be stretched in the cold condition it is normal for this to be done at a temperature above the glass transition, the resulting orientation of the molecular chains being frozen by subsequently chilling the fibre below its glass transition temperature.

As well as additions made for the purpose of improving processability, some additives are aimed at direct influence upon the properties of the product. Plasticizers are added to increase flexibility; there are also particulate additives of a rubbery nature which, forming an incompatible phase within the polymer matrix, are able to absorb energy under conditions of shock loading. This technique is employed to produce high-impact grades of polystyrene and ABS. Additions made for the purpose of mechanical reinforcement are even more important and the effects on properties produced by fillers such as glass beads, glass fibres and carbon fibres have been discussed in Chapter 6.

13.6 The processing of composites

Most composite applications can be divided into:

(1) structural – continuous unidirectional fibres;
(2) semi-structural – continuous multi-directional fibres;
(3) non-structural – discontinuous fibres.

The major processing routes used to fabricate thermoset composites are shown in Figure 13.4, where they are related to the above divisions. There is currently a gap in the top right of Figure 13.4, i.e. no production techniques which address volume markets for structural components, although variants of the resin transfer moulding (RTM), pultrusion and filament winding techniques are continually being developed to try to solve this problem.

Key

RTM   Resin transfer moulding
SMC   Sheet moulding compound
BMC   Bulk moulding compound

Figure 13.4 Composite manufacturing processes. (From Furness[10] after Sheard[11].)

It can be instructive to divide the processing techniques into three types:

(1) open mould (e.g. hand lay-up, filament winding);
(2) closed mould (e.g. vacuum bag moulding, RTM);
(3) continuous (e.g. pultrusion).

Sheet moulding compound

Sheet moulding compounds (SMCs) are usually a filled polyester resin with chopped or continuous glass fibres. The consistency is normally similar to that of thick putty. The sheets of SMC are essentially compression moulded to final shape. SMC can produce both large and small parts with a surface finish generally superior to all other composite processing methods, hence its attraction to the automotive industry for appearance-critical body parts. Tooling costs tend to be high, so production runs should be long. Raw materials suppliers are working with the equipment manufacturers in an attempt to reduce cycle times using faster cure cycles and increasing automation.

Resin transfer moulding

Resin transfer moulding (RTM) involves the injection of a low viscosity resin into a closed mould, usually under vacuum, containing the fibre reinforcement as a preform. The Ford Motor Company recently demonstrated that the entire 90-piece front end of the Escort car could be replaced by a 2-piece RTM structure. Production cycle times were estimated at less than 10 minutes and the RTM structure was lighter than the steel structure it replaced, as well as being stronger and stiffer. As with SMC, low viscosity, fast curing resins along with highly automated equipment are being developed in order to attract the automotive industry to the process.

For low volume production, it is possible to use low cost soft-tooling, referred to earlier in this chapter. For larger components, the tooling is frequently manufactured from composite itself. For larger production runs, aluminium tooling is generally used with built-in heating and cooling channels for faster cycle times. Semi-structural automotive components, with a fibre volume fraction approaching 60%, can be produced in large numbers using modified reaction injection moulding (RIM) equipment. The quality of surface finish with RTM is generally good and parts can be produced to tight tolerances. It is also possible to mould-in inserts and foam cores.

Pultrusion

In the pultrusion process, dry fibre reinforcements are passed through a resin bath and impregnated with resin before being drawn through a heated die, where the thermoset resin is cured. In general terms, pultrusion allows the manufacture of any profile that could normally be produced by extrusion of a thermoplastic. The applications for pultruded sections range from skis and ladders to structural sections for bridges and buildings. As a consequence of the fibre alignment, pultruded sections are stiff along their length but display poor mechanical properties in a transverse direction. This is often compensated for by over-wrapping the pultrusion with a woven reinforcing fabric to be impregnated with resin or prepreg, providing reinforcement in the transverse direction. It is also possible to pultrude woven cloth to prevent the need for further processing.

Polyester resins are generally used, although epoxies and phenolics are available for pultrusion. Pultruding with thermoplastic resins is also extremely attractive and is of considerable current interest.

Pultrusion is unique among the processing techniques for composites as it is capable of producing complex components on a continuous basis.

Prepreg moulding

Prepreg moulding is in many respects the next step up from the impregnation of resin by hand into layers of dry chopped-glass strand mat laid onto the mould. Using prepregs, the resin content and fibre orientation of the finished component can be more accurately controlled. Prepregs command a price premium and require oven-curing and vacuum bag or autoclave moulding to take full advantage of the properties. Prepreg moulding is particularly cost-effective in the manufacture of large one-off composite structures, such as ocean-going racing yachts, where performance requirements outweigh the costs. Moulding by autoclave is used in the aerospace and high-performance racing car industries, where the high temperatures and pressures achieved during moulding ensure that the optimum composite properties are achieved.

For the manufacture of stiff 'sandwich' structures, where two laminate skins are separated by a low density core material such as PVC foam or aluminium or paper honeycomb, filmed adhesive can be first applied to the core material and the prepreg plies for the laminate are then placed onto this film adhesive, which provides the necessary resin to reticulate around the core material, thereby achieving a good skin-to-core bond.

Filament winding

Resin-impregnated fibres are wound onto a mandrel for the manufacture of components with cylindrical symmetry in the filament winding process. Pressure vessels, pipes and automotive drive shafts have all been produced by this technique. One European company manufactures commuter train carriages by a variant of the filament winding process. The mandrel can be any shape that does not possess re-entrant curvature, although it can be possible to remove the component from the mandrel before it has fully cured and reverse curvature can be produced by suitable forming operations.

Filament winding is usually computer-controlled and the reinforcement can be oriented to precisely match the design loads. The fibres may be impregnated with resin before winding (wet winding), pre-impregnated (dry winding) or, more unusually, post-impregnated. High volume fractions of fibre are attained (60–80%). Only the inner surface of a filament wound structure will be smooth unless a secondary operation is performed on the outer surface.

Thermoplastic matrix composites

Many of the processing problems associated with polyester and epoxy thermosetting composites (e.g. limited shelf life of resins, long cure times and labour intensive) can be overcome by the use of thermoplastic resins. Thermoplastics need only to be warmed in order to flow in indirect forming processes such as stamping and hydro-forming. Thermoplastic matrix composites (TMCs) can also be produced into laminates by continuous tape laying techniques, into tubes by filament winding and rod by pultrusion. TMC prepreg material has an infinite shelf life and does not require refrigerated storage, unlike epoxy resin prepregs. The material is dry and therefore easier to handle and therefore processes are easier to automate. Once the part has been formed, it is relatively easy to cool the part to room-temperature rigidity. While early TMC work concentrated on the more exotic thermoplastic resins such as PEEK, there is much interest into the use of commodity plastics such as polypropylene with continuous glass fibre reinforcement, particularly for semi-structural automotive, aerospace and construction components.

13.7 Fabrication from powder

Here we imply a system where the material in powder form is bound together at a temperature below the melting point of the major constituent to form a coherent solid mass. The powders can be pure metals, alloys, intermetallic compounds, metal compounds such as oxides, nitrides, borides, carbides and carbon. It is a way of producing simple and complex components without the requirements for conventional melting and casting processes, and very often without having to revert to working processes and fabrication techniques of any other kind, although in some cases the powder route may be chosen for the initial production of simple forms which are subsequently mechanically worked. Speaking in general terms there are two main reasons why the powder route may be chosen: first, for technical reasons over other fabrication techniques or where there is just no alternative (e.g. non-alloying metallic or metal/non-metal systems) and second, for large-scale bulk production of articles which can easily be made in other ways but where an economy is effected by using powder.

The primary fields of application may be classified as follows:

(1) The manufacture of materials and components which cannot conveniently be produced by conventional fusion and casting, e.g. the high melting point metals (W, Mo, Ta), carbides, nitrides, borides, 'heavy' alloys (W–Cu–Ni and Mo–W), ceramic systems, insulators, metal/metal oxide dual systems.

(2) The manufacture of components from constituents which are not miscible in the fused state, but where a combination of properties from the constituents is required. In metal systems this is true of W–Cu and W–Ag for hard-wearing contacts. Cu–C electric motor brushes and to an extent to Cu–Pb bearings.

(3) Where a component has some specific porosity requirement, as in porous bearings (to be oil-filled) and filters. This is a very big field for both ferrous and non-ferrous parts. An organic binder is frequently added prior to sintering to facilitate the formation of interconnected porosity.

(4) The manufacture of alloys to precise uniform compositions, and thus having uniform physical properties which fall within narrow limits; for example, alloys for sealing to glass and ceramics, magnetic materials and bimetallic elements. Mixed and sintered powders are worked down to whatever form is required and all the oxidation and segregation problems in getting precise compositions in cast material are avoided. In the particular case of the very important field of magnetic materials the powder route also makes possible the enhancement and control of magnetic properties in ways not possible by conventional fabrication. Not only are many of the magnet alloys susceptible to strong segregation in casting, they are also difficult to work and machine.

An interesting application in this context was in the production of very uniform alloy standards for spectrographic analysis, where the fluctuations inherent in cast material from segregation produced an unacceptable background scatter.

The ability to produce close compositional control with only very short-range segregation (i.e. limited to the particle size) has attracted interest in areas that are more directly of an engineering character. Two examples suffice: the hot isostatic pressing and forming of complex alloys such as high-speed steels and certain classes of nickel-base superalloys, both where casting and fabrication to a uniform material can result in a high reject rate.

(5) The manufacture of suitable engineering components direct to finished size and shape without the need for expensive machining operations. In this respect it is directly competitive with die-casting and die-forging and is particularly important for iron and steel systems.

The powders

It is outside the intended scope of this book to discuss the ways in which powders are produced, but the route by which they are obtained will control the shape and the surface characteristics of the powder (particularly the presence of surface oxide). The former is of particular importance in the die-filling and compaction process and, where a full density is achieved by forging after sintering, the cleanliness of the powder surface becomes of paramount importance in determining the level of toughness which can be achieved. The most common methods for producing powders are atomization of liquid by water or gas, the reduction of oxides and salts, electrolysis and the thermal dissociation of metal compounds.

Since the packing and moulding properties of powders are very important it is clear that the size and distribution of sizes is an important aspect of powder control. Shrinkage during consolidation by sintering is also at a minimum where the initial compaction density is high.

Consolidation of powders

Depending on their nature, powders are consolidated by mechanical deformation (giving surface-to-surface pressure welding) and/or thermally activated diffusion processes (with or without the presence of a liquid phase developed) whereby material transport (sintering) reduces the amount of internal surface to give densification and external shrinkage, and where chemical homogeneity is achieved through diffusion over short distances where the powders are fine and mixed (i.e. not prealloyed). The presence of a liquid phase during sintering, either developed temporarily or throughout the process or infiltrated from outside, can accelerate densification, particularly in view of the higher rates of transport involved.

Hot pressing

The fabrication of most powder compacts is carried out by pressing at room temperature and then sintering at elevated temperature. It is possible to combine these two to a one-stage process by pressing at an elevated temperature where considerable plastic deformation is possible. (If the pressing is achieved isostatically the term 'HIPing' is employed.) Plastic deformation and sintering take place simultaneously giving a product having a recrystallized structure and little or no porosity. Since there is easy plastic deformation the shape and size characteristics of the powder are not as critical, and an exact replica of the die will be made without the need for any allowance for shrinkage during sintering. Because of the high density, strength and ductility are maximized.

There are, however, some obvious difficulties. Heating the large initial surface area of uncompacted powder can produce excessive oxidation, with poor mechanical properties, particularly ductility, as the result. An evacuated system, or the provision of a neutral or reducing atmosphere, is essential. In producing simple shapes by hot-pressing for subsequent forging or extrusion, the powders can also be canned and sealed.

When making a finished shape by direct hot-pressing a major problem is in finding a suitable die material for use at elevated temperatures.

Mechanical properties

Mechanical properties are of particular significance in the (5) category of usage in the production of engineering components, particularly in iron and steel. The density of sintered materials is the primary factor determining the mechanical properties,[12] and increases in density by whatever means – purity, soft particles, increased compaction pressures, high compaction velocity, powder sizing and shape control, have much the same effect. The tensile strength of ferrous components is improved by a reduction in porosity and pore and particle size, and by alloying. The toughness also increases with density, most rapidly at high densities[13]. The impact strength of sintered iron is very low until densities of about $7\,g/cm^3$ are reached, when it rises rapidly with density.[12] It is greatly improved by phosphorus additions,[14] probably because the liquid phase formed during sintering reduces the angularity of pores at similar density.

Hardness is mainly controlled by alloying; an effect of density has been claimed[12] but as Arbstedt[14] points out, hardness measurements are unreliable on porous materials.

In a complete review of the literature it becomes clear that whilst empirical relationships may be established with tensile strength the effect on other properties is confused by their sensitivity to pore structure and microstructure.

The porosity can be reduced to very low levels by post-sintering mechanical consolidation treatments,[15] for example powder-forging where full density is achieved by hot-forging sintered preforms made from annealed water-atomized pre-alloyed powders.[16] With these processes capable of producing pore-free components, designed for highly stressed components in plain carbon and alloy steels, with strengths of a similar order to wrought products, the presence of non-metallic inclusions, largely associated with the prior-particle boundaries, controls the fracture resistance and toughness (as measured by crack opening displacement, COD).[17] Fracture occurs by the formation and linking of voids associated with inclusions along the prior particle boundaries and resistance to fracture can then be correlated with oxygen content and with the inter-inclusion spacing, correlating well with earlier work on the influence of inclusions on COD in wrought steels.[18] In the case of powder forgings, however, since the inclusions are not randomly distributed the effect of a given volume fraction of inclusions (i.e. oxygen content) is more pronounced. For toughness, therefore, once full density is achieved it is important to ensure that the oxide on the surface of the powder is minimized in production and reduced to a low level during the sintering operation.

In sintered ferrous components the fatigue limit increases with density in relation to the tensile strength. In comparing fully dense powder forgings with conventionally wrought materials there can be an advantage in fatigue strength associated with the fine grain size and the generally small size of the non-metallic inclusions, which are less prone to initiate fatigue cracks than in wrought steels, where the occasional large inclusion may be encountered. In guaranteeing the advantage in fatigue strength, however, there must never be any entrainment of particle-sized non-metallic material in the powder system.

Powder injection moulding

Powder injection moulding allows the near-net shape forming of metals, cermets, intermetallics and ceramics just as for polymers and composites. A thermoplastic binder is mixed with the powder, which forms a low viscosity slurry when heated. This slurry can be processed by injection moulding and the thermoplastic is removed from the resulting compact which can be sintered to near full density. The ceramic body of the spark plug is manufactured by an injection moulding process, but interest in the process is growing in the manufacture of parts from metals

and intermetallics, as it has a number of advantages over conventional powder-processing techniques. The main advantages are perceived to be the manufacture of relatively complex parts, the use of high performance metals and intermetallics and the relatively low cost of the capital equipment costs.

Materials that have been processed by powder injection moulding include steels, nickel alloys, niobium (columbium), superalloys, cobalt alloys and metal matrix composites. The process is not cost competitive with traditional hot isostatic pressing of powders for shapes with relatively simple or axi-symmetric geometries.[19] Powder injection moulding is best for small, complex shapes.

13.8 Fastening and joining

Methods of fastening and joining can be put into three main groups, *viz.*, mechanical, metallurgical and chemical. The first of these comprises screwed fasteners, rivets, spring clips and similar devices; the second, welding, brazing, soldering and diffusion bonding; and the third, adhesion.

The first criterion in selecting a method of joining is whether or not the joint must be demountable because welded, brazed and glued joints are intended to be permanent, as are also some mechanical joints such as rivets and certain types of spring clips which, once fixed, cannot be disengaged. Screwed fasteners in general are readily demountable and soldered joints in electrical circuits can also be taken apart and rejoined, if necessary.

Mechanical fastening

Screwed fasteners

The clamping force exerted upon a joint by a screwed fastener has its origin in the elastic stress developed in the shank of the fastener as the male and female threads are tightened, one upon the other. The first requirement of the fastener material is therefore a high elastic limit to allow the application of a high clamping force. If the fastener is also required to produce a pressure-tight joint then a high stiffness is also desirable since the lower the stiffness, the greater is the reduction in clamping force when a given pressure (equivalent to an additional tension in the fastener) tends to open the joint. However, under fatigue conditions the stress amplitude in the fastener will be reduced if it is less stiff.[20] Bolting materials must possess sufficient workability for heading in the soft condition either hot or cold, and also for thread-rolling, done cold in the final heat-treated condition.

For most general engineering purposes, mild steel fasteners are entirely satisfactory and can be used within the temperature range from normal room temperature up to 300°C (575°F). Mild steel fasteners for consumer durables are generally electroplated with cadmium, zinc or chromium to prevent rusting (a matter of appearance rather than function).

Mild steel fails as a fastener material when it is inadequate in respect of (1) strength, (2) corrosion resistance, (3) temperature resistance and (4) weight. For moderate increases in strength it is possible to choose from the quenched and tempered, medium-carbon low-alloy steels on the basis of the known relationships between hardenability, ruling section and heat-treatment. In doing so it has to be remembered that, because a screwed fastener is severely notched, notch sensitivity limits the strength to which a given steel can be heat-treated. In general engineering, a popular high tensile bolting steel is 0.4C–1Cr–0.3Mo (709M40 in BS970, formerly En19) but lower- and higher-strength steels are also used. In aerospace, the need for ultra-high strength to save weight favours the use of maraging steels and the precipitation-hardened martensitic stainless steels. Titanium alloys are also used, the most common being Ti–6A1–4V,[21] but the higher strength Ti–11.5Mo–6Zr–4.5Sn (Beta III) has been proposed.[22]

For applications requiring good resistance to corrosion, it is possible to move towards copper-base alloys and stainless steels. The cheapest copper-base alloys are the brasses, because zinc is one-third the cost of copper. Strength increases with increasing zinc but ductility reaches a peak at 30% and decreases thereafter: the ordered body-centred-cubic beta phase is quite brittle at room temperature. The tendency, therefore, is to choose single-phase alpha brasses where good cold headability is required, especially for recessed heads, and alpha-beta alloys for good machinability, even though this may necessitate hot heading. For reasons of economy, the zinc content in the single-phase alloys is set as high as the manufacturing requirements will allow, up to 38%. The alpha-beta structure obtains at about 40% zinc and above this level the alloy becomes too brittle for cold forming. Two to three per cent of lead may be added for additional machinability. Ordinary brasses are not satisfactory in marine applications, for which it is necessary to look to dilute tin bronzes, aluminium bronzes, high-tensile brasses (sometimes known as manganese 'bronzes') or stainless steels.

Copper-base fasteners should not be used in aluminium structures because of electrolytic action: it is better to use aluminium alloy fasteners. The Al–5Mg alloy in the half-hard condition and Al–1Mg–1Si, precipitation-hardened, have excellent corrosion resistance, can be anodized and develop strengths equal to that of mild steel. Aluminium wood screws are made in Al–4.5Cu–0.5Mg–0.7Si–1Mn for additional strength.

Bolts for service at elevated temperatures, as in power generation plant, must be made of creep-resisting material. The 0.4C–1Cr–0.5Mo alloy can be used up to 450°C (840°F), but 565°C (1010°F) can be achieved with higher alloying: Wyatt[23] gives 0.2C–1Cr–1Mo–0.75V–0.1Ti–0.005B as the steel with the highest relaxed stress in common use, and states that the 12Cr steel is no longer used for high-temperature bolts. For the most onerous conditions, Nimonic alloys should be used.

Where weight reduction is important, fastener materials must be judged in terms of the strength–weight ratio and for screwed fasteners the appropriate criterion of strength is the notched tensile strength. On this basis, Ti–6Al–4V with a σ_{UTS}/s.g. value of around 350 MPa

(50 ksi) is slightly superior to the high-strength stainless steel 431 at 300 MPa (43 ksi).

Rivets

Traditionally, steel rivets were formed hot. The final clamping force is then mainly due to the thermal contraction of the rivet as it cools down to ambient temperature. Due to recrystallization, the rivet is left in a fairly soft condition, depending upon how rapidly it is cooled by the tools and surrounding plate material. However, this method of joining steel has been superseded by welding and most riveting of other materials is carried out cold. In cold-riveted joints there is no thermal contraction to contribute to the residual clamping force, which is determined solely by mechanical effects. Due to strain-hardening, cold-formed rivets are stronger than those formed hot, and this results in a stronger joint, but of course the joint is less easily made. In light alloy work it is better, in the interests of productivity, to use a precipitation-hardening material for the rivets. These are supplied in the annealed controlled-drawn condition and then allowed to strengthen after forming by natural ageing. This produces a strong joint but the residual clamping force in the joint will be lower than if an initially stronger rivet were used. Suitable materials are Al–3.5Mg, Al–5Mg, Al–1Mg–1Si and Al–4.5Cu–0.5Mg–0.7Si.

Snap-fits

For thermoplastics with good stiffness, toughness and, for repeated opening and closing, good abrasion resistance and fatigue endurance, snap-fits can be suitable for joints in non-critical, non-load-bearing situations. The springy properties of acetals are very suitable for snap-fits.

Welding

Welding processes join materials in ways which attempt to develop at the joint interface the strength of the basic interatomic, or inter-molecular, bond of the materials concerned and in this respect they differ fundamentally from mechanical or adhesive methods of joining. Energy must be supplied to make a weld and this can be provided in the form of heat, as in fusion welding, which is essentially a localized, small-scale casting operation; or plastic deformation, as in pressure-welding; or kinetic energy, as in friction-welding; or the energy of a beam as in electron-beam or laser welding. Fifty or more distinct variants of the basic welding processes can be listed but the most important are fusion arc welding and electric resistance pressure welding. Since the aim of welding is to produce a joint which is no less strong and tough than the materials to be joined, the more complex these are, the more difficult it is to join them satisfactorily and some materials are regarded as unweldable.

The reason for this lack of weldability is usually the production of a welding defect which is characteristic of a particular combination of material and method of welding, and it follows that the production of a satisfactory welded joint requires a clear understanding of all the factors concerned in the interaction between material and process. Obvious materials factors are chemical composition, compositional heterogeneity, metallurgical and physical condition prior to welding, dimensions and cross-sections, mechanical constraints, and physical properties. Process factors include the nature of the energy source, rate of energy input, welding environment and degree of automation.

The most significant defect encountered in the fusion-welding of steels is hydrogen-induced cracking of the heat-affected zone of the weld. The conditions which must simultaneously be present to cause hydrogen cracking are

(a) hydrogen dissolved in the steel of the joint,
(b) transformation of the steel in the heat-affected zone to a hard microstructure,
(c) stress in the joint which has,
(d) cooled to a sufficiently low temperature.

The hydrogen dissolved in the weld has its origin in the process consumables (stick electrodes or welding wire) or, less often, in the parent plate. The stress in the joint is determined by the geometry and mechanical fit-up of the parts to be welded together with the external restraint imposed by the surrounding metal. The hardness of the microstructure in the heat-affected zone depends upon the composition of the steel in relation to the rate at which it cools from the welding temperature. For a given steel and given welding consumables, process factors such as welding energy input and the heat sink around the weld (determined by the thickness, geometry and thermal properties of the surrounding metal and such auxiliary heating as may be employed to extend the cooling time) will determine whether the hydrogen content of the weld metal or heat-affected zone is sufficient to cause cracking when a sufficiently low temperature is reached. For some lightly alloyed steels it may be necessary to hold the welded joint at a sufficiently high temperature for a time to allow hydrogen to diffuse away from localized concentration, or alternatively to allow the joints to transform to a ductile microstructure less susceptible to embrittlement by hydrogen.

Carbon is the compositional element having the strongest effect on the weldability of steels and modern low-carbon mild steels can be welded without difficulty except in very thick sections or in material from the highly segregated zones of ingots. Higher carbon contents make welding more difficult and beyond 0.2% carbon much care must be taken. In strong low-alloy steels, the carbon content is often around 0.4%, but to make steels of this type weldable the carbon content must be reduced to 0.3%. Medium alloy steels (alloying additions ~5%), such as HY 130/150, intended for deep submersibles, must be highly weldable and the carbon content is reduced to 0.15%. It is, of course, not only the carbon which influences weldability, mainly through the effect on hardenability, and the concept of a carbon-equivalent reflecting the full effect of the composition of the steel on propensity to cracking in the heat-affected zone has been established. In the UK it is possible to purchase weldable structural steels to limiting values of carbon equivalent (CE), which is defined in BS4360 as

$$CE\% = C\% + \frac{Mn\%}{6} + \frac{Cr\% + Mo\% + V\%}{5}$$
$$+ \frac{Ni\% + Cu\%}{15}$$

The establishment of a safe welding procedure is a complex matter but, broadly, it can be said that as the carbon equivalent increases above 0.40%, it becomes increasingly necessary to employ special precautions in welding.

High-alloy steels include such widely different materials as the 9% nickel, the stainless and the maraging steels. Because the carbon content of the 9% nickel cryogenic steel is limited to 0.13%, welding is trouble-free, using nickel-base electrodes, and there is no need for pre-heat or stress-relief. The nickel-free stainless steels with 13% chromium and carbon contents ranging from 0.08 to 0.35%, being air-hardenable, transform to martensite during cooling from the welding temperature, but so long as the carbon content is below 0.1% the weld zone is reasonably ductile. Carbon contents higher than this produce brittle welds and although this brittleness can be reduced by post-weld heat-treatment, or the use of austenitic electrodes, it is probably better to regard such steels as unweldable and restrict welding to the low-carbon varieties. Austenitic steels suffer from solidification cracking in the weld metal, a defect which is generally countered by adjusting the composition of the weld metal so that it contains about 5% delta ferrite. Austenitic steels also suffer from the weld-decay phenomenon, caused by localized precipitation of chromium carbides in the heat-affected zone. This can be remedied by adding titanium or niobium (columbium) to the parent metal, to extents depending on the carbon content. However, this remedy can bring its own problems and it is probably better to reduce the carbon content to 0.03%, at which level the defect is not appreciable. The ferritic stainless steels with chromium contents of 17–20% can also suffer from carbide precipitation adjacent to welds. These steels are intended to be non-transformable but denudation of chromium in the matrix by precipitation of chromium carbide can cause martensitic regions with consequent embrittlement which is exacerbated by large grain size.[24]

Aluminium alloys are subject to four main problems on welding: production of oxide films, weld metal porosity, softening of the weld zone and solidification cracking. The first two of these problems are dealt with by the use of inert-gas shielded processes together with electrode-positive DC (for metal arc processes) or AC (for tungsten arc processes) power sources and high-purity consumables. Softening of the weld zone occurs in strain-hardened non-heat-treatable aluminium alloys and also in the fully heat-treated alloys, but the changes in the latter class of alloys are more complex. Post-weld cold working of weld-softened zones in strain-hardened alloys may produce some recovery of properties but this is not fully effective and the designer should endeavour to locate welds in regions that are lightly loaded. In the heat-treatable alloys, the properties of the weld zone can be restored by full post-weld heat-treatment but in large structures this is hardly practicable. Some recovery of properties may occur in alloys which age naturally, e.g. Al–4.5Zn–1Mg.

Solidification cracking in aluminium alloys is a matter of the solidification temperature range of the alloy concerned and this problem is countered by appropriate choice of electrode. The binary aluminium–magnesium alloys are welded with the overmatching (i.e. higher alloy content) Al–5Mg electrode whilst the heat-treatable Al–Mg–Si alloys require the overmatching Al–5Si or Al–10Si electrodes. Some loss of strength occurs and in the latter case blackening of anodic protective films occurs. The high strength Al–4.5Cu–Mg–Si alloys, using Al–12Si electrodes, can be welded with some chance of success but the properties of the weld are poor and these alloys are not recommended for welding. The Al–4.5Zn–1Mg–Mn alloy can be welded with Al–5Mg–Mn electrodes, again, with some loss of strength. Degradation of weld zone properties will be smaller if welding can be performed with low energy input.

Copper-base materials can be welded satisfactorily but some problems occur. Unalloyed copper must be preheated to allow for its high thermal conductivity and the welding of brasses is complicated by fuming of the zinc – bronze fillers are used to counteract this. Stress relieving is generally carried out after welding to eliminate residual stresses and the associated danger of stress corrosion. In bronzes and cupro-nickel problems of embrittlement and the need for deoxidation are countered by careful process control and the use of specially developed electrodes.

Commercial-purity titanium and the alpha-phase alloys weld very nicely but it is vital to guard against contamination from the atmosphere. The alpha-beta alloys form brittle welds and whilst Ti–6Al–4V can be welded by the electron-beam process, the high-strength alpha-beta alloys such as Ti–4Al–4Mo–2Sn–0.5Si are not normally considered weldable but work carried out by the aerospace industry in recent years has demonstrated that even these materials can be fusion welded under special circumstances.[25]

The welding behaviour of commercial-purity nickel resembles that of mild steel, with the important exception that it is very sensitive to certain contaminants, especially sulphur. Nickel-rich alloys such as Monel, the Inconels and the lower Nimonics can be welded provided appropriate filler metals are used, but the higher-strength Nimonics are unweldable.

The preceding account of the weldability of various materials has assumed that the necessary equipment and expertise are available. This varies greatly. The manual metal arc welding of mild steel non-critical parts can be carried out with little cost and only moderate skill by the domestic householder. In contrast, the equipment for tungsten inert gas welding can cost £10,000 and requires a very high level of operator skill. Robot spot-welding for the manufacture of modern automobile bodies represents a considerable capital investment and the setting up of this equipment must be done by qualified welding engineers. In critical applications, such as pressure vessels, especially for nuclear applications, weld quality must be carefully controlled and tested by non-destructive techniques such as radiography and ultrasonics.

Soldering and brazing

These are methods of joining in which the filler material is fused while parent metal is not, which distinguishes them from welding.

Soldering was originally quite easily distinguished from brazing – soldering referred to low-temperature soft solders of the lead–tin type melting below 250°C, and brazing referred to copper–zinc filler alloys of high melting range – usually above 850°C, based on Cu–40Zn. This classification also coincided with a difference in strength of joints. Soldered joints are relatively weak (38–55 MPa, 5.5–8 ksi) and brazed joints strong (~300 MPa, 44 ksi).

The mechanical strength of soldered and brazed joints

Figures for the strength of a solder or brazing alloy have no direct value in establishing the strength of a soldered or brazed joint. Where the solder or brass is joining stronger materials than itself there is constraint of the solder in the joint by the rigid parent metal interface and the yield and ultimate strength of the assembly can substantially exceed that of the filler material. The constraint developed is a function of the thickness of the joint in relation to its characteristic length,[26] and the thinner the joint, the stronger it is. This is illustrated in Figure 13.5, a phenomenon known as 'contact strengthening'. All brazing and soldering methods depend upon capillary flow of the filler and joint design is therefore critical: butt and fillet joints are permissible in brazing and hard soldering, but only lap or otherwise enclosed joints should be employed for soft solders.

With the development of other alloys, mainly by the addition of silver, the melting range of the soft solders was raised, with corresponding gain in strength of the joint, and that of brazing alloys was progressively lowered without any loss of strength – in fact in most cases with gain in both strength and ductility. Although not now a necessary distinction, 'soldering' is used to describe joining methods employing a filler

Figure 13.5 Relationship of shear strength to brazed joint thickess for pure silver joints in 12.7 mm (0.5 in) diameter steel drill rod (1 psi = 6.894 × 10³ Pa). (From Eagar and Baeslack[27].)

melting below 550°C and 'brazing' a filler melting above that point.

If further subdivision is required it is common practice to describe soldering with a maximum operating temperature below ~350°C (662°F) as 'soft soldering' and between 350 and 550°C (662 and 1022°F) as 'hard soldering'. 'Low-temperature brazing' extends from 550 to 800°C (662 and 1472°F) and includes the use of the well-known 'silver solder' alloys. 'High-temperature brazing' is effected above 800°C, usually with 60/40 brass filler.

The types of alloy are numerous. Soft solders are to BS219, BS441 or BSAU90 (ASTM B32-70) and are divided principally on the basis of whether or not they contain antimony. Antimonial solders are cheaper (with scrap sources of lead containing the antimony, which replaces some tin), stronger and of wide application, but are not suitable for brasses, zinc and galvanized materials and are usually not preferred for radio and electrical assembly. Some examples are shown in Table 13.2.

The higher melting-point solders, or 'hard' solders, mainly introduce silver and antimony in greater quantity to a lead or lead–tin base. They are used where service temperatures are higher than normal soft solders will tolerate under

TABLE 13.2 Typical BS Soft solders

Grade BS 219	Nominal composition (%)			Melting range (°C)		Uses
	Sn	Sb	Pb	Solidus	Liquidus	
K	60	0.5 max	bal	183	188	Tinman's eutectic solder, electrical
F	50	0.5 max	bal	183	212	General bit and blowpipe soldering – 2–5% resin in cored solder
B	50	2.5–3.0	bal	185	204	as F
H	30	0.3 max	bal	183	244	Wiping lead cable and pipe joints and for dipping baths

stress, and temperatures required for brazing cannot be employed. The 4% Ag, 96% Cd alloy, with a melting range of 338–390°C, gives the highest temperature service.

The brazing alloys of the low-temperature type contain silver as an essential constituent. Whilst no flux is necessary when using Types CP1 and CP3 on copper-base materials, these solders should not be used on ferrous or nickel-base materials as they form brittle intermetallic compounds at the interface. A particular feature of the silver solders is their high fluidity and the ease with which penetration of a joint can be effected. The silver–copper eutectic is not frequently applied as such, but assemblies such as radiators made from oxygen-free copper or copper-base alloys may be coated with a thin layer of silver before fabrication. When heated in a reducing atmosphere, the silver–copper eutectic is formed by diffusion, melts at 778°C and effects joining. This is also the basis of the production of 'Sheffield Plate', where silver is bonded to copper by the soldering action of eutectic formed at the interface under the influence of pressure and heat, a process known since Minoan Times[28]

Traditionally brazing, using essentially a 40% Zn brass filler metal, is frequently misnamed 'bronze welding' and is applied to copper, mild steel and cast iron. The tensile strength of copper is about 210 MPa (30 ksi), brass for brazing 390 MPa (56 ksi), with engineering silver solders at 450–500 MPa (65–72 ksi). The special brazing alloy containing silver and phosphorus added to a copper base is even stronger at ~600 MPa (87 ksi). Brazing alloys are covered by BS1845; the US equivalent is AWSA5.8 (Table 13.3).

The metals and alloys which are most difficult to solder and braze are those which are, or contain, metals of high oxygen affinity and which form coherent oxide films, preventing metallic bonding with the solder. Difficulty is therefore experienced with aluminium and its alloys, aluminium bronze, and stainless steels and other chromium-containing alloys. Because of this, aluminium conductors in electrical applications must be fitted with copper end-tags so that soldered connections can be made. The tags are joined to the aluminium by friction welding.

Fluxes which are fluid and sufficiently active at the working temperature may be used to dissolve or reduce such oxides and enable wetting by the solder or braze. Soft solder assembly can be greatly facilitated by preliminary electrotinning (e.g. 'tin' cans, copper heat-exchanger jackets).

Poor joints may result from the formation at the interface of embrittling intermetallic compounds, as occurs during the brazing (silver soldering) of titanium alloys. Although brazing alloys and techniques have been developed for titanium they are not entirely reliable and more interest is currently being shown in diffusion bonding.

TABLE 13.3 Typical BS Brazing Alloys

Grade BS 1845	Nominal composition (%)					Melting range (°C)		Remarks
	Ag	Cu	Zn	Sn	Other	Solidus	Liquidus	
Brazing brasses								
C23	–	60	40	–		885	890	For brazing and
C25	–	60	39	1		880	890	'bronze welding'
C28	–	50	40	–	Ni 10	–	–	
Engineering silver solders								
AG 4	61	28	10	–	–	690	735	Use borax as flux
AG5	43	37	20	–	–	700	775	Use borax as flux
AG 3	38	20	22	–	Cd 20	605	650	Use fluoride base flux
Special brazing alloys								
CP 3	–	93	–	–	P7	705	800	No flux necessary on
CP 1	15	80	–	–	P5	645	700	copper-based materials
Silver–copper eutectic	72	28	–	–	–	780		

Brazing is one of the best methods for joining ceramics. The wetting of the ceramic surface is achieved by first metallizing the ceramic or by modifying the brazing alloy itself to give surface penetration.

Diffusion bonding

This is a form of pressure-welding in which the joint is effected by atomic diffusion across the interface without need for fluxing or significant plastic deformation. It requires time, high temperature and controlled atmosphere or vacuum. Although solid-state bonding is possible, the process is accelerated if liquefaction occurs at the interface. This can be brought about by interposing between the parts to be joined a thin layer of metal which forms a eutectic with them. Roll-diffusion bonding adds to the bonding mechanism plastic deformation by hot-working, but the need for a controlled environment makes the process difficult to control.

Iron, titanium and copper can be diffusion bonded as their oxide will diffuse into the bare metal at the elevated temperatures necessary. Aluminium and magnesium are more difficult to

diffusion bond. The process is only economic when combined with super plastic forming.

Diffusion bonding has been used to bond ceramics but usually requires the use of metallic interlayers.

The welding of thermoplastics

The welding of plastics is restricted to thermoplastics, which can be melted or softened by the action of heat without degradation in the subsequently cooled product. Thermosets, which do not soften on heating, cannot be welded. Three methods are employed to generate the heat at the interface between and two components – (a) processes involving external heating such as a heated tool and hot gas, (b) processes where mechanical forces are employed to generate the necessary heat, and (c) processes involving electromagnetic heating methods, such as resistance, microwave and laser welding.

(a) In heated tool welding the geometry of the tool relates, of course, to the shape of the components. In the simplest form an internally heated flat plate is clamped between the two surfaces to be joined under controlled pressure.

Welding temperatures are usually in the range 180–230°C, and once the surface layers have attained this temperature the tool is removed and the two surfaces are brought together under pressure to produce the weld. The equipment is simple and the method is versatile, giving strong liquid-tight joints, making it ideally suited for containers and pipes for gas and water distribution (p. 162). For mass production items, however, the process is usually too slow.

Hot gas welding is also relatively slow. It has many similar features to the gas welding of metal, for example, edge preparation, matching filler rod, the need for operator skill, but with heated air, nitrogen or other gas substituted for the flame. Compressed air is usually used since it is cheap. The method is used in the fabrication and repair of plant and is particularly suited to the construction of large complex components.

(b) Where mechanical movement is employed to generate localized heating it may take a number of different forms, ultrasonic, friction or vibration. In ultrasonic welding it is necessary to provide ridges or bumps on the mating parts, usually included as parts of the original mouldings. These localize the energy to provide rapid melting and the molten material thus formed is dispersed in the joint. Higher modulus rigid plastics weld most easily by this method, with lower internal energy loss. The method is particularly suited to mass production at a high level of automation since speeds and tooling costs are high.

Plastics may be friction welded in a similar way with similar control variables to metals but with the difference that melting occurs at the interface, but is largely displaced into the flash on the application of the final jointing pressure. Relative motion can be of any form, but rotational welding is perhaps most used, for circular solid or hollow sections.

Vibration welding is again similar except that heat is developed by friction from linear oscillations rather than continuous relative movement, prior to the application of a final welding pressure. It is capable of producing long joints at high production rates and has found its main use in the domestic appliance and automobile industries, for example, in joining two-part thermoplastic car bumpers.

It is a method suitable for almost all thermoplastics, but is particularly good for the crystalline polymers such as acetal, nylon, polyethylene and polypropylene and does not require the degree of rigidity associated with ultrasonic welding.

(c) In resistance and induction implant welding a metal insert is heated by resistance or induction at the joint interface. The local heating causes fusion and the heating element is entrapped in the joint.

Dielectric and microwave heating can also be used to fuse certain plastics. However, microwave welding is not common and dielectric welding is generally restricted to joining thin films of dipolar polymers such as PVC and ABS.

Infrared welding involves heating the joint surfaces with infrared lamps and pressing the parts together once the surfaces have melted, just as in hot-plate welding. Laser welding is very similar except that the heat source has a much greater energy density. This latter technique is not yet fully commercialized.

Adhesives

Adhesion is an important method of joining which can be applied to all classes of materials, with perhaps thermoplastics presenting the most difficulty. Butt, lap, tapered lap, double lap and scarf joints can all be adhesively bonded and the shape of the joint is an important consideration in relation to the universal problem of geometric stress concentration.

Preparation of adherend surfaces is usually necessary, and adhesive is then applied and the joint closed or clamped. Heat and pressure may be applied. Where contraction occurs after the setting reaction, stress concentrations may be set up within the adhesive or at the adherend/adhesive interface, which may seriously weaken the joint. Obviously, adhesives which rely on the loss of a volatile solvent for setting cannot be satisfactorily employed for non-porous adherends in a closed joint.

The requirements for an adhesive can be summarized as:

(i) it must wet the adherend surface, indicated by a small contact angle, generally less than 30°;
(ii) it should be set to a high intrinsic strength;
(iii) it should have the ability to deform during the setting process to release elastic stresses and stress concentrations.

The small contact angle not only reflects the existence and magnitude of bonding forces to the substrate, it also ensures that stress concentration at the end of a joint is at a minimum. Further, assuming that the gas absorbed in surface roughness can be displaced, it reflects the extent to which the adhesive can penetrate the roughness to maximize surface contact area.

The nature of the bonds established between adhesive and substrate depends, of course, on the structure of the two. In the adhesive the existence and orientation of polar groups towards the surface, interacting with the orbiting electrons in a metal or with ions in a substrate such as glass or metallic oxide on the surface of metal produces bonding forces. How far covalency or other forms of chemical bond are formed with surfaces and are necessary for good adhesion is not clear. Certainly chemical bonds have been established in the sulphur vulcaniza-tion process for rubber-metal bonding and have also been suggested for phenol formaldehyde resin with aluminium alloys.

The aerospace industry is currently making extensive use of structural adhesives since the elimination of rivets not only saves weight but produces a smoother profile in the aircraft skin. The load distribution is also over a large area, preventing stress concentrations. Thus there is a growing requirement for adhesives withstanding temperatures in excess of 150°C (300°F), e.g. polymides. Whilst such resins have existed they have usually had poor processing character-istics[29], particularly in relation to solvent removal and the condensation reaction of polya-mic acid to polymide on subsequent heating to a temperature above 150°C during bonding, which results in the liberation of water. Both factors can lead to void formation within either an adhesive bond-line or the matrix phase of a composite. This leads to a need for high processing pres-sures, accurate venting techniques and restric-tions on component size and bond area. It is not a good system, therefore, for aerofoil surfaces.

Developments have been mainly in the areas of improving the processing characteristics of the imide resins, lower boiling point solvents, chain modifications to induce flexibility, and introduc-ing thermoplasticity by fluorination of the poly-mides. By using low molecular weight imide

TABLE 13.4 Typical maximum service temperatures and shear strengths for some adhesives (From Holloway[30])

Adhesive	Maximum service temperature		Shear strength[a]	
	(°C)	(°F)	(MPa)	(ksi)
Epoxy	90	194	36	5.2
Epoxy-phenolic	200	392	25–40	3.6–5.8
Epoxy-polyamide	100	212	25	3.6
Polyimide	400	752	14–18	2.0–2.6
Silicone	340	644	14	2.0
Phenolic-polyamide	150	302	36	5.2
Phenolic-vinyl	100	212	36	5.2
Polyamide	100	212	40–60	5.8–8.7

[a] Estimated at 21°C

prepolymer there is potential for use as a melt processable system using 100% solids with no solvent, or at least minimal low boiling point solvent. They have also been developed to have reactive terminal groups to convert the polymer to a cross-linked product without the evolution of volatiles, which presents such a problem where large areas of non-porous material is to be bonded.

Table 13.4 gives typical shear strength properties for some adhesive resins used for aerospace applications.

The anaerobic adhesives which cure at room temperature and only in the absence of air, are widely used as sealants or locking adhesives in closely mating metal parts. They are single component acrylic systems which cure primarily by a free-radical mechanism in contact with an active metallic surface (i.e. metal ions). Secondary curing may be achieved by the use of an activator or by heating. They are normally applied in liquid form and their good wetting and penetrating characteristics allow the gap between the mating surfaces to be filled. The good penetration also makes the adhesives useful for sealing microporosity in castings.[31]

The strengths developed with an anaerobic adhesive of this type (Permabond A118) are indicated by Table 13.5. Thin, close tolerance joints/fits are required to maximize strength.

Maintenance of strength with temperature varies with the particular grade of adhesive employed. For specialized applications where service temperatures are about 200°C there are adhesives with ring structures or which are highly cross-linked. The combination of penetra-

tion and strength at temperature has led, for example, to use in the fixing of studs on automobile engine castings.

The advantages of adhesion as a method of joining are that the materials to be joined are unaltered by the joining process, either mechanically or metallurgically. Stress concentrations due to fastener holes are absent, as are heat-affected zones due to welding. Dissimilar materials are readily joined without the danger of electrolytic corrosion.

A particular advantage of bonded joints is that they have better fatigue resistance than mechanically fixed assemblies. Since the adhesive makes intimate contact with the metal, it fills all the microroughness in the surface and spreads the load evenly along the bond line, instead of merely at asperities with local stress concentrations as in a mechanical joint, this quite apart from the avoidance of the more major stress concentrations mentioned above. Since air is excluded from the joint, fretting corrosion is eliminated and other corrodents are also excluded (seals). The major disadvantage is that most adhesives cannot be used at temperatures higher than about 200°C (400°F). The joints are also permanent and cannot be dis-assembled for inspection readily. Special surface preparations can also be a disadvantage.

TABLE 13.5 Shear strengths of steel joints with Permabond A118 acrylic adhesive (from Baldwin[32])

Test Condition	Shear strength	
	MPa	ksi
24h at 23°C (73°F)	20.5	3.0
Aged for 1000h at 100°C (212°F)	31.5	4.6
Aged for 1000h at 150°C (234°F)	34.0	4.9

References

1. J. A. REYNOLDS: *Br. Foundryman*, 1973; **66** 245.
2. P. R. BEELEY: *Progress in Cast Metals*. Institution of Metallurgists, 1971.
3. W. J. JACKSON: *Mater. Eng.*; 1981; **2**, 187.
4. J. A. REYNOLDS: *Progress in Cast Metals*. Institution of Metallurgists, 1971.
5. M. C. FLEMINGS and D. PECKNER: *Mater. Des. Eng.*, August 1963.
6. J. CAMPBELL: 'Castings'. Butterworth-Heinemann, Oxford, 1991.
7. F. WILD: *Met. Mater.* November 1969, p. 123.
8. P. L. CLEGG and S. TURNER: *Metallurgist*, December 1975.
9. R. D. DEANIN: *Polymer Structure, Properties and Applications*. Cahners Books, Boston, 1972.

10. J. A. G. FURNESS: *Thermoset Composites, Materials Information Service Profile.* Inst. of Materials 1995.

11. P. SHEARD: Fibre Reinforced Composites, Materials Information Service Profile, Inst. of Materials 1992.

12. A. SQUIRE: *Trans. AIME*, 1947; **171**, 485.

13. G. ZAPF: *Powder Metall.* 1970; **13**, 170.

14. P. G. ARBSTEDT: *Met. Technol.*, 1976; **3**, 214.

15. R. T. CUNDILL, E. MARSH and K. A. RIDAL: *Powder Metall.* 1970; **13**, 165.

16. G. T. BROWN: *Powder Metall.* 1971; **14**, 124.

17. F. L. BASTIAN and J. A. CHARLES: *Powder Metall.*, 1978; **4**, 199.

18. T. J. BAKER and J. A. CHARLES: *Effect of Second-phase Particles on the Mechanical Properties of Steel.* Iron and Steel Institute Publication 145, London, 1971, pp. 79–87.

19. R. M. GERMAN: in *The Encyclopedia of Advanced Materials*, eds D. BLOOR, R. J. BROOK, M. C. FLEMINGS and S. MAHAJAN. Pergamon Press, 1994.

20. D. H. CHADDOCK: *Introduction to Fastening Systems*, Engineering Design Guide 01, Design Council/ OUP, 1974.

21. R. G. SHERMAN, E. K. BUKONT and V. BENSON: *Met. Eng. Q.*, 1970: **10**, 5.

22. J. B. GUERNSEY: *Met Eng. Q.*, 1970; **10**, 10.

23. L. M. WYATT: *Materials of Construction for Steam Power Plant.* Applied Science, 1976.

24. D. SLATER: *Metallurgia*, May 1966; **73**, 212.

25. B. HANSON: *The Selection and Use of Titanium.* A Design Guide. The Institute of Materials, 1995

26. J. R. GRIFFITH and J. A. CHARLES: *Met. Sci.*, 1968; **2**, 89.

27. T. W. EAGAR and W. A. BAESLACK: *Joining Processes*, in: *The Encyclopedia of Advanced Materials*, eds D. BLOOR, R. J. BROOK, M. C. FLEMINGS and S. MAHAJAN. Pergamon Press 1994.

28. J. A. CHARLES: *Antiquity*, 1968, **XLII**, 278–85.

29. S. J. SHAW: *Mat. Sci. Tech.*, 1987; **3** (8), 589.

30. L. HOLLOWAY: 'Space Structure Applications of Polymers', in: *The Encyclopedia of Advanced Materials*, eds D. BLOOR, R. J. BROOK M. C. FLEMINGS and S. MAHAJAN. Pergamon Press, 1994.

31. P. S. CLOUGH and A. S. PARKES: 'Anaerobic Sealing', 51st meeting, Inst. of Gas Engineers, November 1985, London.

32. T. R. BALDWIN: *Mat. Sci. Tech.*, 1986, **2**, 1–7.

14
The formalization of selection procedures

In selecting a material for a given application the materials engineer is faced with an almost endless number of possibilities. If the choice is to be made with economy of time and effort, but also with the assurance that no possibility is overlooked, some systematization of procedures is essential.

The basis for materials selection is a 'shopping list' of design requirements and, as stated in the Introduction, the selection procedure should be as numerate as possible. The extent to which this can be achieved, however, varies from one design requirement to another, extending from attempts to quantify the aesthetics of design and current fashion to the considerable precision attainable in some property parameters.

A useful reduction in the initial number of candidate materials can be obtained by establishing at the outset upper and lower bounds for the various design requirements. On the basis that every design requirement must be present to an acceptable degree, but that costs must inevitably increase if it is present to a greater extent than is strictly necessary, a table can be drawn up (Table 14.1) summarizing the merits and demerits of the contenders so as to permit early elimination of unsuitable materials. Materials M1 to M6 represent generic types of material for this initial materials selection step, for example, cast aluminium alloys, austenitic stainless steels, polyamides, polyacetals, etc. Analysis of individual materials at this stage, e.g. PA 6, 6–30% glass fibre filled, is dangerous, as other families of material may be missed; what is needed is a broad sweep of all genetic materials' families.

Considerable knowledge and experience are required to reject a material at this stage, because materials properties can be varied widely during manufacture and processing, and so also can costs. A material would not necessarily be rejected because it was unsatisfactory in respect of a single secondary design requirement, or even a primary one, if there were scope for ameliorating the disadvantage during design and manufacture. Whether or not over-provision of some property is cause for rejection depends upon the effect upon cost. Excessive cost is always cause for rejection but cost is also a function of processing.

TABLE 14.1.

Materials	Design requirements					Cost	Decision
	Primary			Secondary			
	DR1	DR2	DR3	DR4	DR5		
M1	A	O	A	A	A	E	Reject
M2	A	A	A	O	A	A	
M3	U	A	A	A	A	A	Reject
M4	A	O	A	A	O	A	
M5	A	A	A	A	A	E	Reject
M6	A	A	A	U	A	A	

U = underprovision: O = over-provision: E = excessive. A = acceptable.

TABLE 14.2.

Material	Heat resistance	Rigidity	Resistance to stress cracking	Mouldability	Overall rating (Max = 20)
M1	4	3	3	3	13/20 = 0.65
M2	2	3	4	3	12/20 = 0.60
M3	5	4	1	1	11/20 = 0.55
M4	1	1	4	3	9/20 = 0.45
M5	4	5	1	3	13/20 = 0.65
M6	3	2	5	5	15/20 = 0.75

Clearly, any version of a basically expensive material, such as titanium, will be costly but whereas steels are mostly cheap they become expensive when highly alloyed or manufactured to tight tolerances or compositional limits. It is likely that any class of material which passes this initial stage of selection will produce three or more competing variants of the same material to be considered at a later stage.

Table 14.1 can be refined by replacing the simple go/no-go criteria of satisfactory and unsatisfactory by varying degrees of merit. For properties that are not reliably quantifiable, more-or-less vague terms such as poor, fair, excellent, etc. are best abandoned in favour of numerical ratings of, say, 1 to 5 in ascending order of merit. The individual merit ratings can then be totalled to give an overall numerical rating as in Table 14.2.

Clearly, the overall superiority of M6 derives from its maximum ratings in respect of stress cracking and mouldability, but what if heat resistance and rigidity were the properties more urgently required? This might well be so, since although the life of the component could be determined by its resistance to stress cracking, its resistance to heat and its rigidity could determine whether or not it can do the job at all. (Mouldability is important mainly through its influence on costs.) The relative importance of the various properties therefore depend upon the nature of the application and this can only be assessed by the designer using the tools of function analysis and quality function deployment, as discussed in earlier chapters. For example, he can exercise his judgement in this respect by assigning weighting factors to the various properties, as in Table 14.3.

The choice now moves to M5, which means that a short-life material has been preferred to a long-life material. This emphasizes that weighting factors must be used cautiously, since by

TABLE 14.3.

Material	Heat resistance × 5	Rigidity × 5	Resistance to stress cracking × 2	Mouldability × 3	Overall rating (Max = 75)
M1	20	15	6	9	50/75 = 0.67
M2	10	15	8	9	42/75 = 0.56
M3	25	20	2	3	50/75 = 0.67
M4	5	5	8	9	27/75 = 0.36
M5	20	25	2	9	55/75 = 0.73
M6	15	10	10	15	50/75 = 0.67

TABLE 14.4.

Material	σ YS (MPa)	K_C (MPa m$^{1/2}$)	ρ (tonnes/m^3)	E (GPa)	Cost (£/tonne)
Aluminium alloy 1	350	45	2.7	70	590
Aluminium alloy 2	550	25	2.7	70	700
Titanium alloy	880	60	4.5	110	5500
Stainless steel	900	100	7.8	200	500
Unidirectional high strength carbon fibre epoxy laminate (60% carbon)	1300	50	1.8	130	15000

their use small changes in heavily weighted properties can mask the effects of large changes in more lightly weighted properties. However, the next stage will be to examine specific materials within the M5 family in more detail, although for the example shown, families M1, M3 and M6 should not be ruled out.

As always, materials selection is more effective when it can be carried out in terms of precisely defined quantitative property parameters. A method of dealing with this may be exemplified by means of the data given in Table 14.4, in which are listed materials which might be considered for use in an aeroplane wing. These materials could have been arrived at following a 'sweeping' exercise as illustrated above. The values of cost/tonne given are illustrative only, and must not be taken as definitive. Price

instability will bring change of magnitude and possibly, even, relationships.

These data cannot be used in raw form, first, because the significance of the individual properties varies from one part of the structure to another, and second, because the units are variegated. The first point can be dealt with by combining units into so-called 'merit parameters'[1], formulated by consideration of the target function, as discussed in previous chapters; the second by expressing the data in each column as proportions of the largest figure appearing in that column (see Table 14.5).

The values of the three merit parameters listed in Table 14.5 should be maximized to provide the minimum weight of a flat panel, loaded in bending where the thickness is free. Where strength is important, $\sigma^{1/2}/\rho$ is important; where

TABLE 14.5.

Material	$\sigma_{YS}^{1/2}/\rho$		$K_C^{1/2}/\rho$		$E^{1/3}/\rho$		Cost (£/tonne)		Overall rating $\dfrac{A + B + C + (1 - D)}{4}$
	Abs.	Rel. (=A)	Abs.	Rel. (=B)	Abs.	Rel. (=C)	Abs.	Rel. (=D)	
Aluminium alloy 1	6.93	0.35	2.48	0.63	1.53	0.54	590	0.04	0.62
Aluminium alloy 2	8.69	0.43	1.85	0.47	1.53	0.54	700	0.05	0.60
Titanium alloy	6.59	0.33	1.72	0.44	1.06	0.38	5500	0.37	0.45
Stainless steel	3.85	0.19	1.28	0.33	0.75	0.27	500	0.03	0.44
Unidirectional high strength carbon fibre epoxy laminate (60% carbon)	20.03	1.00	3.93	1.00	2.81	1.00	15,000	1.00	0.75

Abs = Absolute value; Rel. = Value relative to largest quantity.

stiffness is important, $E^{1/3}/\rho$ is important; where toughness is important, $K_C^{1/2}/\rho$ is important. These are all reasonable parameters to apply to aircraft wing skin materials. Despite its high price, the overall rating of the composite is the highest. There is apparently little to choose between the aluminium alloys or, at the bottom of the five candidate materials, titanium alloy and stainless steel. It is hardly surprising that where cost is relatively insignificant, which is generally the case for military supersonic aircraft, composite materials are used (see Figure 15.11). Even in civil aircraft, the use of composites is increasing (see Chapter 15). However, aluminium alloys remain a good choice, partly for conservative reasons because their processing and behaviour is well characterized. The use of stainless steels in wing structures is limited, despite their corrosion resistance, but some titanium alloys have been developed for selected aircraft skin applications, despite the low overall rating shown in Table 14.5, and the reasons are explored in Chapter 15. Illustrating, perhaps, that while merit parameters are useful, in the real world the situation is often more complicated.

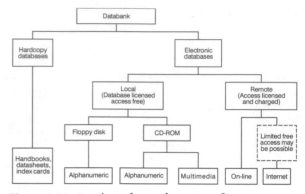

Figure 14.1 Database forms (from Maier[8]).

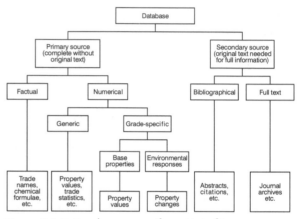

Figure 14.2 Database types (from Maier[8]).

14.1 Materials databases

The importance of integrating materials selection and manufacturing method with the design process has been highlighted in Chapter 2, as has the role of computer-aided design and computer-aided manufacture. Therefore, it is logical that the materials selection should also be computerized[2]. Also, the computer is able to sort through masses of data rapidly, which is invaluable given the ever expanding number of available engineering materials, making a manual search of data very laborious if not impossible. Much of the available materials data does now exist in computerized form, and several information networks now exist[3–7]. However, hardcopy databases should not be ignored, and the relationship between the different database forms is illustrated in Figure 14.1. The different types of database available are shown in Figure 14.2.

The use of a database with a particular selection methodology was an idea developed many years ago by Waterman[9,10] and further discussed by Appoo and Alexander[11]. Gillam[12] also discussed some aspects of computer selection, and Reid and Greenberg[13] described, as an example, how a simple computer program could aid optimal materials selection for a compound beam. The approach has now become well-established, with several systems available commercially.

Generic databases

Generic databases are concerned with materials at the 'family' level. Property values are those considered typical for specific grades within the family, although the materials are often sub-divided into a reasonably narrow band of materials, generally suitable for the conceptual design stage.

The *Cambridge Materials Selector* is a generic database that builds on the graphical methods described by Ashby[14], which are user-friendly and very effective in the initial sorting stages. The selection philosophy involves first identifying the design requirements. These are then expressed in terms of a merit parameter, such as those listed in Table 1.4. The merit parameter can be plotted as a straight line onto the Ashby map, where on logarithmic scales, the properties under consideration are displayed along the axes. Two examples are shown as Figures 14.3 and 14.4, where Young's modulus is plotted against density

and Young's modulus is plotted against strength, respectively. The range of properties displayed by each family of materials is described by a 'balloon' on the map. In Figure 14.3, the lines for E/ρ, $E^{1/2}/\rho$ and $E^{1/3}/\rho = C$, where C is a constant, are drawn. For minimum weight design, the value of C should be maximized, which represents moving the lines from right to left. As already noted, when selecting materials for a light, stiff column, $E^{1/2}/\rho$ should be maximized, and, as can be seen from Figure 14.3, wood and composite CFRP are good candidate materials, residing in the unshaded portion of the graph.

In Figure 14.4, the lines for σ_γ/E and $\sigma_\gamma^2/E = C$ are drawn. To minimize the volume of a spring with a specified energy storage value, for example, the parameter σ_γ^2/E should be

Figure 14.3 Materials selection chart: Young's modulus/density indicating relationships of specific stiffness.[14]

Figure 14.4 Materials selection chart: Young's modulus/strength indicating relationships for consideration as springs.[14]

maximized, regardless of spring shape or loading method. Moving the line for σ_y^2/E to increase C (N.B. left to right), materials with maximum C values include rubber and spring steels, which might have been anticipated. Note, however, that CFRP and glass make excellent springs in the right circumstances, hence the use of glass galvanometer suspensions and CFRP vehicle springs.

The method is very powerful and makes it possible to sweep across the properties of each of the generic materials families, useful in design innovation exercises. The *Cambridge Materials Selector (CMS)* allows these maps to be plotted from the comprehensive database, with continuous properties such as strength plotted as

shown above; discontinuous properties such as corrosion resistance in certain environments are ranked on a scale of 1 to 5. After the design requirements are finalized, it is possible to perform several sequential selection steps as well as combine properties on one axis (e.g. cost multiplied by density) and the materials passing each and all of the steps can be displayed. Although included in the generic databases, *CMS* uses several specific databases as well as a front-end generic database, including polymers, light alloys and metal matrix composites. Once the generic database has highlighted, for example, that a number of polymer families are candidate materials, it is possible to refine the selection further by interrogating the polymers database, although there is no data on specific commercial grades, and the selection of 'short fibre reinforced polymers' with the generic database may be refined only to 'polyamide 6/6–30% short glass fibre reinforced' using the polymers database. The generic database contains 143 materials (15 of which are thermoplastics), the polymers database contains 178 materials (54 of these are thermoplastics).

Plascams is a generic database for plastics, with 351 material types included, arranged under 84 generic family titles. The search philosophy is more basic than the approach used in CMS. Initially, an elimination search is conducted to ascertain those materials meeting certain essential design criteria, e.g. 'tensile strength greater than 60 MPa'. Materials passing these elimination searches are then ranked by a combined weighting search of desirable properties. Each of the 351 materials has been given an 'expert' ranking between 0 and 9 for each property. These rankings are simply multiplied by the weightings assigned to the property by the designer. There can be a danger in using the weightings to arrive at the answer that might have been anticipated from past experience. *Rubacams* is a sister database to *Plascams*, containing data on 38 generic types of rubbers and elastomers, utilizing a broadly similar selection philosophy to *Plascams*. The producers of *Plascams* and *Rubacams* have used this software in recently released knowledge-based systems for plastics and rubbers. Using CD-ROM technology, the producers have been able to combine several software programs and databases into one system. This enables it to cut across the boundaries shown in Figure 14.2, containing abstracts, encyclopaedic information, materials processing analysis software as well as materials selection systems. The knowledge-base allows all of these components to be 'hot-linked' so that a line of enquiry can be followed through the system.

Grade-specific databases

These databases hold property values for identified commercially available material grades, so they work well in conjunction with a generic database, refining the materials selection to specific grades and suppliers. For plastics, *Campus* is the world-leader in grade-specific databases and emerged from an initiative by the major materials producers. Now more than 30 materials suppliers produce *Campus* disks, holding data on their own material grades. Compatibility of data between suppliers is ensured by utilizing the ISO standards for single- and multi-point data. It is possible to search the grades of all suppliers simultaneously using a program developed by the producers of the database engine, M-Base. This allows access to more than 6000 grades of plastic materials. Materials selection is made by defining upper and lower bounds to the desired property.

Other grade-specific database for plastics include *Selector II*, giving a total coverage of some 21,500 grades of thermoplastic and thermoset. *Plaspec* contains data on more than 11,000 grades. The selector program can be enhanced with the purchase of other database modules containing bibliographic information. For metals, a number of disks are available in the UK that originated from a recent government initiative. Databases for copper alloys, magnesium alloys, stainless steels and titanium alloys have been produced to run under DOS. The Aluminium Federation produce a database of aluminium alloys, *Aluselect*.

There are also a number of databases providing grade-specific information on a range of materials. *Mat. DB* is a modular database that can be built to suit the user. Ten modules are currently available, including thermoplastics, composites and steels. The data are derived largely from standards and critical assessment of the scientific literature. It is, however, expensive and somewhat cumbersome to use. *Cen BASE/Materials* is another grade-specific database covering about 30,000 plastics, metals, composites and other materials. All property fields are searchable and the data are backed-up by full-text descriptions and supplier contact details. In a recent development, *CenBASE/Materials*, has become available on the Internet, with unlimited access for a small annual subscription (<http://www.icentor.com>, see below). *M-Vision* is an expensive system, primarily containing data for aerospace alloys and composites, but the selection procedures are very flexible, although it needs a workstation to operate.

The Plastics Design Library have developed *Rover PDL Electronic Handbooks*, which are a modular database system, building on the PDL hardcopy data on the environmental properties of plastics, and will include data on metals and ceramics.

Bibliographical databases

Bibliographical databases are designed to rapidly assemble a series of references relevant to a particular enquiry. They normally comprise abstracts but sometimes include numerical data and statistical information. *Rapra Abstracts* is the world's biggest store of plastics information, containing over 400,000 references, all classified by keywords. *Engineered Materials Abstracts* is a bibliographical database of over 75,000 references concerned with plastics, ceramics and composites. The abstracts are derived from technical journals, books, patents, conference proceedings and suppliers' literature. *Metadex* is the equivalent database for metals. Each of these bibliographical systems is available on-line or via CD-ROM.

The Internet

The Internet population is growing rapidly and most information providers now have a presence on the world-wide web. There is a good deal of information made available without charge, but several sites are used to act as 'tasters' for a subscription service, although the charges for these can be modest. Useful Internet addresses are listed below at the end of this chapter.

Data quality

In all these selection systems, whether computerized or not, much hinges on the current recognition of the service requirements and correct instructions in terms of weightings given to various factors and the significance of constraints. Several important properties are not easily quantifiable and data may not be obtainable, particularly in relation to wear and various forms of corrosion. The GIGO principle applies, namely garbage in – garbage out. And no database has been developed to truly deal with synergistics, making the testing of materials in their anticipated service conditions just as essential before final selection.

The quality of the data held in computerized and hard copy systems is always of concern[15] and it is important to ensure that data used for comparative purposes results not only from equivalent testing but that its accuracy is also validated.

At an advanced stage in the selection process, whether or not computerized handling of the data is envisaged, it becomes convenient to combine cost and property parameters because processing accounts for a large part of final costs, and it may happen that the properties exhibited by a material processed in one way are different from the properties of the same material processed in another way. It is best then to employ compound parameters such as $(C_R)\rho/\sigma_{YS}$, where (C_R) is the price per unit mass, and where the parameter takes different values for different materials variants and different manufacturing processes. An overall rating can then be obtained in terms of these parameters in the manner previously described, using weighting factors and lower bounds, as appropriate. A list of compound parameters of this sort is given in the Introduction (Table 1.4).

It may be noted that where space-filling is the major requirement, as applies for example to a pressure-activated device, then the only criteria of choice are surface stability and price per unit volume since mechanical properties are irrelevant.

PC-based materials databases

Cambridge Materials Selector

Granta Design Ltd
20 Trumpington Street
Cambridge CB2 1QA, UK
Tel: + 44 (0)1223 334755
Fax: + 44 (0)1223 332797
WWW:http://www.granta.co.uk

Plascams, Rubacams, Rapra Abstracts, Plastics and Rubber Knowledge-Based Systems

Rapra Technology Ltd
Shawbury
Shrewsbury
Shropshire SY4 4NR, UK
Tel: + 44(0) 1939 250383
Fax: + 44(0) 1939 251118
WWW:http://www.rapra.net

Campus
(disks available from plastics material manufacturers)

M-Base GmbH
Dennewart Strasse 27
DS-2068 Aachen, Germany
Tel: + 49 241 963 1450
Fax: + 49 241 963 1469
WWW:http//www.m-base.de

Selector II

Data Business Publishing
15 Inverness Way East
PO Box 6510
Englewood
CO 801556510, USA
Tel: + 1 303 799 0381
Fax: + 1 303 799 4082

Aluselect

Aluminium Federation
Broadway House
Calthorpe, Fiveways
Birmingham B15 1TW, UK
Tel: + 44 (0)121 456 1103
Fax: + 44 (0)121 456 2274

Mat.DB, Engineered Materials Abstracts, Metadex

ASM International
Materials Park
OH 44073, USA
Tel: + 1 216 338 5151
Fax: + 1 216 338 4634
WWW:http://www.asm-intl.org

CenBASE/Materials

Infodex
12782 Valley View Street
Garden Grove
CA 92645, USA
Tel: +1 714 893 2471
Fax: +1 714 893 4856
WWW:http://www.centor.com/cbmat/index.html

M-Vision

PDA Engineering
2975 Redhill Avenue
Costa Mesa
CA 92626, USA
Tel: +1 714 540 8900
Fax: +1 714 979 2990

PDL Electronic Handbooks

Plastics Design Library
13 Eaton Avenue
Norwich
NY 13815, USA
Tel: +1 607 337 5000
Fax: +1 607 337 5090
WWW:http://www.norwich.net/~wai001

On-line materials databases

Plaspec

D & S Data Resources, Inc
PO Box H
Yardley
PA 19067, USA
Tel: +1 215 428 1060
Fax: +1 215 428 1069

Engineered Materials Abstracts, Metadex
ASM International
as above

Rapra Abstracts
as above

On-line service providers:

ERA-IRS Dialtech
British Library
25 Southampton Buildings
London WC2A IAW, UK
Tel: +44 (0)171 323 7951
Fax: +44 (0)171 323 7954

Datastar Dialog Europe
Knight-Ridder Information Ltd
Haymarket House
1 Oxenden Street
London SWIY 4EE, UK
Tel: +44 (0)171 930 7646
Fax: +44 (0)171 930 2581

Questel Orbit
18 Parkshot
Richmond
Surrey TW9 2RG, UK
Tel: +44 (0)181 332 7888
Fax: +44 (0)181 332 7449

STN International
FIZ Karlsruhe
PO Box 2465
76012 Karlsruhe
Germany
Tel: +49 7247 808555
Fax: +49 7247 808131

Internet sites

General materials

Cen BASE/Materials
as above

Cambridge Materials Selector
as above

The Institute of Materials
http://www.instmat.co.uk

University of Cambridge, Department of Materials Science and Metallurgy
http://www.msm.cam.ac.uk

Index to WWW Materials Engineering
http://www.materials.drexel.edn/others/industry.html

Metals

The Copper Page
http://www.copper.org

The Aluminum Industry WWW Server
http://www.euro.net/concepts/industry.html

Metal Powder Industries Federation
http://www.well.com/user/css/mpif.htm

Ceramics

Technical ceramics – WWW virtual library
http://www.ikts.fhg.de/ceramics.html

Plastics

Rapra Technology Ltd
as above

PDL/Plastics Design Library
as above

Polymer DotCom
http://www.polymers.com/dotcom/subscribe.html

Plastics News
http://www.plasticsnews.com

Allied Signal Plastics
http://www.asresin.com

BASF
http://www.basf.de/basf/html/e/home.htm

GE Plastics
http://www.ge.com/plastics/index.htm

Hoescht
http://www.hoescht.com

IDES Databases
http://www.odesinc.com

IRC in Polymer Science
University of Leeds
http://www.irc.leeds.ac.uk

Composites

Turner Moss Co
http://www.advmat.com/links.html

References

1. R. SANDSTROM: Merit parameters in materials optimization, *Materials and Design*, **13**, 1992, 131–137.
2. J. A. CHARLES: Paper 1 Materials' 88, Materials and Engineering Design, Institute of Metals, London, May 1988.
3. V. E. HAMPEL, W. A. BOLLINGER, C. A. GAYNER and J.J. OLDANI: UCRL Report No.90276, May 1984.
4. V. E. HAMPEL, C. A. GAYNER, B. E. HEGEMANN, R. D. SANNER and Y. WOLMAN: UCRL Report No. 90942, June 1984.
5. National Bureau of Standards, *Computerized Data Systems*, Fairfield Glade Workshop, TN, Nov. 7–11, 1982.
6. CEC Factual Material Data Banks Workshop, JRC Petten 14–16 Nov. 1984, ECSC-EEE-EAEC Brussels, Luxembourg, 1985.
7. CODATA, *Materials Data for Engineering*, Schluchsee Workshop FRG Sept. 22–27, 1985, Fach-Informations Zentrum 1986.
8. C. MAIER: *Materials Databases 1, Materials Information Service Profile*. The Institute of Materials, UK, 1995.
9. N. A. WATERMAN: I. Mech. Conference Publication No. 22, 1973, p. 15.
10. *Fulmer Materials Optimizer*, Fulmer Research Institute, Stoke Poges, Bucks.
11. P. M. APPOO and W. O. ALEXANDER: *Met. Mater.*, July/August 1976, p. 42.
12. E. GILLAM: *Metallurgist*, 1979, **11** (9), 521.
13. C. N. REID and J. GREENBERG: *Metallurgist*, 1980, **12** (7), 385–387.
14. M. F. ASHBY: Materials Selection Charts in *The Encyclopedia of Advanced Materials*, eds D. BLOOR, R. J. BROOK, M. C. FLEMINGS and S. MAHAJAN. Pergamon Press, 1994
15. P. SARGENT: *Data quality in materials information systems, Computer-aided Design*, **24** (1992) 477–90.

Case studies in materials selection

15
Materials for airframes

The aerospace industry is in business to provide a means of transport, and the broad service requirement to convey maximum cargo at minimum cost is the same as for other forms of transport. However, the materials problems are greatly intensified by the fact that failure in the air is much more likely to involve catastrophic losses, including loss of life, than it is on the surface.

Lifting a payload against gravity in order to transport it by air is an expensive process, so that designs must be as efficient and light as possible. The aerospace industry therefore makes great demands upon its materials. Although always stringent, these demands vary according to the nature of the intended service, e.g. civil or military. Civil aircraft vary in size from the very small single-seater plane, intended for pleasure or small-scale commercial operations such as crop-spraying, through a wide range of increasingly large executive-type planes to the very large jumbo transports, capable of carrying 300 or more passengers or the equivalent quantity of freight. Military aircraft, once described either as fighters or bombers, now include a wide variety of craft ranging from the high-flying reconnaissance or 'spy' plane through operational transports and helicopters to the supersonic or terrain-following combat aircraft.

These different types of aircraft cover a wide range of design requirements. In any given case the important design-determinant factors will include range, speed and altitude, manoeuvrability (as in air superiority fighters), and load.

The life of an aircraft consists of a repeated sequence of operations which is made up of four phases: (1) ground, (2) take-off, (3) cruise (civil liners) or operational (military), and (4) landing.

For civil planes, take-off is the most demanding phase. It requires engines to be operated close to full power and the angle of climb must be sufficient to clear all obstructions in the vicinity of the airport with a sufficient margin of safety and also to comply with stringent noise abatement procedures. In contrast the cruise phase, involving steady-level flight, is relatively undemanding, since modern weather forecasting methods, employing radar to detect high-moisture concentrations, enable storms to be avoided and with jet propulsion most weather problems can be avoided (although not, perhaps, clear air turbulence) by flying in the stratosphere at, say 12,000 m (7.4 miles), where the air has only one-quarter its normal density.[1] Compared with a short-range machine, a long-range aircraft spends a much higher proportion of its life in the cruise mode. On the other hand a military combat jet must not only be capable of a fast take-off, but must also be prepared for the succeeding operational phase to consist of a succession of high-speed manoeuvres with possibly an almost total absence of steady level flight.

In general, the more arduous the service the shorter is the design life, although efforts to increase life are always continuing. Thus, the lifetimes of civil aircraft are now stretching upwards into the range 50,000–100,000 hours (although the supersonic airliner Concorde was designed for a life requirement of as little as 20,000–30,000 hours). As could be expected, the design lives of military aircraft are much shorter – possibly little more than 6000 hours (Figure 15.1), although many remain operational for long periods (e.g., 25 years), so, apart from fatigue considerations, selection of materials for inherent corrosion and stress corrosion resistance is still important, together with high integrity corrosion protection schemes.

As with all transport systems, the aim is to maximize payload in relation to cost. Costs are

Figure 15.1 (a) Anglo-French Jaguar, 6000 hours +; (b) European Airbus, 60,000 hours +[2]. (c) Boeing 777, 40,000 cycles + (courtesy the Boeing Company)

determined by what has to be paid for the carrier and its propulsion. Thus, any transport system consists of three components. These are (1) payload, consisting of the goods and/or passengers carried; (2) the carrier, which includes the hull structure, control systems and crew; (3) the means of propulsion, made up of the power-plant and the fuel.

Increasing the payload means reducing one or both of the other two in relation to the total weight of the aircraft. According to Page,[3] the biggest contribution in this direction so far has been in increased efficiency of power-plant: whereas engine weight may formerly have been equal in weight to that of the payload it now more commonly amounts to little more than half. Unfortunately, in long-range aircraft there remains the problem arising from the need to carry large quantities of fuel. Table 15.1 shows that the payload as a proportion of total weight has actually decreased as a direct result of increases in the corresponding figure for fuel weight.

The materials engineer is involved in contributions to the improvement of fuel efficiency in two ways. First, the thrust to weight ratio in gas turbines has been increased significantly over the years. This has been as a result of the efforts to improve the elevated temperature properties of materials (as in the Nimonic series of alloys, (p. 280), the development of directionally solidified and single crystal turbine blades and the application of blade cooling, requiring the development of new manufacturing techniques). In the future the development of new engine materials, – metal matrix composites, oxidation resistant coatings and reinforced ceramics may see a further impact of the materials engineer on this thrust to weight ratio aspect and fuel efficiency.

Second, the materials engineer must direct his attention towards the aircraft structure, and because of the small payload he finds the greatest return for his efforts in the long-range transport. This is because if a given reduction in structural weight is considered to be transferred directly to payload then the smaller the payload as a percentage of structural weight the greater will be the percentage increase in payload. Engineers will therefore consider carefully the cost per unit weight saved, as new material might save a lot of weight, but will be expensive. This extra expense, divided by the weight saved, is a useful design parameter to evaluate new materials. At one extreme, it costs about $20,000 to lift 1 kg into a geosynchronous orbit, demonstrating the importance of weight in spacecraft. The argument for aircraft is similar, where fuel costs can account for about one-third of the total operating costs of a large civil aircraft and the value of weight savings can be as high as $1,000 per kg.

Improvements in the properties of any structural material enable less of it to be used and thereby reduce structural weight. However, Page[3] has made the point that different criteria

TABLE 15.1. Percentage of total take-off weight (including information from Page[3] and Edwards[4])

	Vimy Commercial 1920	Vickers Viscount	Modern short-range subsonic	Modern long-range subsonic Boeing 747–400	Concorde supersonic
Payload	17	14	24	15	9
Fuel	25	23	18	40	48
Systems, crew, etc.	11	25	18	13	10
Power-plant	18	12	11	6	10
Structure	29	26	29	26	23

for materials selection apply in different parts of an aircraft so that improvements in any one particular property may only be effective in part of the structure. Increases in static strength are only useful in regions of high loading intensity, such as wing sections, whereas in regions of intermediate loading, stiffness is the major design criterion. In lightly loaded regions the sizing of parts is dominated by the minimum-gauge criterion, so that in these areas the resistance to thinning by superficial corrosion becomes an important consideration. However, it is self-evident that one important property, density, exerts its influence throughout the whole of the aircraft. Page[3] estimates that a reduction in density of one-third has a greater effect in reducing structural weight than an improvement of 50% in either strength or stiffness. Any preliminary assessment of material suitability for structural purposes in aircraft must take this into account.

As well as aircraft weight, designers are continually minimizing the drag coefficient, which can have a dramatic effect on fuel efficiency. For example, it was calculated that with the Boeing 777 a badly painted square foot of fuselage on each of ten aircraft, producing a rough surface, could result in an extra fuel consumption of 21 gallons per year. Increase that area to two doors, and over 750 extra gallons would be burnt in a year.[5]

15.1 Principal characteristics of aircraft structures

The significant parts of an aircraft structure for the present study are the wings, the fuselage or hull, the landing gear and the moving operational parts, and control surfaces such as flaps, rudder, elevators and ailerons.

The wings are subjected to the highest levels, and also the most complex variation, of stresses. When the plane is on the ground the wings hang down due to self-weight, the weight of fuel stored inside them and the weight of the engines if these are wing-mounted. The upper wing surfaces are then in tension and the lower surfaces in compression. This loading pattern continues whilst the plane is taxiing, which represents a significant contribution to the service life of the plane. However, the largest forces on the wings occur when the plane is airborne. Since the wings must then support the whole weight of the aircraft the steady stresses are high, and with the wings bending upwards the upper surfaces are in compression and the underside in tension. Each wing therefore acts as a cantilever with the maximum bending moment occurring at the wing roots. If the engines are mounted on the wings, then engine weight, together with weight of undercarriage and fuel, oppose the lift force and in a small way reduce its effects. Superimposed on to the steady stresses are fluctuating stresses which are complex in form and origin. They occur mainly at low altitude where the air is dense, as a result of manoeuvring or gusty weather conditions, but may also be due to clear air turbulence at higher altitudes. In general, however, aircraft cruising at high altitude should not be subject to extreme stress fluctuations: these will tend to be restricted to take-off, climb, approach and landing. In military combat aircraft fluctuations of stress are liable to be higher due to the requirement for frequent and fast manoeuvring. In contrast, a high-altitude military reconnaissance plane should be subject to few fluctuations. Since fatigue crack growth is favoured by tensile, but less so by compressive, stresses the upper and lower wing surfaces have different materials requirements. The main requirement of the upper wing surface is for resistance to compression. In the lower wing surface, whilst there is clearly a requirement for static tensile strength, the more critical need is for resistance to fatigue. Additional properties are, of course, required in both surfaces but these depend to some extent upon the method of manufacture. The earlier method for manufacturing wings is to thread two wing spars, or girders, through the fuselage. These may stretch from wing-tip to wing-tip, but in larger aircraft they may be attached, as separate spars for each wing, to the fuselage which is then strengthened in the appropriate position by a special bulkhead. The aerofoil surface is provided by a skin which is attached by rivets to the spars and ribs. Modern methods of construction aim at diffuse load paths which are more easily achieved when the wings are formed by machining from a solid plate. In this case there is an additional requirement for good resistance to stress corrosion since the parent plate must be thick and this, together with the fact that it has to be heat-treated, means that complete freedom from residual stresses cannot be guaranteed. The tensile nature of the stresses also introduces a requirement for fracture toughness. There is also a need for stiffness to resist bending and buckling.

The fuselage, or body, is a long approximately cylindrical shell, closed at its ends, which carries the whole of the payload. The effect of the payload, acting vertically downwards and supported by the wings at a nearly mid-length position, is to subject the shell to considerable bending. The lower part of the body, especially beneath the wings, is therefore subject to compression whilst the upper fuselage correspondingly experiences tension. In addition, when the aircraft rolls, this applies torsion to the fuselage. In aircraft which fly at altitude the cabin must be pressurized, and this subjects the shell to additional longitudinal and circumferential tension. Since the fuselage then becomes a pressure vessel there is not only a requirement for static tensile strength but also for fracture toughness and damage tolerance. Furthermore, the need for pressurization and de-pressurization once per flight establishes a critical requirement for resistance to low-cycle fatigue.

In aircraft designed to fly at very high speeds, airframe surfaces and especially leading edges are heated by interaction with the air: these effects are discussed in Section 15.3.

Stresses in landing gears are high on take-off when the aircraft has a full load of fuel. Fatigue loads can be generated during taxiing and there are high static loads on the undercarriage components of military aircraft standing on an airfield at a high state of readiness with a full fuel load. The greatest stress, however, occurs on landing as the weight of the whole aircraft hits the ground on touchdown. Whilst the vertical descent velocity of more than $1\,m/sec$ ($>2\,mph$)[1] is not high enough to rate as shock loading, the stresses are very high. During flight, the landing gear must be retracted into the wings or body so as not to spoil the aerodynamic performance of the aircraft, and it follows that these parts must take up as little space as possible. Performance criteria must therefore be referred to the minimum volume condition. Because density-compensated property values are not then appropriate, landing-gear components must be manufactured from materials with the highest commercially available levels of static strength, low-cycle fatigue strength and fracture toughness. Since the critical components are often heat-treated forgings there is also a need for good resistance to stress corrosion.

The control surfaces consist of the rudder, elevators, ailerons and flaps. These parts are, in general, lightly loaded so that static strength is not a major requirement although the flaps must be sufficiently robust to withstand flying debris from the runway. Control surfaces are rather thin components which are still commonly of skin and stringer construction, and in view of the function they have to perform must be provided with adequate stiffness. Any control surface in the vicinity of an engine exhaust will also need to possess temperature resistance. Further, if there is any direct influence from engine noise the resulting acoustic loading of structural panels will demand good resistance to high cycle fatigue.

15.2 Property requirements of aircraft structures

Appropriate materials selection criteria vary according to the loading of the part concerned, i.e. whether it is a simple tension member, a beam, a strut, or a panel. Applying these ideas to aircraft structures, and remembering that density is always important, the properties of the candidate materials can be assessed in terms of the following parameters:[6]

Wing, upper surface: $\dfrac{\sigma_{YS}}{\rho}$ (compr.);

$\dfrac{E^{1/3}}{\rho}$; K_{ISCC}

Wing, lower surface: $\dfrac{\sigma_{YS}}{\rho}$ (tension);

$\dfrac{E}{\rho}$; K_{ISCC}; σ_{FS}; $\dfrac{da}{dN}$; K_C

Fuselage: $\dfrac{\sigma_{YS}}{\rho}$ (tension); $\dfrac{E^{1/3}}{\rho}$;

σ_{FS} (low cycle); $\dfrac{da}{dN}$;

K_C; exfoliation corrosion resistance

Spars, frames, ribs: $\dfrac{\sigma_{YS}}{\rho}$; $\dfrac{E}{\rho}$; K_{ISCC}; σ_{FS}; $\dfrac{da}{dN}$

Landing gear: σ_{YS}; E; K_{ISCC}; $\dfrac{da}{dN}$; K_{IC}

It is worthwhile considering the principal differences between compression-loaded and tension-loaded structures, as follows.

Compression-loaded structures

Consider a panel in the upper surface of a wing. As previously discussed, this is stressed predominantly in compression during flight and will

therefore be subject to failure either by buckling or plastic crushing.

The failure stress by buckling of a solid panel in compression is given by

$$\sigma_B = \frac{\pi^2 E}{3(1 - v^2)} \left(\frac{t}{b}\right)^2 \qquad (15.1)$$

E is the stiffness modulus; v = Poisson's ratio; t is the thickness of the sheet; and b is the width, i.e. the distance between stiffeners (see Figure 15.3b).

Now, the efficiency of a structure can be measured by dividing the load which the structure can support by the weight of the structure. Re-arranging equation (15.1) and taking $v = 0.3$: $t^3 = Pb/3.62E$ where P is the total load supported by the panel and W, the weight of the panel, is $tb\rho$, where ρ = density. Hence,

$$\frac{P}{W} = 1.54 \frac{E^{1/3}}{\rho} \left(\frac{P}{b^2}\right)^{2/3}$$

$E^{1/3}/\rho$ is the materials selection criterion whilst P/b^2 is the structural loading index.

Table 15.2 gives some relevant data for three materials, any of which could be used for an aeroplane wing in the right circumstances. The quantity $E^{1/3}/\rho$ is seen to be highest for aluminium and lowest for steel, whilst the value for titanium is intermediate.

However, this simple basis of comparison tells us nothing of the limiting conditions which may apply to the various materials as the loading intensity increases. Higher end loading develops higher stresses in the panel, and if the loading intensity is high enough the elastic limit of the material will be exceeded and the panel will start

TABLE 15.2.

	Yield stress (MPa)	E (GPa)	Density (tonnes/m³)	$\dfrac{E^{1/3}}{\rho}$
Stainless steel, FV 520	1081 (0.1% PS)	215	7.83	0.765
Titanium alloy, Ti–6Al–4V	830 (0.2% PS)	110	4.43	1.08
Aluminium alloy, 7075-T76	470 (0.2% PS)	72	2.80	1.48

to go plastic.[7] Now, although an aircraft structure must not be allowed to enter the plastic regime in service, regions of plasticity can be envisaged when considering the ultimate strength of the structure, since a small amount of plasticity will not endanger its ultimate integrity. On entering the elastic–plastic regime, the panel buckling strength is diminished, because once the elastic limit is exceeded the effective stiffness criterion is the tangent modulus rather than Young's modulus and the former is not constant but decreases with increasing stress. Because of this, as the panel starts to go plastic its weight increases more rapidly with respect to end load and eventually, if the end load is high enough, aluminium panels become heavier than titanium panels due to the fact that titanium, being an intrinsically stronger material, remains elastic at higher loads than does aluminium (Figure 15.2a). If the panel width is increased the buckling

Figure 15.2 Panel weights for steel, titanium and aluminium. (Taken from Webster[7].)

stress is lower and for a given end load the panel must be thicker and therefore heavier (Figure 15.2b).

Harpur[8] has compared several aerospace materials in terms of buckling stress and structural loading index. He writes:

$$\sigma = 3.62E \left(\frac{t}{b}\right)^2 \quad \text{But } P = \sigma tb \quad \therefore t = \frac{P}{\sigma b}$$

$$\therefore \sigma = 3.62 \frac{E}{b^2} \cdot \frac{P^2}{\sigma^2 b^2}$$

$$\therefore \sigma^3 = 3.62E \left(\frac{P}{b}\right)^2$$

$$\therefore \sigma = 1.5355E^{1/3} \left(\frac{P}{b^2}\right)^{2/3}$$

$$\text{and} \quad \frac{\sigma}{\rho} = 1.5355 \frac{E^{1/3}}{\rho} \left(\frac{P}{b^2}\right)^{2/3}$$

Figure 15.3 shows that the intersection point of the curves for the aluminium alloy 7075-T6 and the titanium alloy Ti–6Al–4V occurs at a structural loading index of about $50\,\text{MN/m}^2$. This figure must be exceeded if titanium alloys are to be competitive with aluminium alloys for wing skins. There are two ways of achieving this:

(1) the width, b, of the titanium panel must be reduced as compared with an aluminium panel; or

(2) the titanium panel must be subjected to a higher loading intensity than the aluminium panel.

Reducing the width of the panel allows titanium to be employed in thinner gauges and the panel is therefore lighter. There is, of course, a limit to the benefits obtainable thereby since reduction in weight is eventually offset by increased weight of stringers and supporting structure. Now a fairly typical end-load, P', for an aluminium alloy panel structure would be about $3.5\,\text{MN/m}$, i.e. $P' = P/b = 3.5\,\text{MN/m}$.[9]

From Figure 15.3 the requirement for Ti–6Al–4V alloy to be competitive with aluminium alloy 7075-T6 is $P/b^2 > 50$, i.e. $3.5/b > 50$ and b must therefore be less than 70 mm. This is hardly a practical condition for conventional skin-and-stringer construction since, with this spacing, riveting is impractical and although narrow panels could be achieved by integral machining from solid plate, the incidental rejection of 80% of the starting material as low-grade swarf is not acceptable with a material as expensive as titanium. Thus, if titanium is to be used, some novel method of construction has to be developed.

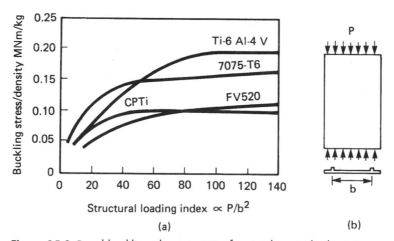

Figure 15.3 Panel buckling characteristics for stainless steel, aluminium and titanium. (After Harpur;[8] reproduced by permission of Royal Society of Arts.)

If end-loading is regarded as a variable, taking 100 mm as a practical minimum value for panel width then $P'/0.100 > 50\,\text{MN/m}^2$ and the end-load P' must be greater than $5\,\text{MN/m}$. This is a higher loading intensity than could normally be justified in civil aircraft.

On the basis of these simple criteria, therefore, titanium would seem to be ruled out as a major aircraft structural material and since similar arguments apply with even greater force to steels, it seems that these ought to be ruled out also.

However, the solid panel construction considered so far is a very simple form of design and it is necessary to consider whether more sophisticated methods of construction would return a different balance of merit. Sandwich structures are prime contenders since these can produce panels which are inherently much more stiff on a weight-for-weight basis than solid panels.[7] Figure 15.4 shows that the higher efficiency of the honeycomb structure can make even stainless steels competitive, and a good deal of honeycomb panel in controlled transformation stainless steels has been used in airframes. The recent development of processes which combine superplastic forming with diffusion bonding has given a fresh impetus to the use of titanium alloys.

Titanium responds particularly well to the diffusion bonding process, since the oxide film which would normally be expected to act as a barrier to bonding appears to dissociate and then dissolve into the base metal at the temperatures appropriate to this process. Stiffened panels in Ti–6Al–4V can be manufactured by diffusion bonding thin titanium parts which are shaped by superplastic forming, and it is estimated that weight savings of around 30% can be achieved in this way (see also p. 252).

One important conclusion to which we are led by the foregoing arguments is that attempts to substitute one material directly for another in a static design are rarely successful. Some element of redesign is always desirable and for best results the suitability of a material should only be considered in terms of designs and manufacturing methods that exploit to the full the potential of that material.

Despite the foregoing, it is unlikely that titanium would ever be preferred to aluminium for skinning purposes if it were not for its superior resistance to temperature. Figure 15.5 shows that titanium maintains its properties to higher temperatures than does aluminium, giving it sufficient temperature resistance for use

Figure 15.4 Weights of honeycomb panels in stainless steel and titanium alloy. (After Webster[7].)

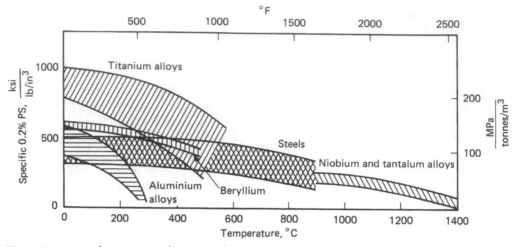

Figure 15.5 Specific 0.2% proof stresses of various alloys at room and elevated temperatures. (Stubbington[10].)

at flight speeds of Mach 3. This is true also of certain of the stainless steels, but, as described later, these are more difficult to form and are also less competitive when density is taken into account.

The present position is that whilst attempts have been made to manufacture whole aircraft skins from stainless steels, they have not been successful, and these materials are now used only locally, e.g. in engine-heated areas, where there is special need for temperature resistance. Similar comment applies to titanium, except that titanium alloys have been used more extensively in certain supersonic military aircraft. Aluminium continues to hold its position as the dominant material for the surfaces of civil aircraft, but is coming under increasing competition from composites. Composite materials already make up large parts of the structures of military aircraft (see p. 254) and are also used in control surfaces in some civil aircraft.

Tension-loaded structures

The compressive loading discussed so far occurs principally in the upper wing surface and the lower regions of the fuselage, especially beneath the wings. In contrast, the lower wing surface and major regions of the fuselage are subject to fluctuating tension and thus prone to fatigue. In the lower wing surface the fatigue stresses are of high frequency whilst the amplitude varies in a random manner. In the fuselage the stresses arise partly from bending but also from repeated pressurization and depressurization, i.e. low-cycle fatigue.

The failure of the two de Havilland Comets, which crashed early in 1954, was shown to be due to low-cycle fatigue. A crack growing outwards from a stress concentration caused by a bolt-hole near the corner of a window caused the passengers' pressure cabin to explode. Because the fracture toughness of the alloy used in the Comet was low the critical crack length at failure was only a few millimetres. The likelihood of failure should have been revealed by testing in the development stage when a prototype cabin was inflated cyclically between zero and service pressure for 18,000 cycles before failure. Unfortunately, the same cabin had previously been pressurized about 30 times to pressures approaching double the cruising pressure and this had artificially raised the fatigue strength of the material. When, after the catastrophes had occurred, testing was repeated under directly valid conditions failure occurred after only 1830 cycles. These disastrous failures established that in fatigue we are not only concerned with the high-cycle end of the endurance curve, where the life is estimated in millions of cycles, but it can also occur in the low-cycle fatigue region after only a few hundred cycles.

After the obvious requirement for minimization of stress concentrations has been dealt with, it is clearly essential to use a material which exhibits a low rate of crack propagation together with a large critical crack length for final fracture and these requirements apply as much to the lower wing surfaces as they do to the fuselage. They apply, in fact, to any structural part loaded in variable tension.

Unfortunately, high fracture toughness is not compatible with high static strength – they are generally inversely related – and it is also true that the increases in strength that have been achieved in highly developed materials have not been accompanied by corresponding increases in fatigue resistance. The modern development of alloys therefore consists of a juggling act in which several different properties must be manipulated so as to achieve an overall optimization in which further improvement in one property can often only be achieved when it is accompanied by deterioration in another. Modifications were made to aluminium 2024–T351 plate that resulted in 2324–T39 plate and 2224–T3 extrusions finding application in the lower wing structure of Boeing aircraft.

One property requirement which frequently sets a limit on the improvement of other properties is corrosion resistance. Corrosion can take many forms, some of which are more difficult to deal with than others. A frequently used method of protecting aircraft against general corrosion is to clad the structural material with a thin layer of corrosion-resistant material. This practice is used for fuselage and wing skins and control surfaces (except, of course, where non-metallic materials are involved). However, the wing planks of many military and civil aircraft are now fabricated by the integral machining of unclad plate and the corrosion resistance and stress corrosion resistance of the alloy used and its protection by effective corrosion protection schemes, e.g. chromate films and paints, becomes again of great importance.

Unfortunately, the mechanical properties of corrosion-resistant materials used for cladding are generally rather poor. For example, aluminium of about 99.3% purity, being soft, has a seriously adverse effect on fatigue resistance when used for cladding. Concorde uses a binary Al–1% Zn alloy but Simenz and Steinberg[11] refer to an aluminium–zinc–magnesium alloy (7008) which, whilst providing a corrosion-resistant surface, can be heat-treated to approach the strength of the base material.

Two other corrosion phenomena of critical importance in aircraft are exfoliation corrosion and stress corrosion. Exfoliation corrosion is corrosion that starts internally at preferred sites, such as grain boundaries, and proceeds along planes parallel to the surface. Corrosion products then exert a mechanical delaminating action, forcing surface metal away from the body of the material. In stress corrosion, a crack is initiated at a corrosive defect in the surface of a metallic part and penetrates into the bulk of the material under the combined action of chemical corrosion and stress. Residual stresses derived from previous processing often provide much of the driving force for stress corrosion. As a broad generalization it can be said that exfoliation corrosion is the main corrosion problem in thin sheet, whereas stress corrosion is a major source of failure in thick plate and forgings. Forgings in heat-treatable materials are especially vulnerable to the development of residual stresses. When a substantial forging of variable section is quenched after solution treatment, the thin sections cool faster than the thick sections and the latter are subject to compressive forces. These, still being at high temperature, yield plastically and in their turn, as they cool, try to contract away from the thin sections. There is a final tendency, therefore, for the thick sections to be in tension and the thin sections, where they embrace the thick sections, to be in compression. Residual stresses are profoundly modified by any asymmetrical machining operations performed subsequent to heat treatment. It should be noted that skin material with integral stiffeners machined from thick plate inherits residual stresses from the parent plate, and must therefore be regarded as subject to stress corrosion. Parts of appreciable section thickness should be designed so as to eliminate as far as possible differential contractions during heat treatment; alloy compositions and heat treatment procedures should be established with a view to minimizing the quench rates necessary; machining should be performed before heat treatment as far as possible; and alloys should be developed with a maximum resistance to crack propagation under conditions where stress corrosion is to be expected.

It is worth emphasizing that aluminium alloy forgings for aircraft can be very large. Turley and Gassner[12] state that in the DC10 fuselage, side frames are machined from basic forgings weighing up to 317 lb (144 kg). Floor beams in the centre fuselage are machined from a basic forging weighing 400 lb (182 kg), whilst the spar frames which provide the load path from the upper vertical stabilizer spar around the tail-mounted engine inlet into the fuselage structure are each machined from a forging of weight 5200 lb (2360 kg).

Landing gear materials

Landing gear assemblies, hydraulic jacks, etc., combine the need for extremely high mechanical properties and total reliability with the ability to stow into small space. The materials selection criterion should then properly be σ_{YS} rather than σ_{YS}/ρ and this means that compensation by low density for deficiencies in other properties is less useful. Since, however, there is always an incentive to minimize the overall weight of the aircraft, the selection of landing gear materials becomes a matter of fine judgment. Table 15.3, from Imrie[13] gives data relating to possible contenders for landing gear forgings. The ultra-high-strength steels are most likely to be selected, aluminium alloys having been largely abandoned in large civil craft. The highest-strength titanium alloys exhibit a σ_{YS}/ρ value only about 20% higher than that for 300M steel, a difference

that cannot justify the increased cost in civil aircraft. Cost may be a less important factor in military craft and titanium alloys were selected for landing gear components in the American B1 bomber.

Given that the necessary static strength can be provided – and this means yield strengths in excess of 1500 MPa (218 ksi) – the principal requirement is to provide adequate resistance to cracking. Selection then requires a proper combination of static strength with fracture toughness and resistance to stress corrosion, hydrogen embrittlement and fatigue, including fatigue crack initiation and propagation.

Service failures are most likely to occur at points of stress concentration where the stress concentration factor, K_t, typically takes values around 3.[13] Because reductions in fatigue crack propagation rates are difficult to achieve in steels it is probably more fruitful to aim at prevention of crack initiation.

Control surfaces

Control surfaces such as the rudder and elevators are rather lightly loaded, and where metals are still used the construction is of the conventional skin and stringer type. Large deflections cannot be tolerated in control surfaces so that the materials employed in these positions must have adequate stiffness. Reinforced plastics allow significant weight savings to be achieved and carbon-fibre-reinforced plastics (CFRP) are prime contenders. CFRP is found in the elevators and rudder in the tail section of the Boeing 777, for example, where it is used with Nomex honeycomb, resulting in a light, stiff structure. The resins used are toughened epoxies. The use of multi-directional lay-up allows control of anisotropy, which can thus be matched to function,[2] and there is also the possibility of using dual reinforcement ('hybrids') in which glass fibres and carbon fibres can be mixed in the proportions necessary to give the stiffness required at any particular point in a structure. In 1994, 11×10^6 kg of composites produced in the US were sold to the aerospace industry.

TABLE 15.3. Nominal composition and tensile properties of undercarriage materials (from Imrie[13])*

Alloy	Nominal chemical composition							Mechanical props			
								0.2% PS		σ_{TS}	
								MPa	ksi	MPa	ksi
Magnesium	Al	Zn	Mn	Zr	Ag	r.e.					
A8 (L121)	8.0	0.5	0.3	–	–	–		77	11	200	29
Z5Z (L127)	–	4.5	–	0.7	–	–		143	21	230	33
MSR–B (DTD.5035)	–	–	–	0.7	2.5	2.5		178	26	245	35
ZE63A (DTD.5045)	–	5.8	–	0.7	–	2.5		188	27	275	40
Aluminium	Zn	Mg	Cu	Cr	Mn	Ag	Zr				
L77	–	0.8	4	–	–	–	–	325	47	417	60
DTD.5024	5.5	2.5	0.5	–	0.5	–	–	420	61	495	72
7075-T6	5.5	2.5	1.5	0.3	–	–	–	430	62	510	74
-T73	5.5	2.5	1.5	0.3	–	–	–	385	56	455	66
AZ74	6.0	2.5	1.0	0.2	–	0.4	–	450	65	510	74
DTD.5120	6.2	2.5	1.7	–	–	–	0.14	450	65	520	75
7050-T736	6.2	2.3	2.4	–	–	–	0.12	420	61	490	71
Titanium	Al	Mn	V	Mo	Si	Sn					
IMI 314	4	4	–	–	–	–		927	134	1004	146
IMI 318	6	–	4	–	–	–		927	134	1004	146
6–6-2	6	–	6	–	–	2		980	142	1080	157
IMI 551	4	–	–	4	0.5	4		1130	164	1270	184
Steel	C	Si	Ni	Cr	Mo	V					
S99	0.40	–	2.5	0.65	0.55	–		1080	157	1240	180
4340	0.40	–	1.8	0.80	0.25	–		1500	218	1790	260
300M	0.42	1.6	1.8	0.80	0.40	0.1		1560	226	1900	276

*Reproduced by courtesy of The Metals Society

The various types of reinforced plastics composites and their properties have been discussed in Chapter 6.

15.3 Requirements for high-speed flight

In flight an aircraft's skin is heated due to viscous shear in the stagnant layer of air contiguous with the surface. In the case of leading surfaces on the wings, and especially at the nose, the heating effect is greater since just ahead of these regions there is a steep pressure gradient. Successive pockets of air become sharply compressed as the plane flies into them, and since this compression occurs under essentially adiabatic conditions the air is heated and the rise in temperature is transmitted to the aircraft skin.

Doyle[14] refers to the unpublished results of Hartshorn, who gives the saturation skin temperatures for an aircraft having an emissivity factor of 0.9 flying at 75,000 ft (23,000 m) where the ambient temperature is –56°C.

TABLE 15.4.

Mach no.	Saturation temperature	
	(°C)	(°F)
2.0	100	212
2.5	150	302
3.0	200	392
3.5	300	572
4.0	370	698

All aluminium alloys lose strength rapidly at temperatures above 150°C (302°F) and the speed of Concorde, the civil supersonic transport, is therefore limited to Mach 2.2. This corresponds to a saturation skin temperature of about 125°C (257°F). For many metals this would normally be regarded as a low service temperature but for the aluminium alloy required to give a service life of 20,000–30,000 hours creep resistance is a major requirement. Clearly, to achieve the speeds envisaged for an Orient Express, a Mach 8 aircraft that could fly from New York to Tokyo in three hours the skin temperatures will be very

high and demand the development of new skin materials.

Another factor which becomes important when structures are subjected to variable heating is thermal stress. If the outside skin of the aircraft heats up whilst the internal structure stays cool, the skin will be put into compression whilst the interior is in tension. It is possible for thermal stresses to be comparable in magnitude to the primary stresses to which they are additive.

15.4 Candidate materials for aircraft structures

The data given in Table 15.5 allow a preliminary assessment to be made of several materials of interest. The figures given are, of course, selected since most materials can be manipulated to give a range of properties.

Of the materials given it is notable that those with the lowest densities are not used – either because, as with wood, other properties are not adequate, or because certain inherent disabilities have been found resistant to improvement.

TABLE 15.5.

Material	Density (tonnes/m^3)	Strength (MPa)	E (GPa)	$\dfrac{\text{Strength}}{\text{Density}}$	$\dfrac{E^{1/3}}{\text{Density}}$	Ref.
Sitka spruce	0.42	39 parallel	9.4 parallel	93	5.0	15
Plywood		23.5 perpendicular	0.7 perpendicular	56	2.0	
Glass-reinforced polyester						
50% glass cloth 0°	1.7	275	17	162	1.5	17
50% glass cloth 45°	1.7	882	9	48	1.2	
70% glass cloth 0°	1.9		24		0.7	
Carbon-fibre-reinforced-epoxy:						
Unidirect. HT	1.5	1600	129	1070	3.4	18
Unidirect. HM	1.6	1280	192	800	3.6	
Aluminium alloys:						
2024–T3	2.77	$\sigma_{YS} = 345$	73	124	1.5	
7075–T6	2.8	0.2% PS = 500	71.6	178	1.48	
Ti–6Al–4V	4.43	0.2% PS = 830	110	187	1.08	
Steel FV520	7.83	0.1% PS = 1081	215	138	0.77	
Maraging steel	7.8	0.2% PS = 1750	190	224	0.74	

Wood

Spruce and birch were widely used for the airframes of the early powered aircraft and even as late as the Second World War plywood construction was giving effective service in the de Havilland Mosquito, a highly developed fighter/bomber capable of flying at 400 mph (650 kph). Murphy[15] states that in 1945, 17% of the entire aircraft production of the United Kingdom was still of wood. Table 15.5 shows that in terms of density-moderated property parameters wood is extremely competitive. However, certain basic characteristics, such as anisotropy of properties, moisture absorption, dimensional instability and vulnerability to various kinds of decay means that effective maintenance is always tiresome and frequently impossible. Gordon[16] has entertainingly described some of the problems encountered with the Mosquito in the Second World War.

Magnesium alloys

Although magnesium alloys are attractive by reason of their low density, applications have been generally limited by cost and by concern over the relatively poor corrosion resistance associated with this metal. However, new high purity alloys have improved inherent corrosion resistance and the necessary protective coatings are available which, together with good design in relation to such aspects as avoiding contact with dissimilar metals, should enable satisfactory service life.

Most castings are in the various Mg-Al alloys containing small amounts of zinc (e.g. A8 [US: AZ81], AZ91 [AZ91]) but the higher strength systems are based on zinc additions (up to 15%) with zirconium (e.g. Z5Z [ZK51]) and, in particular, creep strength, where alloys are of the Mg-Zn-RE-Zr (RZ5 [ZE41]) and Mg-Zn-Th-Zr (ZT1 [HZ32]) types. These alloys also have greater uniformity of properties, freedom from stress corrosion cracking and less microporosity from casting. Room temperature mechanical properties have been further improved to levels approaching those of the high strength aluminium-based casting alloys by alloying with silver, again with rare earths and zirconium. A casting alloy of this type, MSR-B (QE22) was selected for the nosewheel fork in the Anglo-French Jaguar fighter. Heat treatment is, of course, necessary to achieve the optimum properties for many of those sophisticated alloys.

As mentioned, the relatively poor corrosion resistance of magnesium alloys necessitating careful protection systems has always been a disincentive to its use. Recently it has been found that additions of yttrium greatly improve corrosion resistance. The new casting alloy WE54 combines yttrium with rare earths and zirconium to give also good room temperature and elevated temperature properties, good castability and weldability.

Many of the alloy systems used in castings are also available in wrought form, usually as extrusions, and in some cases they are amenable to post-extrusion heat treatment to achieve attractive properties. A recent development has been the weldable Mg-Zn-Cu alloy (ZCM711). The room temperature properties for this alloy are listed in Table 15.6.

These various developments suggest that the use of magnesium in airframe components may increase in future years.

TABLE 15.6. Properties of Mg–Zn–Cu alloy (ZCM 711)[10]

Condition	Yield Stress MPu	Tensile Strength MPa	Elongation %	Density
As extruded	185	275	12	1.83 g/cm^3
As extruded, artificially aged	235	280	9	
As extruded and fully heat-treated	340	365	6	

Steels

The high density of steel – nearly 20 times that of plywood – would seem to make it a poor prospect for use in airframes and yet it has been used quite extensively. In the interwar years it was employed successfully for both spars and skins. However, to minimize weight, sections must be very thin. To provide the necessary stiffness thin spars had to be tubular and often of complex cross-section to resist buckling. This gave rise to considerable manufacturing problems and steel has been little used for spars in the post-war period. For skins, steel has to be corrosion-resistant and the steels with the best corrosion resistance and also the best formability are the straight stainless austenitics of the 18Cr–8Ni type. Unfortunately, these are inherently low-strength materials and the only method of strengthening them is by cold work which severely reduces their formability. At the present time there does not seem to be any proposal to use steels in aircraft skins except in localized regions, commonly near the engines where the combination of engine heat, kinetic heat and acoustic loading is too severe for aluminium alloys. In Concorde, the secondary exhaust system mounted behind each engine nacelle consists of honeycomb panels manufactured from PH15–7Mo stainless steel mounted on a substructure of the same material.[17] Non-structural parts, such as ducts, are made of niobium-stabilized 18/8 type steel.

The small usage of stainless steels in airframes is undoubtedly due to the increased strength of competition from titanium alloys, especially the Ti–6Al–4V alloy fabricated by superplastic forming and diffusion bonding.

Steel continues to be used for undercarriage forgings and these have generally been the high-strength varieties of the conventional low-alloy engineering steels. The steel chosen for the main undercarriage forging in Concorde has the composition and properties:[15]

C	Ni	Cr	Mo	0.1% PS (MPa)	TS (MPa)	%El
0.35	3.8	1.7	0.3	1670	1850	11

This steel is heat-treated by air cooling from 875°C, refrigerated to −70°C to eliminate retained austenite, and then tempered at 200°C.

In this sort of application there is a clear need for the strength to be supplemented with high levels of fracture toughness and resistance to stress corrosion. Simenz and Steinberg[11] consider that advanced alloys should aim at tensile strengths above 1900 MPa together with a fracture toughness around 88 MPa m$^{1/2}$ and a threshold K_{ISCC} around 23 MPa m$^{1/2}$. They are also concerned that available expertise in non-destructive testing is not compatible with the critical crack sizes likely to be encountered in steels at this strength level.

For high-strength forgings, maraging steels possess many advantages including a better resistance to hydrogen embrittlement than is exhibited by most low-alloy steels, good fracture toughness and increased convenience in processing. The strength of maraging steels is developed by age-hardening a low-carbon ductile martensite at 450–500°C. The softness of the martensite

TABLE 15.7. Maraging steels

	Grade of steel		
	140	175	200
0.2% Proof stress (MPa)	1312–1451	1605–1822	1775–2085
Tensile strength (MPa)	1374–1513	1652–1884	1822–2131
Percentage elongation on 4.5 $\sqrt{A}$	14–16	10–12	10–12
Modulus of elasticity E (GPa)	180	186	189
Fracture toughness K_{IC} (MPa m$^{1/2}$)	110–176	99–165	88–143

prior to ageing, and the absence of distortion on ageing, means that most, sometimes all, of the machining can be done prior to final heat treatment, with a consequent reduction in processing costs which goes a long way to offset the increased materials costs.

Haynes[18] describes three grades of maraging steel with properties as listed in Table 15.7. The middle grade has been used extensively in many undercarriage components of the Hawker–Siddeley VTOL Harrier support aircraft. Maraging steel has also been used for the main landing gear axles of the BAC Super VC10 civil airliner. Critical components such as wing-fitting forgings and wing-hinge fittings in swing-wing aircraft are also suitable applications for this material.

Unfortunately, the fatigue properties of as-machined maraging steel components are no better, and sometimes somewhat poorer, than those of comparable low-alloy steels. Surface peening is usually carried out to remedy this deficiency. Again, although the resistance of maraging steels to stress corrosion is superior to that of comparable low-alloy steels, their resistance to generalized corrosion is relatively poorer and they usually need to be protected by cadmium plating followed by baking for 24–48 hours at 200°C.

Maraging steel is however, frequently judged to be too expensive and the ultra-high strength steels, such as 300M, the quenched and double-tempered high-silicon variant of 4340 (see p. 55), continue to dominate the field for landing-gear components. Steels of this type must be processed by vacuum-arc or electro-slag remelting to establish low inclusion counts and high transverse properties. Notched fatigue life may be improved by shot-peening.

Aluminium alloys

Aluminium alloys are still the major materials for airframe construction, at least for civil applications, and seem likely to remain so for the foreseeable future, although the proportion of total take-off weight which they represent will no doubt progressively decrease due mainly to competition from composites. As shown in Table 15.8 in terms of strength/density ratio aluminium alloys can be superior to steel, though not to titanium alloys: in respect of stiffness criteria they are better than both steel and titanium. Nevertheless, the competition between materials is fierce, and the aircraft designer continues to call for higher levels of fatigue resistance, fracture toughness and resistance to stress corrosion and exfoliation corrosion.

Apart from a small number of special cases there are two main groups of candidate aluminium alloys for aircraft. One of these groups consists of the descendants of the first precipitation-hardening alloy, discovered by Wilm in 1911, which contained 3.5% copper, 0.5% magnesium together with silicon as an adventitious but important impurity. Alloys in this group have often been referred to as duralumins. They are grouped together in the 2xxx-series of the Aluminium Association designations and the best known member of the group is the high-magnesium version known as 2024. Generalized compositions and properties, together with approximate UK specifications equivalents, are given in Table 15.8. The aluminium alloy 4-digit designation system developed by the Aluminium Association indicates the aluminium content or main alloying elements by the first number (1 – >99%Al, 2–Cu, 3–Mn, 4–Si, 5–Mg, 6–Mg–Si, 7–Zn, 8–others). The second indicates compositional modifications and the last two the aluminium alloy. Following the alloy designation, a letter denotes the basic treatment or condition (F–as fabricated, O–annealed, H–cold-worked, T–heat treated) with numbers following which relate to details of the treatment or condition. For example, on Table 15.8, 2024–T3 relates to the alloy in which copper is the main alloying constituent which has been solution-treated, cold-worked and naturally aged; 2024–T6 relates to the same alloy solution treated and artificially aged.

The other group comprises the more recently developed Al–Zn–Mg–Cu alloys in which are to be found the strongest aluminium alloys of all: in the American designations this group comprises

TABLE 15.8.

US		UK
2014	4.4Cu–0.4Mg–0.8Si–0.8Mn	L76.L77
2017	4.0Cu–0.6Mg–0.5Si–0.7Mn	L65,L70,L71
2024	4.4Cu–1.5Mg–0.6Mn	
2124	4.4Cu–1.5Mg–0.6Mn (higher purity version of 2024)	
7010	6.2Zn–2.5Mg–1.7Cu–0.14Zr	L95,L96
7075	5.6Zn–2.5Mg–1.6Cu	
7079	4.3Zn–3.3Mg–0.6Cu–0.20Mn	
7178	6.8Zn–2.7Mg–2.0Cu–0.30Cr	DTD 5074
7001	7.4Zn–3.0Mg–2.1Cu	
7050	6.2Zn–2.25Mg–2.3Cu–0.1Zr	
7475	5.7Zn–2.3Mg–1.6Cu–0.2Cr	

	σ_{YS}		σ_{TS}		%El	
	MPa	ksi	MPa	ksi		
2014–T4	290	42	427	62	20	Solution-treated; naturally aged
2014–T6	414	60	483	70	13	Solution-treated; artificially aged
2014–T3 Alclad	276	40	434	63	20	Solution-treated; cold-worked; naturally aged
2017–T4	276	40	427	62	22	
2024–T4	324	47	469	68	20	
2024–T3	345	50	483	70	18	Solution-treated; cold-worked; naturally aged
2024–T6	395	57	475	69	10	
7001–T6	627	91	676	98	9	
7075–T6	503	73	572	83	11	
7079–T6	469	68	538	78	14	
7178–T6	538	78	607	88	10	
7010–T76	484	70	544	79	12	
7050–T76	476	69	545	79	12	
7475–T651	560	81	590	86	12	

the 7xxx series of alloys and the quintessential member of this group is 7075. Data relating to important examples of this class are given in Table 15.8.

The story of the development of aluminium alloys over the greater part of the last few decades has been one of early, and quite successful attempts to increase static strength followed by the realization that this property is not necessarily the one of most importance. A high value of static strength is, of course, always eminently desirable and the development procedures applied to any alloy cannot be allowed to sacrifice it to any considerable degree: never-

theless, almost all of the research effort which has been directed towards aircraft alloys in recent years has been aimed at the enhancement of toughness, fatigue resistance and resistance to the various types of corrosion, even where this has involved some concomitant loss of strength.

It was the perceived need for high static strength that led to the development of the high-strength Al–Zn–Mg–Cu alloys for use in aircraft. These alloys were first investigated in the 1920s but metallurgical complexities inhibited their serious candidature as aircraft alloys until the Second World War. Even after that, when

Figure 15.6 L1011 basic materials and design features. (Taken from Simenz and Steinberg[11].)

their limitations in respect of fatigue strength and resistance to stress corrosion became evident, designers felt obliged to give preference for a time to the older 2xxx series. Then, following intense research and development the pendulum swung again and designers returned to the 7xxx-type alloys but not to the total exclusion of the 2xxx-type alloys so that both types of alloys are still in use (Figure 15.6).

The 2xxx-series alloys

When Wilm discovered age-hardening, the property changes that he observed were the result of natural ageing at room temperature. It was subsequently discovered that significant further increases in strength could be induced by artificial ageing at elevated temperatures. Compare, for example, 2024–T4 and 2024–T6 in Table 15.8. These differences in strength were, of course, accompanied by changes in other properties. In the naturally aged tempers the ductility is satisfactory and the fatigue performance is, for

an aluminium alloy, quite good. However, there is a high susceptibility to stress corrosion and exfoliation corrosion. In contrast, in the T6 and T8 tempers the 2xxx alloys are highly resistant to both stress corrosion and exfoliation corrosion. The changes in properties are even more marked when silicon is present as a deliberate addition, as in 2014.

Unfortunately, 2014 is also subject to stress corrosion in the T6 temper. Despite this, 2xxx alloys in general are more likely to be used in artificially aged tempers. However, the greater strength associated with artificial ageing is not accompanied by any increase in fatigue strength – indeed there may even be a decrease. Similar remarks apply to fracture toughness, because this property is lowered by the presence of second-phase constituents, some of which may derive from the elements added to provide age-hardening strength. Thus, 2024 sheet which has been under-aged will exhibit higher toughness than similar sheet over-aged to the same strength level. In respect of fatigue resistance and fracture toughness, therefore, the naturally aged versions of the standard 2xxx alloys are superior to the same alloys in the artificially aged condition. Considerations of materials selection are therefore complicated by a clear conflict between corrosion resistance and mechanical properties, and efforts have been devoted to optimization of these properties. Thus, in the alloy 2048 the copper content has been lowered so as to eliminate most of the copper-rich intermetallics which are present in 2024.[19] In the T851 condition this alloy exhibits 50% increase in toughness with only a slight decrease in strength as compared with 2024–T8. Hyatt[19] refers to a composition of 3.9Cu–1.55Mg–0.5Mn for achieving highest strength consistent with least amount of copper-rich second-phase particles.

Alloy development has also been directed towards lowering the impurity levels in all alloys with a view to increasing fracture toughness, particularly in regard to anisotropy of this property. The high-purity version of 2024 is 2124, and plate in the T851 condition develops excellent short-transverse properties as well as improved toughness in other directions. According to Staley[20] it is used extensively of the F111 and in the space shuttle. Hyatt considers that in the naturally aged temper, reduction of iron and silicon each to around 0.05 will yield adequate fracture toughness and also decrease the rate of fatigue crack propagation at the higher ΔK levels. Further compositional and processing developments of 2024 have led to the materials 2324–T39 and 2224–T3. By controlling the iron and silicon to low levels, the fracture toughness and fatigue crack growth rate at high levels of stress intensity were improved in 2324, cold worked to 9%. By controlling processing conditions to retain the deformation crystallographic texture after solution heat treatment, 2224–T3 extrusions achieve higher strength. Both these materials are used by Boeing for lower wing structures.

The 7xxx alloys

In the 1940s and early 1950s the aluminium–zinc–magnesium–copper alloys were welcomed by aircraft designers on account of their high strength and the prospect of significant weight savings. Their use was rapidly bedevilled by failures due to stress corrosion, mainly in the short-transverse direction, and also the recognition that their fatigue properties left much to be desired. Stress corrosion is less of a problem in sheet than in thick plate and forgings, whereas fatigue failure can present a problem in all types of product. It was common for a time, therefore, to use 7xxx alloys for upper wing surfaces and 2xxx alloys for lower wing surfaces. For forgings there was a clear need for alloy development. The first step towards solution of this problem was taken with the observation that greatly improved resistance to both stress corrosion and exfoliation corrosion could be obtained by the use of temperatures in excess of 150°C for precipitation treatment.[20] The over-ageing that this represents results in a considerable loss of strength as compared to the T6 condition. This loss in strength was reduced to 10–15% by the subsequently developed T73 temper in which the 150°C over-age is preceded by ageing at 132°C. Forgings and plate 7075–T73 have been in use since the 1960s, and no stress corrosion failures have been reported. Resistance to stress corrosion requires a higher degree of over-ageing than is the case with exfoliation corrosion, and the lower over-ageing given in the T76 temper is aimed mainly at the latter. Higher strength with slightly reduced resistance to stress corrosion can be obtained by taking an intrinsically stronger alloy, 7175 and over-ageing to the strength level of 7075-T6, giving the 7175–T736 material. Some manufacturers have used silver additions of around 0.4% to increase strength but this procedure has not been universally adopted. More recently the problem has been tackled by optimization of composition, leading to alloy 7050 in the USA and 7010 (British Alcan) in the UK, in which advantage is taken of the known effect of copper in increasing resistance to short-transverse stress corrosion and the beneficial effect on quench sensitivity of using zirconium or manganese rather than chromium to inhibit recrystallization. Staley[20] gives the optimum composition as 6.2Zn–2.3Cu–2.25Mg–0.1Zr (7050) and observes that since it exhibits high resistance to exfoliation corrosion it can also be used for sheet – a high-strength cladding alloy has been developed for this purpose. Indeed, thick plate and forged products of 7050–T74 are able to match the tensile properties of 7075–T6, with stress corrosion resistance similar to that of 7075–T73, as well as displaying superior fracture toughness and fatigue characteristics. Sheet and extruded 7050–T76 products also match the strength of 7075–T6 and the exfoliation corrosion resistance of 7075–T76. Hence, 7050–T74 plate and forged products are standard for thick sections in airframe applications, with 7050–T76 also finding application. In fact both 7010 and 7050 are now used extensively in airframe applications and for the Tornado the alloys have equivalent status.

As in the 2xxx alloys, fracture toughness is enhanced by reductions in impurity levels. Alloy 7475 is an example of this effect, and in all tempers this alloy will develop strengths comparable to those of 7075 in similar tempers, but with increased toughness. It has been possible to use alloy 7475 in various T7xx tempers for lower wing skins and fuselages instead of 2024.[21]

Fatigue resistance remains a major problem, especially in the presence of notches: Forsyth[22] quotes the notched fatigue strength of 7050–T73651 as 75.8 MPa (11 ksi) compared with a yield strength of 441.3 MPa (64 ksi). Problems are also encountered with the use of soft cladding materials since the Stage I contribution to fatigue life is then small. This means that interest centres on fatigue crack propagation (FCP) rates. FCP rates in 7xxx alloys are much faster than those in 2xxx alloys (Figure 15.7)[23]

The requirement for high strength materials to reduce weight along with increased fracture toughness and resistance to fatigue cracks led to the development of 7150 alloy plate and extrusions, primarily for use in upper wing structures. While alloy 7150–T6 was used successfully by Boeing, the temper T61 was developed to improve the exfoliation corrosion resistance, and 7150–T61 plate and extrusions are used on the McDonnell-Douglas MD–11 airliner. More recently, the temper T77 has been developed for 7150 in order to maintain the

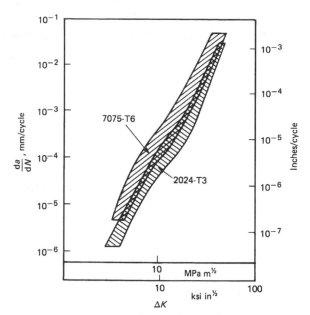

Figure 15.7 Fatigue crack growth rates for 2024–T3 and 7075–T6 (From Hahn and Simon[23])

strength and fracture toughness at the levels of 7150–T6, but to enhance the resistance to exfoliation corrosion and stress corrosion up to that of 7075–T76.

The most recent developments with the 7xxx series of aluminium alloys has resulted in the alloy 7055. Increased understanding of the metallurgical variables controlling fracture toughness focused research on higher solute alloys, where strict attention to the amounts and ratios of the major alloying elements and the thermo–mechanical processing combined to produce 7055–T77 plate and extrusions[24], with nominal composition 8.0Zn–2.3Cu–2.05Mg–0.1Zr. This alloy offers greater compressive strength than its forebears, along with improved fatigue resistance, fracture toughness, exfoliation corrosion resistance and stress corrosion resistance. Boeing have selected 7055–T77 plate and extrusions for the main compressive load bearing components in the upper wing structure of the 777 airliner. Alloy 7055–T77 forgings are also thought to be promising. The progress in aircraft upper wing skin alloys is plotted in Figure 15.8. It may be

Figure 15.8 Upper wing skin plate alloy/temper chronology. (From Staley[24].)

that significant further progress in fatigue resistance cannot be achieved solely from alloy development but that advanced design procedures employing crack stoppers and load-shedding techniques are also required.

Aluminium–lithium alloys and metal matrix composites

There is at present a great deal of interest in the development of alloys in the aluminium-lithium system for use in aircraft, taking advantage of the reduced density to obtain high specific stiffness, but not increasing the lithium content to a level where casting difficulties and embrittlement result. To obtain the required levels of strength with toughness, additions of one or more of the elements copper, magnesium, zinc and silicon are made to give conventional precipitation hardening[25].

With the objectives of a density reduction of 10% as compared to the current 2000 and 7000 series (i.e. a target density of $2.52\,\mathrm{gcm^{-3}}$ as opposed to $2.8\,\mathrm{gcm^{-3}}$) and with a target stiffness of $80\,\mathrm{GPa}$ (10% greater than 2000 series, 15% greater than 7000 series), the most promising alloys commercially available would seem to be 8090 and 8091, with the compositions given in

Table 15.9. 8091 is the higher strength variant and the two alloys now cover a wide range of strengths and tempers[26].

It had been anticipated that the introduction of lithium would reduce the corrosion resistance, but provided the copper content is kept reasonably low (the guide of 1.7% as used in the 7000 series seems appropriate), and the material is homogenized and wrought, the corrosion resistance seems to be equivalent.

Undoubtedly the development of the aluminium-lithium system as a conventional alloy, seen with potential for replacing existing aluminium and other alloys now in use in the aerospace industry, would also lead to its incorporation into fibre-reinforced laminated materials and particulate and fibre-reinforced metal matrix composites. The alloy X2080, containing 0.2 wt%

TABLE 15.9. Composition of selected commercial aluminium-lithium alloy

	Li	Cu	Mg	Zr	Fe	Si	Specific gravity
8090	2.5	1.3	0.8	0.12	0.10	0.05	2.53
8091	2.6	1.8	0.9	0.12	0.10	0.05	2.54

Li, has been developed for use with particulate SiC reinforcement. The composite $X2080/SiC_p$ offers greater stiffness than conventional aluminium alloys and enhanced resistance to fatigue failure compared with other MMC systems. This particular MMC is a candidate for compressively loaded structures in commercial airliners.

Powder metallurgy and rapid solidification

Processing by powder metallurgy and rapid solidification offers the potential for improved material properties, as it can result in a fine stable grain structure, often dispersion strengthened, as well as allow the use of novel alloying elements. The materials available commercially are[24]: forgings and extrusions produced from 7093 powder (9.0Zn–2.5Mg–1.5Cu–0.2O–0.1Zr), displaying high strength and stress corrosion resistance; forgings and extrusions produced by mechanical alloying of alloy 905–XL powder (4.0Mg–1.3Li–1.1C–0.4O–0.1Zr); and sheet, extrusions and forgings for use at elevated temperatures produced from X8019 powder (8.3Fe–4.0Ce–0.2O) or particles of rapidly solidified ribbons of 8009 alloy (8.65Fe–1.8Si–1.3V–0.3O), relying on dispersion strengthening. According to Staley, these aluminium–iron alloys could find application in supersonic commercial airliners, with attractive elevated temperature properties.

Aluminium laminates

In an attempt to combine the attractive properties of both aluminium alloys and polymer matrix composites, researchers at the University of Twente, The Netherlands, developed two laminates, ARALL and GLARE. Layers of aircraft alloy sheet are bonded by layers of adhesive containing either aramid (ARALL) or glass fibres (GLARE). These laminates display superior specific properties to aluminium alloy sheet and are very damage-tolerant, providing excellent resistance to fatigue cracks. Fabrication techniques are similar to those employed conventionally with metals. ARALL has found application in the cargo doors of the C–17 military transporter, while GLARE is a candidate material for jumbo jet fuselages. These laminates are, however, much more expensive than standard aluminium alloys and not as stiff.

Titanium alloys

Titanium alloys were first used in aircraft as engine alloys, but they have subsequently found a progressively increasing use in the aircraft structure. Titanium alloys constitute ~20% of the airframe mass in the Tornado and Grumman F–14 Tomcat, as compared to 5–7% in most aircraft. The supreme virtues offered by titanium alloys for use in aerospace are their abilities, first, to develop high strength-to-weight ratios, and second, to maintain their properties at moderately elevated temperatures. It is the second of these virtues that dominates when titanium alloys are used in gas turbine engines and the alpha or near-alpha alloys have been chosen, since only they possess adequate creep resistance in high-temperature use. The presence of beta phase rapidly degrades creep properties so that alpha–beta alloys can only be used for parts in which resistance to creep is less important.

However, the alpha alloys are non-heat-treatable and solid-solution strengthening by the alpha-stabilizing elements, aluminium, silicon and oxygen, is the only way of increasing tensile or compressive strength. Higher-strength alloys must therefore contain the heat-treatable beta-phase and there are three possibilities: alpha–beta, metastable-beta and stable-beta. By definition, a stable-beta alloy cannot be strengthened by transformation to alpha-phase: hardening could therefore be accomplished only by some other form of precipitation. This has proved elusive and there appears to be no immediate prospect of an established high strength commercial stable-beta alloy.

The metastable-beta alloys offer the highest strength possibilities and should be able to combine this with other significant advantages, the most important being high fracture toughness and deep hardenability. The high-strength metastable-beta alloy Ti–13V–11Cr–3Al has been most widely used – it is capable of strengths in excess of 1500 MPa (218 ksi) and made up 93% of the airframe weight of the military YF12 aircraft.[27] Beta–titanium alloys also have the significant advantages of excellent formability in the solution-annealed, low-strength condition. The principal disadvantage arises from the fact that the beta-phase has an intrinsically low stiffness modulus and this, taken together with the necessarily rich alloying with dense beta-stabilizing elements, means that the stiffness modulus–density ratio is rather low, thus limiting the usefulness of beta-titanium alloys for stiffness-critical components. The tensile ductility is also poor, except in thin sections.[28] It is thus, perhaps, not surprising that the only well-established application of beta-titanium alloys at present is the use of Ti–11.5Mo–6Zr–4.5Sn (Beta III) and Ti–8Mo–8V–2Fe–3Al for fasteners. Even here, the well-known propensity of titanium for galling dictates that nuts running on titanium bolts must be of monel or stainless steel[10] thereby cancelling some of the saving in weight.

However, the great bulk of titanium alloys is made up of alpha–beta alloys and of these the most widely used is the Ti–6Al–4V alloy, variously described as the 'general-purpose' or 'workhorse' titanium alloy which has been in use since its development 40 years ago, with the aluminium offering both strengthening and a density benefit and the vanadium making hot working of the material easier. It is used for aircraft engine discs and compressor blades up to a temperature of about 350°C.[29] Alpha–beta alloys at the lower strength levels are used in the annealed condition but the strength can be enhanced by solution treatment followed by ageing (Table 15.10).

An important advantage for some alpha–beta titanium alloys such as Ti–6Al–4V is the fact that they can be joined by electron beam welding, thus enabling complex structures to be built up with high quality joints over a wide range of thicknesses. The high level of titanium usage in the Tornado (~20%) already mentioned is because the wing carry-through box is an electron beam welded fabrication.

The compositions of the higher-strength duplex alloys must be carefully balanced so that each phase makes its proper contribution to the overall properties. It is, of course, possible to influence the relative proportions of alpha and beta phases by heat-treatment procedures. Alpha-stabilizing elements are usually present to provide solid solution strengthening of the alpha phase. This strengthening mechanism is not available to any significant extent in the beta phase but the composition must contain sufficient beta-stabilization to provide strength from transformations within the beta.

The main disadvantages of the heat-treatable alpha–beta alloys are, first, their poor deep-hardening characteristics (in the high-strength heat-treated condition Ti–6Al–4V is limited to a maximum section of only 25 mm) and, second, their poor fracture toughness, which can be low enough for critical cracks to be only marginally detectable by non-destructive methods[27] Optimization of properties by microstructural control is difficult and careful control of processing is

TABLE 15.10[30]

Ti–6Al–4V	0.2% PS		TS		%El
	MPa	ksi	MPa	ksi	
Annealed, 700°C	925	134	990	144	4
Solution treated and aged, 500°C	1100	159	1170	170	10

TABLE 15.11.

	0.2% PS		TS		%El
	MPa	ksi	MPa	ksi	
Ti–4Al–2Sn–4Mo–0.5Si (IMI 550), Solution-treated and aged	1000	145	1100	159	14
Ti–4Al–4Sn–4Mo–0.5Si (IMI 551), Solution-treated and aged	1200	174	1310	190	13

necessary if unacceptable scatter in properties is not to result. Even different specimens from the same batch of material may yield widely different test results. These problems have led to the development of specialized processing techniques, such as isothermal forging, to allow better control of properties.

Nevertheless, certain generalizations are possible. Thus, according to Morton,[28] the fracture toughness can be improved by solution treatment high in the alpha-plus-beta field to give a low proportion (10–25%) of primary alpha. On the other hand, the best fatigue properties are obtained when the alpha-plus-beta structure is fine-grained but with a relatively high proportion of primary alpha.

Titanium alloys in airframes can be used in two main forms: forgings and sheet. There are two principal ways of producing a bulky component of complex shape. One is by machining from a solid billet the other is by forging. The emergence of numerically controlled machining, by which highly complex parts could be produced rapidly at a low level of labour intensity, has ensured a degree of popularity for the first of these options in many applications. However, as materials costs increase there arrives a break-even point which is determined by the balance of costs between, on the one hand, the material buying-in cost, the scrap value of the swarf, machine-time costs, machine tool depreciation, costs of tapes, and on the other hand die costs, costs of final machining and rectification.[3] Titanium is intrinsically an expensive material which also suffers an exceptionally large depreciation in value upon conversion into swarf. The break-even point, above which forging is preferred, therefore occurs

at quite small order sizes (Figure 15.9). There is thus considerable incentive for high-strength forging alloys and improved forging methods to be developed, and there has been considerable success. Thus, Ti–4Al–4Mo–2Sn–0.5Si (IMI 550) and Ti–6Al–4V (IMI 318) are used in the Tornado, Jaguar and Harrier combat aircraft, for applications such as flap tracks and brackets[30] High-strength forgings are also used for engine brackets and landing-gear components.

The situation with sheet is rather more difficult, because the decision to manufacture the entire skin of an aircraft from titanium must be taken in more of a pioneering spirit than fitting in a few forgings – the incentive to develop high-strength titanium sheet alloys has therefore been correspondingly less intense. The early applications for titanium sheet involved regions of the

Figure 15.9 Relative costs for producing flap tracks by three different methods. (From Page[3].)

aircraft subject to heat, such as exhaust shrouds and firewalls, and the materials used were titanium–oxygen alloys, often misleadingly referred to as commercial-purity titanium (Table 15.12).

For more comprehensive skinning the alpha alloy Ti–8Al–1Mo–1V was widely canvassed at one time and was used for much of the skin of the A11 supersonic aircraft[31] It derives its strength solely from solid-solution hardening. The alloy Ti–2.5Cu (IMI230), the only titanium alloy based on a classical precipitation-hardening reaction, has also been produced in sheet form, and has been used in the Tornado and Jaguar combat aircraft and also in Concorde (Table 15.13).

The principal disadvantage of these sheet materials (and also of Ti–6Al–4V when it is used in sheet form) is the very limited formability which they exhibit at room temperature. For the production of any shape of more than moderate complexity hot pressing at temperatures around or above 600°C is called for.

Attempts have been made to develop the metastable beta alloys as high-strength sheet alloys. These have the advantages of good cold formability in the solution-treated condition and high strength, developed by subsequent ageing.

The beta sheet alloy, Ti–15V–3Cr–3Al–3Sn, originated for the US Air Force by Battelle and Lockheed, is said to be highly formable. Unfortunately, the metastable beta alloys suffer from a low elastic modulus, high density and are difficult to weld (although the older Ti–13V–11Cr–3Al, one of the strongest alloys available, is said to be reasonably weldable and was used for the combat YF12 aircraft). Nevertheless, because of the considerable technical difficulties involved, and above all the expense, it seems unlikely that high-strength sheet alloys will be developed for large-scale use in aeroplane skins.

Superplastic forming (SPF) has become an important means of fabrication for appropriate titanium alloys and, coupled with diffusion bonding for joining, has led to a considerable reduction in the number of separate pieces with many less fasteners in constructions and thus significant saving in weight.

The excellent superplastic forming and solid state diffusion bonding capability of Ti–6Al–4V have made it particularly suitable for the RR Trent 800 hollow wide chord fan blade, where two external skins are separated by either a honeycomb filler, or in the second generation, by a superplastically formed corrugation, which carries a share of the centrifugal load. Both

TABLE 15.12.

	0.2% PS		TS		%El
	MPa	ksi	MPa	ksi	
Ti–0.13 O$_2$ (IMI 125), annealed	360	52	460	67	28
Ti–0.28 O$_2$ (IMI 155), annealed	540	78	640	93	24

TABLE 15.13.

	0.2% PS		TS		%El
	MPa	ksi	MPa	ksi	
Ti–2.5 Cu (IMI 230), solution-treated and aged	620	90	790	115	24
Ti–8Al–1Mo–1V	870	126	965	140	Low

panel–panel and cone–panel bonded joints must achieve parent material properties in terms of impact resistance and fatigue.[29]

Polymer matrix composites

Composites are being used in increasing quantities in airframe applications. These generally consist of a matrix of epoxy-resin reinforced by fibres of boron, carbon or aramid. Sometimes more than one type of fibre may be incorporated into the same composite, producing a so-called hybrid. The properties of advanced composites are discussed in Chapter 6. Composites were first used in aircraft for lightly loaded parts. Thus, for cabin floors it is possible to use aluminium alloy facing sheets over a honeycomb core but a plastics composite core between glass-fibre facing sheets is said to provide better

Figure 15.10 Sections 46 and 47 of the Boeing 777 fuselage with floor beams in place. (Courtesy The Boeing Company.)

resistance to trolleys and stiletto heels[32] Control surfaces were the next areas to which composites were applied and the success of epoxy-carbon materials in these locations gave encouragement to their application to more demanding situations such as wing-fuselage fairings, engine doors and eventually to major structural parts. The L1011 Tristar employs composites for fairings and trailing edges of rudder and elevators (Figure 15.6) As pointed out previously, the elevators and rudder in the empennage of the Boeing 777 are manufactured from CFRP, where weight savings of up to 20% over conventional aluminium alloys are achieved. The floor beams, which need to resist corrosion due to water exposure from galleys and toilets, are also manufactured from CFRP on the 777. A toughened epoxy resin was necessary to provide the required damage tolerance (see Figure 15.10). Other composite applications on the Boeing 777 include secondary structures such as aerodynamic fairings. In fact, including resins and adhesives, composites account for 9% of the structural weight of the 777, which compares to about 3% on the 757 or 767.

Extensive use of polymer matrix composites is made in the McDonnell Douglas AV8B Advanced Harrier. Except for leading edges, tips, pylon, under-carriage attachments and centreline rib, the wing is manufactured entirely from carbon-fibre-epoxy composite giving a weight saving of 330 lb (150 kg)[33] Composites are also used in the tailpiece and front fuselage as carbon-fibre-epoxy sandwich (Figure 15.11). This developing use of composites continues in even more recent military aircraft.

Figure 15.11 Usage of carbon-fibre composite material in the AV-8B Advanced Harrier. (From Warwick[33])

Hybrids are employed sometimes where local variation of properties is required, perhaps in the vicinity of attachments or stress concentrations. The impact properties of carbon-fibre composites can be improved by addition of aramid and glass-fibres and the stiffness of GRP may be increased by the addition of carbon-fibres.[34] The horizontal stabilizer of the B1 bomber is principally a graphite structure but boron fibres have been added to the skins to provide additional stiffening and strength.

An interesting point made by Kelly[35] is that the natural anisotropy once regarded as a disadvantage is now being very imaginatively used to control dynamic and aeroelastic response. Helicopter rotor blades suffer from forced vibrations at particular rotor frequencies. The transmission of vibrations to the rest of the structure can be modified by varying the material's stiffness along the blade by altering the composite structure. The main rotor blade in the Westland EH 101 helicopter consists of a hybrid carbon–glass fibre spar followed by a Nomex honeycomb support for the trailing edge, with glass woven skins at ±45°. Compared to standard blades, the composite structure has improved damage tolerance, with longer life and fewer unscheduled repairs.

Major developments are likely to come in the future from the use of thermoplastic matrices for composites. These have some ductility and give more reproducible strength and toughness. They are also resistant to water and solvents whereas epoxy resin absorbs moisture over long periods of time. They are at present being researched for both fixed wing and helicopter applications – a typical example is PEEK (polyether ether ketone).

The principal cause for caution associated with the use of composites for primary structures was the poor understanding of crack initiation and propagation so far achieved in these materials, so that it was quite difficult to design for them realistically in damage-tolerant terms.[2] However, experience to date has engendered confidence in their fatigue resistance and there has been very marked growth in the rate at which their employment in aerospace has increased.

References

1. I. L. G. BAILLIE and W. P. C. SOPER: Proc. Conf. 'Forging and Properties of Aerospace Materials'. The Metals Society, 1977.

2. W. T. KIRKBY, P. J. E. FORSYTH and R. D. J. MAXWELL: *Aeronaut. J.*, January 1980.

3. F. W. PAGE: *J. Inst. Metals*, 1967; **95**, 65.

4. SIR G. EDWARDS: *Met. Mater.*, 1970; **4**, 323.

5. K. SABBAGH: *21st Century Jet*. MacMillan, London, 1995.

6. R. J. H.WANHILL and G. F. A. VAN GESTEL: Proc. Conf. 'Aluminium Alloys in the Aircraft Industry.' Associazione Italiana Metallurgica, Turin, 1976.

7. D. WEBSTER: *Met. Mater.*, August 1971.

8. N. F. HARPUR: 1978 Chester Beatty Lecture, Royal Society of Arts.

9. T. W.COOMBE: Private communication.

10. C. A. STUBBINGTON: private communication.

11. R. F. SIMENZ and M. A. STEINBERG: in *Fundamental Aspects of Alloy Design* (eds R. L. JAFFEE and B. A. WILCOX). Plenum Press, 1975.

12. R. V. TURLEY and R. H. GASSNER: *Met. Prog.*, March 1972.

13. W. M. IMRIE: Proc. Rosenhain Centenary Conference. The Royal Society/The Metals Society, 1975.

14. W. M. DOYLE: *J. Roy. Aeronaut. Soc.*, 1960; **64**, 535.

15. A. J. MURPHY: 1972 Chester-Beatty Lecture. *J. Roy. Soc. Arts.*, October 1972.

16. J. E. GORDON: *The New Science of Strong Materials*. Penguin Books, 1968.

17. E. A. HYDE: private communication.

18. A. G. HAYNES: *J. Roy. Aeronaut. Soc.*, 1966; **70**, 766.

19. M. V. HYATT: Proc.Conf. 'Aluminium Alloys in the Aircraft Industry'. Associazione Italiana Metallurgica, Turin, 1976.

20. J. T. STALEY: *Met. Eng. Q.*, May 1976.

21. R. J. H. WANHILL and G. F. J. A. VAN GESTEL: Proc Conf. 'Aluminium Alloys in the Aircraft Industry'. Associazione Italiana Metallurgica, Turin, 1976.

22. P. J. E. FORSYTH: Proc. Rosenhain Centenary Conference. The Royal Society/The Metals Society, 1975.

23. G. T. HAHN and R. SIMON: *Eng. Fract. Mech.* 1973; **5**, 523.

24. J. T. STALEY: Aluminium Alloys in: *Encyclopedia of Advanced Materials.*, eds D. BLOOR, R. J. BROOK, M. C. FLEMING and S. MAHAJAN. Pergamon Press, 1994.

25. C. J. PEEL, B. EVANS and D. S. MCDARMAID: *Met. Mater.*, August 1987, 3(8), 449–455.

26. C. J. PEEL, B. EVANS and D. S. MCDARMAID: D. Westec Symposium on Al–Li Alloys, Los Angeles, March 1987 ASM.

27. F. H. FROES, R. F. MALONE, J. C. WILLIAMS, M. A. GREENFIELD and J. P. HIRTH: Proc. Conf. 'Forging and Properties of Aerospace Materials'. The Metals Society, 1977.

28. P. H. MORTON: Proc. Rosenhain Centenary Conference. The Royal Society/The Metals Society, 1975.

29. S. MILLER: *Interdisciplinary Science Reviews*, 1996, **21**, 2.

30. T. W. FARTHING: *Metallurg. Mater. Technol.*, August 1977.

31. M. G. MANZONE: *Titanium Alloys with Molybdenum*. Climax Molybdenum Company, 1966

32. J. M. RAMSDEN: *Flight Int.*, 19 July 1980.

33. G. WARWICK: *Flight Int.*, 29 December 1979.

34. J. SUMMERSCALES and D. SHORT: *Composites*, July 1978.

35. A. KELLY: Phil Trans, Series A, 1987, 322, 102

16
Materials for ship structures

A ship is a carrier, constructed as a hollow shell designed to move its payload at suitable speeds from one place to another. The shape and characteristics of the ship are determined mainly by the nature of the materials to be carried (which include liquid and solid cargo, passengers, or the means to conduct warfare), but also to some extent by the temperament of the waterway involved.

For maximum structural efficiency a vessel should be spherical, and this shape is commonly used for static containers. For transport purposes a more convenient shape is the cylinder and moving vessels which are totally submerged in the fluid medium which supports them, such as airships and submarines, approximate to a cylindrical shape.

However, surface ships operate at the interface between two media of differing density (water and air), and are therefore subject to the interaction between them – determination of the most suitable shape is therefore a complex matter for which there is no comprehensive theory. The shape in the forward part of a ship must take account of the need to cut through rough water and yet have sufficient buoyancy to cope with the action of waves. The major part of the hull must have good carrying capacity but must also be shaped to provide inherent resistance to rotational instabilities (rolling and pitching) since ancillary stabilizing devices not only reduce speed through the water but also have limited effectiveness. The shape can be assessed approximately in terms of the block coefficient, the ratio of the actual volume of the hull below the waterline to that of the rectangular box having the same principal dimensions of length, depth and breadth. Passenger liners may have a block coefficient as low as 0.55, whereas the corresponding figure for a tanker could be 0.85. Thus, for the same principal dimensions, faster

ships tend to have a smaller volume available for the payload. In some circumstances this can be acceptable because, although it is normally desirable to maximize payload in relation to overall weight, it is sometimes economic to carry a smaller load if it can be carried faster and thereby catch a market ahead of competitors. This was the case with the tea clippers, sailing ships designed to carry a rather small cargo of tea as fast as the wind could be made to take them. However, sailing ships pay nothing for their motive power; in contrast, powering modern commercial cargo vessels requires expensive fuels, where maintaining high speeds lowers efficiency of fuel use. Speed therefore gives way in importance to carrying capacity and hull design differs accordingly.

16.1 The ship girder

For present purposes a ship can be considered in two parts, hull and superstructure. Of these two the hull is immeasurably the more important.

The hollow shell making up the hull of a ship is constructed from areas of plate, some flat and some curved, joined together so as to define an enclosed volume which is large enough to provide the required buoyancy and of a shape to ensure good stability and handling. Attached to the plating are long steel sections which confer stiffness and resistance to buckling.

The moulded hull, together with its decks, stiffeners and bulkheads, makes a structure which can be likened to a box girder and the whole structure is referred to by naval architects as the 'ship girder'.[1]

The ship when afloat is considered to be in equilibrium under the action of two sorts of forces: (1) the downward-acting forces of gravity due to the total mass of the ship, and, opposing these, (2) the upthrusts from the water due to the buoyancy of the immersed parts of the hull. In total, these two forces will be equal but their distributions along the length of the hull are not. Thus a floating hull sustains a bending moment and bends longitudinally like a beam.

The major longitudinal bending stresses are resisted by the deck and bottom structures, performing the same functions as the flanges of an I-beam. The sides of the hull hold the top and bottom sections together in much the same way that the web of an I-beam connects the two flanges, thus completing the analogy (Figure 16.1).

Figure 16.1 The hull of a ship regarded as an I-beam (after Watson and Gilfillan[2]). A_D = area of deck plating plus deck longitudinals plus other longitudinal material above 0.9D. A_B = Area of bottom plating plus bottom longitudinals plus other longitudinal material below 0.1D. A_S = Area of shell plus area of longitudinal bulkheads plus longitudinals between 0.1D and 0.9D.

The longitudinal stresses in the top and bottom structures may be calculated from the standard elastic beam formula $\sigma = My/I$ where M is the bending moment sustained by the hull at the point in question, I is the second moment of area of the section resisting bending and y is the distance of that section from the neutral axis.

As a shell, the hull is liable to buckle. Flexure in a seaway loads one or other of the top and bottom structures of the ship girder in compression. Such a plated structure provided with periodic stiffeners and subject to in-plane compressive loading can be considered as an assembly of panels loaded at their ends. As described in Chapter 8, the critical load for failure by buckling of a panel is determined by the thickness of the plating, the spacing of the stiffeners, and Young's modulus for the materials. The thinner the plating, and the further apart the stiffeners, the lower will be the failure load. The likelihood of buckling occurring in top or bottom sections varies according as the hull is longitudinally or transversely framed. Muckle[3] states that in steel ships longitudinal buckling has not presented a problem where longitudinal framing is employed. Transversely framed ships are much more vulnerable to longitudinal buckling and with materials of low elastic modulus, such as glass-reinforced plastic (GFRP), the design of the main structure may be dominated by the need to prevent buckling.

Due to the mismatch between the longitudinal distributions of gravity forces and buoyancy, the ship girder is subject to considerable stress even when the ship is at rest in still water. Frequently, especially with cargo ships, the hull is then subject to a 'hogging' bending moment; i.e. the deckplates are in tension and the bottom is in compression. This is demonstrated in the well-known photograph of MV *Schenectady* – the wide separation of the fractured deckplates indicates that the vessel was hogging just prior to failure. On the other hand, tankers tend to sag.

In a seaway the bending stresses vary as the ship passes through successive waves due to progressive changes in the buoyancy distribution. The naval architect assumes that the worst situation occurs when the ship meets a wave of length equal to that of the ship. Two extreme conditions of loading can then exist when either (1) the crest of the wave is amidships, or (2) there is a crest fore and aft and a hollow amidships (Figure 16.2). Condition (1) produces maximum hogging moment, whilst condition (2) produces maximum sagging moment. This transient effect due to the wave is algebraically additive to the still-water condition so that a ship which hogs strongly in still water may never attain the sagging condition, although it would be unwise to rely on this. This subject is covered in detail in *Muckle's Naval Architecture*.[3]

Figure 16.2 Loading of a hull in a seaway: (a) hogging: (b) sagging on a wave. (From Rawson and Tupper[1]. with permission.)

16.2 Factors influencing materials selection for ship hulls

Deflections

The deflections that occur in a ship's hull must be kept within bounds and Classification Society Rules provide for this. One of the parameters used in ship design is the ratio L/D, where L is the length of the ship and D is the depth, i.e. the distance between top and bottom structures. For a given bending moment, the greater the L/D ratio, the larger the deflections. If the L/D ratio becomes very large then the dimensions of top and bottom sections may be determined by the need to provide stiffness rather than strength.

Muckle[3] has given a method for estimating the effect on deflections of substituting one material for another. If a load W acts on a beam of length L then the deflection of the beam, δ, is proportional to WL^3/EI, where E is Young's modulus

and I is the second moment of area of the cross-section. Then comparing a ship made of a material of Young's modulus E_1, with another ship of which Young's modulus is E_2:

$$\frac{\delta_1}{\delta_2} = \frac{W_1L_1^3}{W_2L_2^3} \cdot \frac{E_2I_2}{E_1I_1}$$

If the two ships have the same principal dimensions, the same load distribution and any difference in hull weight is transformed directly to the cargo, then

$$\frac{\delta_1}{\delta_2} = \frac{E_2I_2}{E_1I_1}$$

and since the total load and its distribution are the same it follows also that the longitudinal bending moment is the same and therefore

$$\frac{I_2}{y_2} = \frac{\sigma_1}{\sigma_2} \cdot \frac{I_1}{y_1}$$

where σ is the longitudinal fibre stress and y is the distance of the fibre stress from the neutral axis. Since y is the same for both ships $(I_2/I_1) = (\sigma_1/\sigma_2)$ and therefore

$$\frac{\delta_1}{\delta_2} = \frac{E_2}{E_1} \cdot \frac{\sigma_1}{\sigma_2}$$

Table 16.1 shows the deflection for some materials taking ordinary mild steel as the standard for comparison. The 20% increase in deflections of the ship in higher tensile steel is acceptable, and this material is commonly used

TABLE 16.1.

Material	E_1/E_2	σ_2/σ_1	δ_2/δ_1
Higher tensile steel	1	1.2	1.2
Aluminium alloy – N8	3	0.6	1.8
GRP Chopped mat 30% glass	20	0.5	10.0
GRP Woven cloth 50% glass	15	1.0	15.0
CFRP (estimated figures)	2	2.7	5.4

within the limits set by Classification Society Rules. The determination of the scantlings of ships' structures has been traditionally based on the use of mild steel. If higher tensile steel is employed, the calculated section modulus must be multiplied by a correction factor. Lloyd's Rules give this as $k = 245/\sigma$ (where σ is the yield stress of the higher tensile steel in MPa units) k is equal to unity for mild steel and is not permitted to be less than 0.72. The lower limit for k also shows that there is no benefit to be obtained from using steel of yield stress higher than 340 MPa, since the Rules would not allow its full strength to be utilized. Although the Rules provide for L/D ratios of 16.95 or above, these do not seem to be employed. According to Watson and Gilfillan[2] tankers allow the largest L/D values because of their favourable structure, with longitudinal bottom framing, longitudinal bulkheads and minimum of hatch openings. However, their data do not include any examples of L/D ratios significantly greater than 14.

Aluminium alloy hulls must be expected to exhibit greater deflections than steel ships, although not impossibly so. Up to the present the cost of aluminium has prevented its use in the main ship girder of large cargo vessels, so the point has not been tested. Aluminium has been used for smaller vessels and special-purpose craft such as hydrofoils in which the low density confers a special advantage.

The deflections shown in Table 16.1 for GRP hulls are impossibly large. Although these could be reduced by increasing the glass content of the GRP it is difficult to obtain satisfactory resin distribution when laying up by hand composites with glass contents much higher than 50%, and even if this were done the deflections would still be too high. It is clear therefore that if deflections in glass fibre hulls are to be reduced to acceptable values, material thicknesses must be far greater than are required by considerations of strength if buckling and excessive deformations are to be avoided. As with aluminium, GRP has not been used for large cargo vessels (although there have been design studies), but it is the most important boat-building material for craft up to 12 m (40 ft) in length and has been used for fishing boats up

to 30 m (100 ft).[4] The extent to which additional material is used to reduce deflections in GRP craft depends upon the field of application. The desire to reduce weight in pleasure craft often leads to local deflections in single-skin GRP structures that are large enough to induce sinking feelings in the inexperienced mariner even though strength may be adequate for the purpose. Larger commercial and naval craft, subject to more arduous service, must be much more substantially skinned. For stiffness and impact resistance, a sandwich construction, for example multi-axial glass skins over a balsa core, is more appropriate if using GRP. In the UK, the Royal National Lifeboat Institution (RNLI) Severn Class lifeboat is manufactured from a single-skin prepreg epoxy laminate for the hull bottom, with top sides, deck and superstructure constructed from stiffer epoxy prepreg/PVC foam core/ epoxy prepreg sandwich.

It is likely that the longitudinal structure of large GRP craft will be determined by buckling criteria because of the low elastic modulus of GRP composites. In transversely framed ships the most practical way of meeting this problem is to thicken up the plating,[5] but in longitudinally framed vessels the possible failure modes are strut-type and panel-type buckling and these may be countered by enlarging the longitudinal stiffeners or decreasing the spacing between them, with the penalty of increased weight.

Weight

The total weight of a ship is made up of four principal components:

> structure;
> propulsive machinery;
> fuel, stores, crew, etc.;
> cargo.

The first two of these components represent capital expenditure which must be repaid from the profits earned from carrying the cargo. It is therefore desirable to maximize the payload as a proportion of total weight since this will reduce the number of voyages needed to repay the initial investment. Economic advantages therefore result if reduction in hull weight can be transferred directly to cargo. Cole and Colvin[6] state that, as compared with mild steel, the use of higher tensile steel in a tanker of deadweight 250,000 tons can reduce the hull weight by 1000 tons (deadweight comprises weight of cargo together with fuel, stores, fresh water, etc.).

A quantitative basis for estimating the weight savings potentially available from the substitution of one material for another can be obtained from the method given by Muckle.[3] Assuming that the bending moment sustained by the ship girder is unchanged by the material substitution, then

$$\frac{I_2}{y_2} = \frac{\sigma_1}{\sigma_2} \cdot \frac{I_1}{y_1}$$

For a box girder type of structure the section modulus can be taken as proportional to the cross-sectional area of the longitudinal material which can be averaged out over the beam of the ship and expressed as an average thickness, t. Then $t_2 = t_1(\sigma_1/\sigma_2)$. The weight of material in the longitudinals, W, is given by

$$\frac{W_2}{W_1} = \frac{\sigma_1}{\sigma_2} \cdot \frac{\rho_2}{\rho_1}$$

where ρ denotes density.

Table 16.2 shows potential weight savings for some materials as compared with mild steel. These savings could not, however, be obtained in practice. The figure for higher tensile steel is the most realistic, provided the material and the L/D ratio of the ship are compatible within the requirements of Classification Society Rules. Even in this case some of the potential weight saving would be lost because a smaller spacing between stiffeners would be required to prevent buckling of the thinner plating.[3]

The use of aluminium alloy for the hulls of large cargo ships is technically feasible, but probably uneconomic. Buckling should not be a problem because of the thicker plating needed to keep the longitudinal fibre stress within the capability of a low-strength material. As previously stated, whether or not large deflections could be accepted in practice is for future determination, since so far no large aluminium hulls have been built, even though the American Bureau of Shipping published Rules for aluminium hulls in 1975.

The weight savings indicated in Table 16.2 for composite hulls are unattainable: the low modulus of elasticity ensures that in hulls designed purely on the basis of strength, overall deflections would be too high and resistance to buckling too low. Correction by larger sections would result in heavier hulls; modern design methods are capable of providing numerical guidance regarding buckling resistance but the level of tolerance towards overall deflections is entirely a matter of experienced judgement. It is likely that the lightest GRP ship capable of satisfying the Rules would be heavier than the equivalent steel ship. The situation is technically better with CFRP because of the much higher modulus. If 50% increase in deflections were permitted as compared with the equivalent mild steel ship then

$$\frac{W_2}{W_1} = \frac{\sigma_1}{\sigma_2} \cdot \frac{\rho_2}{\rho_1} = \frac{1}{1.5} \cdot \frac{E_1}{E_2} \cdot \frac{\rho_2}{\rho_1} = 0.3$$

Further development and cost reduction will, however, be needed before the carbon fibre reinforced material can be envisaged as a major hull material.

Not all types of ships necessarily benefit from reduction in hull weight–it depends upon whether the ship is a weight carrier or a capacity carrier. The latter carries a cargo that stows at

TABLE 16.2.

Material	σ_1/σ_2	ρ_2/ρ_1	W_2/W_1
Higher tensile steel	0.8	1.0	0.8
Aluminium alloy – N8	1.7	0.35	0.6
GRP Chopped mat 30% glass	2.2	0.19	0.4
GRP Woven cloth 50% glass	1.0	0.2	0.2
CFRP (estimated figures)	0.37	0.2	0.07

low density so that the ship has to cope with volume rather than weight. Little benefit is then obtained from reduction in hull weight since the working draught (depth of immersion) can only vary within fairly narrow limits if seakeeping qualities are to be maintained; there is little point in reducing hull weight if the additional cargo required to pull the hull down to a satisfactory working draught cannot be accommodated within the hull volume (cargo can be stowed on deck above the waterline, but there is a limit to this from decreased stability and increased wind resistance). An LNG (liquefied natural gas) carrier, with a cargo which stows at a specific gravity of about 0.4, is a typical capacity carrier,[7] as is a container ship, which may carry much of its cargo on deck. For container ships the ratio of deadweight to displacement is about 0.6 but for passenger liners the much lower figure of 0.35 is typical.[1] At the other extreme, a typical weight carrier such as a tanker or an ore carrier would have a deadweight/displacement ratio of about 0.8. Reduction in structural weight of weight carriers will always be beneficial, the more so as these tend to be the larger types of ships.

Although reduction in structural weight is the commonest justification for using higher tensile steel in ship structures it is not the only consideration. In very large ships it may be difficult to provide the required longitudinal strength with mild steel since the scantlings become very large and, as previously stated, the reduction in thickness which results from the use of higher tensile steel is beneficial in relation to brittle fracture.[8]

Maintenance and repair

A serious disadvantage of all ship steels is their poor resistance to corrosion. Because of this it is necessary to make an allowance in the original construction for progressive loss of thickness during service life and it is also necessary to take the ship out of service for periodic repainting. Below the water line corrosion can be virtually eliminated by cathodic protection (see Chapter 11). Repair of modern steel ships involves fusion welding which can be done only by skilled personnel, although most shipyards can comply.

The most suitable aluminium alloy for shipbuilding is the binary aluminium – 4.5 Mg alloy (N8 in BS1477) and, since this is quite corrosion-resistant, painting should not be essential although it may be done for purposes of decoration. Aluminium alloys are more easily workable than steel and, since the N8 alloy is non-heat-treatable, temporary repairs are more readily carried out. However, the remarks made previously about fusion welding apply.

Composite materials are not normally regarded as vulnerable to corrosion in the usual sense. However, GRP is known to absorb water during prolonged immersion, especially if the gel coat is softened with paint removers – significant loss of strength and stiffness can then occur. By an osmotic process, water builds up in the laminate, resulting in blisters. Repairs to GRP structures are much more readily carried out than is the case with metallic materials and a high level of operator skill is not required. However, Gardner[9] has referred to health and safety problems and the difficulty of providing control of environmental temperature and humidity.

The principal advantage of concrete and ferrocement for ships' structures is the total absence of any need for maintenance – painting is quite unnecessary and repairs are also easily effected. According to Stamford,[7] additional to the reduced cost of maintenance is the corollary that concrete ships may be expected to operate for 353 days per year compared with 340 days for the equivalent steel ship.

Fire risk

The yield strength of steel is highly temperature-sensitive, falling to one-half its room-temperature value at 500°C. On-board fires could therefore be expected to cause considerable buckling and distortion of load-bearing members but would probably not pose a real threat to the ultimate integrity of the ship. With slightly different detail, a similar conclusion could be drawn about concrete.

Aluminium presents a more serious fire hazard, being of much lower melting point. Although not used for large hulls, aluminium has found widespread use in superstructures, not only of passenger liners, for which it became the standard material, but also in warships. It has always been appreciated that aluminium superstructures required insulation to bring their fire resistance to the equivalent of that of steel. The simple melting of aluminium might not be regarded as an event of catastrophic proportions in itself, but the failure of walkways, etc., in this way can prevent any hope of access for firefighting.

There is a potential fire risk with GRP since although glass is not flammable the resin is, so that the actual fire risk in any given case depends on the proportion of resin and the nature of the reinforcement. Not only are the resins combustible, but the amount of smoke given off on burning and the toxicity of the smoke can be unacceptable. Regulations concerning the fire retardant and smoke and toxicity properties of materials for marine craft are becoming increasingly stringent. Phenolic resins have been used traditionally where fire retardant properties are required. Reinforcement in the form of heavy woven cloths as a barrier, intumescent coatings, which foam and self-insulate on burning, can enhance its fire, smoke and toxicity properties. Nevertheless no composite can match steel or even aluminium in this regard.

16.3 Materials of construction

Wood

Although wood has not been used as a major shipbuilding material for over a century, it retained importance for small-boat manufacture until the last decade or so and it is only perhaps with reluctance that it has finally been abandoned in favour of man-made materials. It has adequate strength and toughness provided proper account is taken of its strong anisotropy, it is non-toxic and inert to most cargoes and

retains its properties indefinitely provided it is kept dry. Unfortunately, as a material for shipbuilding it has three main disadvantages:

(1) There is a limitation on size since any tree grows only to a height and girth characteristic of its particular genus. Of course, wood can be joined by gluing, dowelling, bolting and nailing, but only the mechanical methods are appropriate for large structures and such joints are sites of potential weakness.
(2) At above ~18% moisture, wood may rot through attack by various micro-organisms. Resistance may be conferred by permeation with certain chemicals but this increases costs and large sections are difficult to treat.
(3) Wood is subject to attack by certain creatures which enter during their larval stage and, boring into it, produce a multiplicity of holes and channels that destroy the structural integrity. A bivalve mollusc, the teredo worm, or ship-worm, made it necessary to sheath the entire bottom of timber ships with copper sheet for waters in which the teredo were active.

Of the hardwoods, teak and iroko are the most durable for general ship structures. Oak is also much used for frames.

Softwoods such as larch and pitchpine are used for planking, whilst spruce is used for spars and oars. Ash, being straight-grained, is also used for oars. Balsa and cedarwoods are used as core materials in composite sandwich construction.

Iron and steel

Although wrought iron superseded wood for shipbuilding it soon gave way to steel, and mild steel has been the conventional material for large ship construction since the latter half of the last century. As discussed earlier, the desire to reduce weight and cost has directed attention towards the development of higher tensile steels and these are now widely used.

The original classifications of ship steels by the major Classification Societies (Lloyd's Register, Det Norske Veritas, Bureau Veritas, American Bureau of Shipping) varied from one society to another, but following the work of the International Association of Classification Societies there is now a considerable degree of unanimity. In the UK, normal-strength ship steels with carbon contents in the range 0.18–0.23% are in four grades, A,B,D and E, in ascending order of toughness. Higher tensile strength steels are in three grades of toughness, AH, DH and EH, and each grade is produced in three levels of strength, designated 32, 34S and 36. These steels contain 0.18% carbon with manganese in the range of 0.9–1.6%. All grades can be used for longitudinal and strength members under the limitations prescribed by the rules, but the more notch-ductile grades are mostly reserved for strategic localities to act as crack-arresters. Toughness and strength are controlled by steel-works fine-grain practice employing micro-additions of one or more of aluminium, niobium or vanadium. Steels may be employed in the as-rolled, controlled-rolled or normalized condition and impact testing may or may not be mandatory, all of these factors being specified in the rules according to grade and thickness. For example, A grades need not be impact-tested, but the same is true of AH grades only when they are normalized, or else of thickness less than 20 mm. All DH and EH grades must be impact-tested.

Ship steels must be adequately weldable under shipyard conditions, described by Chadbund and Salter[10] as 'no preheat and no complex procedures'. This restricts compositional choice because it is not permissible to use higher carbon contents as a means of enhancing strength and the need to avoid heat-affected zone cracking makes it necessary to restrict the carbon equivalent values of the steel, calculated as C% + (Mn%/6) for normal tensile grades or

$$C\% + \frac{Mn\%}{6} + \frac{Cr\% + Mo\% + V\%}{5} + \frac{Ni\% + Cu\%}{15}$$

for higher tensile grades. Where higher tensile steel is used Lloyd's Rules allow any type of approved higher tensile electrodes to be used provided the carbon equivalent is not greater than 0.41% – preheating is not generally required. When the carbon equivalent is between 0.41 and 0.45% low-hydrogen higher tensile electrodes must be used, but preheating is required only under conditions of high restraint or low ambient temperature. When the carbon equivalent is higher than 0.45%, low hydrogen electrodes and preheating must be employed. However, typically the carbon equivalent is around 0.3% and can be as low as 0.1% for steels now in use in the marine industry.

The situation is different with submarines (submersibles are, of course, pressure vessels), for which the HY80-HY 180 series of quenched and tempered steels were developed (in each case, the number in the designation gives the yield strength of the steel in units of ksi). The welding of these steels is relatively difficult (Table 16.3), as reflecting their high carbon equivalent values.

TABLE 16.3.

	C	Mn	Si	Cr	Ni	Mo
HY80:	0.18	0.3	0.3	1.7	3	0.5
HY100:	0.2	0.3	0.3	1.4	3	0.3

A number of ships have been constructed especially for the carrying of liquefied gases at low temperatures, e.g. LNG (liquefied natural gas) at −162°C and liquefied propane at −51°C. This requirement does not *necessarily* influence the choice of material for the hull but, because the insulation around the liquefied gas tanks must not be disturbed in service, flexing of the hull must be restricted and this does not favour the use of higher tensile steel. However, structural members closely associated with the liquefied gas tanks need to be constructed of cryogenic material such as 9% nickel steel and 18/8 stainless steel.

Reinforced plastic

Glass reinforced polymers have obvious advantages for the construction of marine craft and have been used for many years. They are light, corrosion resistant in the usual sense (although they may need protection against water absorption, for which osmosis-resistant vinyl–ester resins have been developed) and they have a high damping capacity. The problems in their use in terms of deflection and buckling resistance for larger ship's hulls have been outlined already, but the ease with which seamless hulls of complex design can be formed continues to expand their use for fast craft generally less than 10 m long. The use of stronger and stiffer composites than GRP is being introduced widely, based on a variety of polymer matrices and even to knitted as opposed to woven fabric stiffening, in spite of higher costs and more demanding fabrication requirements.

When *HMS Wilton*, a 47 m (154 ft) Royal Navy minesweeper was launched in 1972 it was the world's largest GRP ship and the choice of material for this ship was greatly influenced by the need for it to be non-magnetic and non-conducting. GRP craft with less onerous duties are designed largely on the basis of experience but the construction of *HMS Wilton* was preceded by an extensive design study which fortunately has been described in considerable detail.[5]

The hull was transversely framed with top-hat stiffeners in single-skin construction. Borosilicate E-glass (cheaper than the higher modulus S-glass) in the form of woven roving was used for the fibre reinforcement, hand-laid in polyester resin. This vessel was designed for a service life of 20 years and operated successfully with no serious problems. Similar construction was later used for the 60 m (197 ft) Hunt class mine countermeasure vessels (MCMVs). Sheet moulding compound (SMC) also usually employs a polyester resin, but with chopped E-glass fibre at a glass content normally between 30–60% and can be used for fabrication to covers and structural panels[11].

Kevlar prepreg has a primary use for ballistic protection panels, although it may also be employed for skins over honeycomb structures for floors and bulkheads. CFRP and hybrid composites containing carbon fibres are generally too expensive for marine application, although they are used for wholly performance-oriented craft such as racing yachts and special power craft such as lifeboats and fast patrol boats.

For low volume manufacture, the production of composite structures by hand lay-up of chopped strand mat in polyester resin is ideal, as tooling costs are low. However, a vacuum-assisted resin transfer moulding process known as Seemann Composites Resin Infusion Moulding Process (SCRIMP) is proving popular for the manufacture of large marine structures, such as mine hunters. Parts produced are of consistent quality and low-cost tooling can be used. Ribs, stiffeners and cores can be incorporated in one stage.

Concrete

Concrete has often been used for the hulls of ships, but although having significant advantages for this purpose it has never been truly competitive with steel for general-purpose shipping. Most concrete ships have been built in wartime when there has been a special need to conserve steel, and a lack of a continuous development effort in concrete ship design has caused most concrete hulls to be clumsy and overweight. According to Tosswill[7] such vessels can be 2½–3 times the weight of an equivalent steel ship. This can be reduced by efficient design but Morgan[7] considers that the concrete ship will always be 10–15% heavier than the steel ship unless a high-strength low-density concrete can be produced at low cost.

Concrete hulls are considered to be exceptionally sea-kindly (Stanford[7]) and also have good freedom from vibration. There is also almost complete absence of any need for periodic maintenance or painting, and repairs to a damaged structure are cheaper and require less time out of commission than would be the case for a comparable steel vessel: the ability of the steel reinforcement to hold the fractured concrete in place renders such damage much less catastrophic.

The disadvantage of high weight in a concrete hull can be offset to some extent by the lower first cost, which may be as little as a half of the equivalent steel ship (Morgan[7]).

Although early constructions employed reinforced concrete, hulls of greater than 2000 tonnes deadweight are now more likely to employ prestressed concrete – this allows an element of prefabrication and large ships can be made in sections and joined afloat (Gerwick[7]). Ferrocement is also important, and is widely used for fishing boats and the larger types of pleasure craft.

To avoid excessive hull weight, concrete cube strength should not be less than 60 MPa (8.7 ksi) and preferably higher: Gerwick[7] refers to designs specifying strengths of 100 MPa (14.5 ksi). Hardy and Reynaud[12] consider that the best type of cement is ordinary Portland cement with C_3A (3 $CaO \cdot Al_2O_3$) less than 8%. The cement content should be greater than 400 kg/m^3 (25 lb/ft^3) and the water–cement ratio should be down to 0.40. Good compaction with vibration is recommended. If weight is not important, reasons of economy may dictate the use of crushed stone or gravel aggregates but more generally lightweight aggregates are preferred. Gjerde[12] refers to the use of the German aggregate Liapor, which allows a specific weight of 1.85 tonnes/m^3 (115 lb/ft^3).

Cupro-nickel

At first sight, copper–nickel alloys would seem to be absurdly expensive materials for ships hulls. The factor which makes this usage sometimes possible for particular environments is that other ships are out of commission periodically for removal of marine organisms by scraping and repainting with antifouling paint. The principal toxin in many of these paints is copper – the use of copper-based hulls therefore makes de-commissioning for scraping and antifouling unnecessary and in commercial craft the economic savings can outweigh the increased capital cost.

Early reports referred to the use of 70Cu–30Ni as the structural material but that currently in use seems to be Cu–10Ni–1.5Fe–0.8Mn, which has better resistance to bio-fouling. The ships known to be built so far of through-thickness Cu–10Ni are relatively small with lengths up to 25 m (82 ft) and plate thicknesses not greater than 6 mm (0.24 in).

There are practical difficulties in producing cupro-nickel plate in thicknesses greater than this, and to get the advantages of this material in larger craft it will be necessary to use steel clad with the cupro-nickel alloy by roll-bonding. Trials using steel plate clad with 2–3 mm (0.08–0.12 in) thickness of cupro-nickel have given encouraging results.[13]

Materials for superstructure, non-structural bulkheads and doors

Whilst it has been shown that not all types of ship will necessarily benefit from reduction in hull weight, the replacement of heavy structures above the water-line using lighter materials is desirable, lowering the ship's centre of gravity for improved stability whilst decreasing the overall weight. Thus there is an increasing use of both aluminium alloys (usually aluminium – 4.5% magnesium) and reinforced plastic. Two types of advanced non-structural bulkheads are in service with US Navy ships: aluminium honeycomb with aluminium face sheets for use within the hull, and an aramid honeycomb with E-glass reinforced phenolic skins for ship superstructures[11]. It is considered that ballistic protection can also be improved in superstructures by the use of aramid and ceramic-fibre composites to replace aluminium. In all vessels, of course, resistance to fire penetration is important, as is gas and smoke emission, where all these materials are inferior to steel (see p. 261).

References

1. K. J. RAWSON and E. C. TUPPER: *Basic Ship Theory*. Longman, 1968.
2. D. G. M. WATSON and A. W. GILFILLAN: *Trans. RINA*, 1977; **119**, 279.
3. W. MUCKLE: *Trans. NECIES*, 1976; **92**, 149, and *Muckle's Naval Architecture*, Butterworths.
4. C. S. SMITH: Proc. Conf. on Composites. NPL, Teddington, April 1974.
5. R. H. DIXON, B. W. RAMSEY and P. J. USHER: Symp. on GRP Ship Construction. RINA, London, October 1972.
6. W. COLE and P. COLVIN: *Met. Const. B. W. J.*, April 1971, **3**, 131.
7. Proc. Conf. 'Concrete Afloat'. RINA/Concrete Society, London, 1977.
8. E. E. CLARK: *Trans. NECIES*, 1966–67; **83**, 137.
9. E. W. GARDNER: Discussion to A. J. Harris, *Proc. RINA*, 1980; **123**, 485.
10. J. E. CHADBUND and G. R. SALTER: *Proc. RINA*, 1973; **116**, 13.
11. R. L. PEGG and H. REYES: *Adv. Mater. and Processes*, March 1987, 131 (3), 35.
12. Proc. Conf. *Concrete Ships and Floating Structures* (ed. F. A. Turner), Rotterdam, 1979. Thomas Reed Publications.
13. B. B. MORETON: *Metallurg. Mater. Technol.*, May 1981.

17
Materials for engines and power generation

There are many different sources of energy and a wide range of methods and machinery is employed harnessing them to produce power for propulsion, light and heat. These are summarized in Table 17.1.

It is not surprising that a great diversity of materials problems have had, or need, to be solved. They are frequently concerned either with temperature of service or with size, and sometimes both. For example, the critical parts of internal combustion engines operate at very high temperatures but are mostly small, whereas a water turbine or a tide mill are large but operate at normal atmospheric temperatures. On the other hand, a steam turbine, with its associated equipment, combines both hazards since it is large and hot. The problems of developing materials with properties suitable for operation at elevated temperatures have been discussed in Chapter 10 and some additional considerations will be presented in this chapter.

Where energy from combustion is concerned it is thermodynamically favourable to operate over as large a temperature range as possible. The thermal efficiency of an engine is at a theoretical maximum when it operates on the Carnot cycle and is given by $(T_1 - T_2)/T_1$ where T_1 and T_2 are the temperatures at entry and exhaust, respectively. There are usually practical difficulties which prevent the lowering of exhaust temperatures below certain levels characteristic of the equipment concerned; increased thermal efficiency can therefore only be bought with increased entry temperatures. It is the frailties of available materials which set the upper limits of attainable thermal efficiencies. For example, in the reciprocating internal combustion engine the limiting component is the exhaust valve; in the gas turbine it is the turbine blade (combustion at constant pressure eliminates the need for inlet and exhaust valves); in the steam turbine it is the steam piping. Sometimes the limitation is as much economic as technical. The need to keep capital expenditure within bounds is at least partly responsible for some steam plants operating at 540°C (1004°F), even where 600°C (1112°F)

TABLE 17.1

Source of energy	Method of energy release	Output machinery	
		Reciprocating	Rotary
Fossil fuel	Internal combustion	Petrol engine Oil engine (diesel) Gas engine	Gas turbine Wankel engine
	External combustion	Steam engine	Steam turbine
Nuclear fuel	Nuclear fission	–	Steam turbine
Hydraulic power	Capture of potential energy	Wave engine	Tide mill Water turbine
Wind power	Capture of kinetic energy	–	Windmill
Solar energy	Capture of thermal energy	Not available for direct power generation at present	

might be technically attainable.[1] The operating temperature of the light water reactor is limited to about 300°C (572°F) because of the need for a pressure vessel.[1]

Because the working fluid of an internal combustion engine is air (together with the products of combustion) rather than steam, maximum operating temperatures are much higher. Gas turbine entry temperatures now approach 1350°C (2462°F)[2] and materials development has been largely responsible for this improvement over the 700°C (1292°F) of the Whittle W1 engine in 1941. Combustion within a cylinder closed by a piston allows the reciprocating internal combustion engine to develop even higher temperatures – in excess of 2000°C (3632°F) – but since it is difficult in practice to lower exhaust temperatures much below 500°C (932°F), Carnot efficiencies cannot greatly exceed 70%. Further, such high temperatures mean that external engine cooling is necessary so that actual efficiencies may be nearer 40%. In contrast, every effort is made to reduce heat losses from steam machines; even so, measured efficiencies are still much lower than those of the internal combustion engine.

The difficulties posed by large size are of two sorts. First, it is frequently not easy to find methods of manufacture and treatment that will confer adequate and uniform properties on pieces of very large size. Where cross-sections of castings are large there is always the risk of unsoundness and compositional heterogeneity, whilst efforts to minimize section thickness inevitably increase complexity. Large wrought pieces require still larger forging presses capable of imposing the forces required to effect the required deformation and compaction. This alone has sometimes been sufficient to limit designed sizes of steam turbine and alternator rotors.

According to Sully,[3] the reduction in area required to achieve good consolidation during forging is not less than 2.5:1 and this is most conveniently achieved by forging an approximately cylindrical ingot (octagonal, say) down to smaller diameters. Unfortunately, modern rotors are so large that the ingots needed to give the required reduction by simple diametral forging would be too large to cast with existing plant. It is therefore necessary for the ingot first to be upset (i.e. compressed endwise) prior to diametral forging, but with increasing ingot size, plant for this operation may not be available. Where there is doubt concerning the soundness of rotor forgings it is common practice to bore out the central region – this allows the condition of the bore to be assessed by direct observation. Unfortunately, other things being equal, the presence of a bore doubles the stress at the bore surface as compared with the unbored rotor[3] and power engineers would gladly dispense with the bore if this were possible.

There is also the matter of heat treatment. In the present context the materials to be considered are mainly steels, and the aim is to produce a proper combination of properties: proof strength and fracture-appearance transition temperature (FATT) where temperatures are low, creep strength and rupture ductility where they are high. This implies high hardenability, which in turn requires the addition of significant quantities of alloying elements. But large ingots of alloy steel inevitably exhibit compositional heterogeneity (segregation) – the larger the ingot and the higher the level of alloying the worse this is, and, unfortunately, it cannot be removed by homogenization treatments or mechanical working. If the heterogeneity is unacceptable it then becomes necessary to adopt a lower level of alloying – even so, careful optimization of chemical composition and heat treatment to the best combination of properties remains essential.

The second problem of large size is that it can exacerbate the conditions of service, especially where inertial forces and self-weight are concerned. Design limitations are imposed on moving parts in reciprocating engines by increasing inertial stresses as speed increases: centrifugal stresses in turbine discs and rotors represent significant constraints on design.

One possible solution to the problem of large size is to build up a large component such as a crankshaft or turbine rotor from smaller parts which have been manufactured separately. There is then the problem of joining. The two main methods which have been employed are shrink-fitting and welding. Shrink-fitting cannot be employed on components required to operate in the creep range because at operating temperatures the binding forces disappear. However, the method has been used for low-pressure rotors but there have been many problems, including fatigue of the shaft and stress corrosion at the interfaces.[1] Welded rotors have been more successful but there are materials limitations in respect of weldability of the high-strength creep-resisting materials, and there may also be inspection difficulties.

17.1 Internal combustion

In internal combustion (i.c.) engines the working fluid is produced within the engine by the combustion of fuel with air. In the reciprocating i.c. engine combustion occurs periodically in a chamber (the cylinder) which is sealed temporarily by valves whilst combustion occurs, whereas in the rotary i.c. turbine (the gas turbine), combustion occurs continuously within a chamber which is always open to the gases which, after combustion, provide the working fluid. Both types of engine must be cooled; in i.c. engines by water passages around the cylinder or by air passing over fins projecting from the cylinder, and in gas turbines by taking in more air than is required for combustion and passing some of this air around the outside of the combustion chamber.

For thermodynamic efficiency the air–fuel mixture must be compressed prior to ignition. The compression ratio is often around 10:1; for many spark ignition engines it is less than this whilst for compression ignition engines 15:1 is more typical. In reciprocating i.c. engines the compression is effected by the piston within the cylinder but gas turbines employ a rotary compressor and the compression ratio then varies from 3:1 to 25:1 depending upon the engine.

Because of their greater simplicity of mechanical operation, gas turbines occupy much less space than a reciprocating engine of comparable power output but the gas turbine requires a much greater supply of air. For propulsive machinery in warships gas turbines have widely supplanted steam turbines, but this is not true of marine use generally. In merchant ships the diesel engine is still widely used. Reciprocating i.c. engines cannot be built large enough for large-scale power generation but are used locally for electric power in remote areas or for stand-by or emergency use.

Gas turbines are not favoured for large-scale power generation because of their need for refined fuel.

The reciprocating internal combustion engine

The essential working parts of the reciprocating engine are shown diagrammatically in Figure 17.1.

Figure 17.1 Diagram of reciprocating internal combustion engine.

In engines of short stroke the piston rod is generally dispensed with, the small end of the connecting rod then rocking on the gudgeon pin. In early engines the cylinder head was integral with the cylinder block in a monobloc arrangement which is still occasionally employed, but mostly the cylinder head is bolted to the cylinder block with the joint sealed by a gasket. A spark-plug is not required in engines which operate by compression-ignition (mainly diesel engines).

Most engines operate on the four-stroke Otto cycle in which the induction stroke is followed by compression of the fuel–air mixture (compression stroke). Firing of the mixture then produces the power stroke in which work is done on the piston by the expansion of the exploding mixture. Finally, the products of combustion are expelled from the cylinder during the exhaust stroke. There is thus one working stroke for every two revolutions of the crank shaft. Some small engines use the two-stroke cycle which, by employing crankcase compression from the underside of the piston, completes the same theoretical cycle in one revolution, thereby doubling the rate of working.

Materials for reciprocating internal combustion engines

Although the i.c. engine is best known as a power unit for road vehicles, its uses for propulsion range from the thumb-sized engines of model aircraft to large diesels in cargo ships (it is still used for light aircraft but no longer for airliners). Other applications include lawn mowers, compressors and other outdoor power tools as well as local electricity generation. The high level of technical development which has been achieved in this type of engine partially explains why a smaller range of materials is employed than might be expected from its diversity of applications. Such variations as do occur relate to the severity of the required duty and the design life of the engine.

The cylinder block

This is the heart of the engine containing the cylinders in which the power is generated. The necessity for cooling, as described, means that cylinder blocks are of quite complex shape and are therefore invariably manufactured by casting using alloys of good fluidity and small solidification range. Other general requirements are strength and rigidity, good thermal conductivity and low density. Still more importantly there are the special requirements of good resistance to abrasion, wear and corrosion and low thermal expansion made necessary by the action of the piston as it reciprocates at high speed within the cylinder. Grey cast iron is satisfactory in respect of all these properties except density, and is the traditional material for the purpose: a good choice would be flake graphite cast iron to Grade 17 in BS1452, giving 262 MPa (38 ksi) tensile strength in a standard 30 mm (1.2 in) diameter test bar.

The need to increase power–weight ratio has led to the replacement of cast iron in some engines by aluminium. Since cylinder blocks are lightly stressed it is easily possible for aluminium alloys to satisfy the structural requirements, and as well as low density they offer the advantage of die-casting as a method of manufacture. The thermal conductivity values of aluminium alloys are also attractive (see Table 17.2). However, it is necessary to take steps to overcome their low resistance to abrasion and wear in the cylinder. Such wear is at its worst when similar materials rub together, so that an aluminium piston sliding in

TABLE 17.2

Property	High duty grey cast iron	Aluminium casting alloys
Thermal Conductivity (W/m per K)	50	150
Density (Mg/m³)	7.2	2.7

an aluminium bore of similar hardness would be unsatisfactory. (Although a cast iron piston or ring can run in a cast iron cylinder, this is because graphite in the microstructure of grey cast iron has a lubricating effect – tin performs a similar function in aluminium bearings but this type of alloy is too soft for cylinder blocks). The solution to the wear problem will depend upon the duty required of the engine and the level of additional cost that is tolerable (aluminium is in any case more expensive than cast iron). Accepting that maximizing the difference in hardness between piston and bore will minimize wear, then either the bore must be harder than the piston or vice-versa. The simplest and cheapest solution is to chromium-plate the piston and allow it to run in an untreated aluminium alloy bore. The bore is then soft enough to embed foreign particles sufficiently to prevent scoring. This procedure would be satisfactory for a low-duty engine as used, say, in a domestic lawn mower, the working life of which might not exceed 200 hours. In such an engine, which would be air-cooled, the block could be die-cast from a general purpose non-heat-treatable alloy such as LM2 or LM24 in BS1490 (Table 17.3). For more demanding applications, the reverse solution is more likely to be adopted so that the bore is harder than the piston. Chromium plating of the bore is widely practised and this is facilitated by incorporating pre-finished cylinder liners into the cylinder block; of course, if the liners are of cast iron or steel it may be considered that plating is unnecessary. Suitable block materials for a lined engine would be LM9, containing 12% silicon with a small addition of magnesium to make it heat-treatable or

LM25 solution treated and artificially aged, as in the Rover K series engine. The use of a hyper-eutectic aluminum–silicon alloy such as LM30 may allow the use of an unlined cylinder block. This alloy must be phosphorus-refined to avoid the formation of coarse primary silicon, which makes machining difficult. However, allowing the properties required in the bore to dictate the alloy used for the whole block is possibly not a very elegant solution to the problem, and it is preferable to restrict the hyper-eutectic aluminum–silicon alloy to the liner: the block can then be manufactured from the more easily cast eutectic alloy. Attempts have been made to incorporate a graphitic lubricating phase into the aluminium liner. Obviously, the problems of differential expansion are greatly eased when an aluminium piston slides in an aluminium liner set in an aluminium block and smaller cold clearances are allowable. The hardness difference between piston and bore is sometimes increased by plating the piston with a soft substance such as tin or cadmium.[4]

In order to produce an all-aluminium alloy cylinder block, Honda have built and tested aluminium alloy blocks containing aluminium–metal matrix composite liners[5]. The MMC has been shown to display improved resistance to fretting, abrasive and erosive wear compared with bulk aluminium. The aluminium–MMC block was produced using a new pressure die-casting process and consists of a hyper-eutectic Al–Si alloy incorporating hybrid preforms of short alumina and carbon fibres. Tests reported by Ebisawa et al indicate that these blocks result in a significant improvement in the power-weight ratio compared to cast iron and

TABLE 17.3

Alloy	Cu%	Si%	Al%	Alloy	Cu%	Mg%	Si%	Al
LM2	0.7 2.5	9.0–11.5	rem	LM9	–	0.2–0.6	10.0–13.0	rem
LM24	3.0–4.0	7.5–9.5	rem	LM30	4.0–5.0	0.4–0.7	16.0–18.0	rem
				LM25	–	0.2–0.45	6.5–7.5	rem

Figure 17.2 Test data for the maximum power output of various engine blocks, plotted against the mass of the block, for conventional cast iron blocks, for Al alloy with cast iron liners and for Al alloy with integral carbon/alumina fibre reinforced liners. (From Clyne and Withers[16] from Ebisawa et al.[5].)

aluminium with cast iron liner blocks (see Figure 17.2). However, the increased cost of these components is cause for concern in a cost-sensitive industry.

The cylinder head

Normally a casting, the cylinder head may be of cast iron, aluminium alloy (e.g. LM25 TF) or magnesium alloy. It is less easily diecast than the cylinder block because of complex internal coring and may therefore be sand cast. The Cosworth casting process has been developed for the manufacture of complex cylinder heads and other engine components for high performance cars (see p. 191).

Low-carbon steel castings are used for the cylinder covers in large diesel engines.[6]

The inlet manifold

The admission of the air/petrol mixture to the cylinders via the inlet valves from the mixing carburettor is by way of the inlet manifold which is designed for uniform and smooth flow of the mixture to the individual valves. Normally such manifolds are aluminium alloy (e.g. LM20) die castings with some requirement for post-casting finishing before use as an engine part. A design study[7] has taken the trend towards substituting thermoplastic components for those of metals to this component. A thermoplastic manifold offers weight reduction, reduced machining, good surface-finish and automated production using fusible metal core injection-moulding technology.

The thermoplastic inlet manifold must have complete long-term resistance to oils, greases and the petrol/air mixtures it is carrying, requirements met by a nylon 66 compound containing 33% glass fibre reinforcement. The Ford Endura–E 1.3 litre engine air inlet manifold is injection moulded in one piece from nylon 66 reinforced with 35% glass fibre. It weighs about 2 kg, approximately half the weight of its aluminium predecessor, and reportedly costs less to produce. Using rapid prototyping, the development process was reduced in time from 6 to 3 months. An added benefit was found to result from the smoother interior surface, enabling the engine to develop more power.

The crankcase

Structural support for the crankshaft and sometimes also the camshafts is provided by the crankcase. The mechanical requirements are therefore for strength and rigidity. The design complexity generally stipulates casting as a production method: sometimes a single casting can incorporate both cylinder block and crankcase, either cast iron or aluminium alloy. If the cylinder block is of cast iron then an aluminium alloy crankcase can be bolted to it. A popular alloy for this purpose is the heat-treatable Al–7Si–0.35Mg alloy – LM25 in BS1490.

The sump

The simplest way of making an engine sump is to press it out from a low-carbon steel sheet and this is often done, but sometimes castings are employed. These may be of non-heat-treatable aluminium alloys although magnesium castings have occasionally been used.

The piston

Pistons may be of cast iron, steel or aluminium, but piston technology is very complex and frequently pistons are multi-metallic. For example, an aluminium piston might contain steel inserts to limit thermal expansion. Further, the upper part of a piston is fitted with piston rings and is therefore less likely to rub the cylinder wall than the skirt so that a piston in a slow-running marine diesel might therefore employ a forged 0.3% carbon steel for the upper part (say, 080M30 in BS970 – formerly En5) and cast iron to Grade 14 or 17 in BS1452 for the lower parts.[6]

More generally, pistons are made of aluminium, either cast or forged. The choice of aluminium is not purely a matter of reducing weight and inertial forces, important though this is in fast-running engines. The higher thermal conductivity of aluminium allows a piston of this material to run at temperatures of 200–250°C, some 200°C lower than would be the case with a ferrous piston. Apart from the smaller influence on mechanical properties, the thermal expansion problem is eased – in fact, some lightness of weight in an aluminium piston may be sacrificed to ensure good heat flow. Expansion is still a problem when an aluminium piston runs in a ferrous bore – the coefficient of thermal expansion of aluminium alloys varies from 19 to 25 $\times$ 10^{-6} per °C, whereas that for cast iron is 11 $\times$ 10^{-6} per °C. The compromise between excessive piston slap and total seizure requires an alloy with a low expansion. The proprietary casting alloy Lo-Ex (LM13 in BS1490) was developed with this in mind, and is widely used for cast pistons. Here the nickel in particular improves high temperature properties by forming stable intermetallic compounds, providing dispersion hardening. Still lower thermal expansions are obtained with alloys LM28 and LM29, which are used for high-duty pistons, but since these are hyper-eutectic alloys, more advanced casting techniques are required and machining is more difficult. All of these alloys are heat-treatable (Table 17.4).

Wrought aluminium alloys are frequently used in high-duty engines, e.g. Al–2.25Cu–1.4Mg–0.9Fe–1Ni (RR59 or HF 18 in BS1472) which has been used in Formula racing engines.

Squeeze casting

In order to improve the reliability of pistons in terms of fatigue life, particularly in the pin-boss region, squeeze-casting of Lo-Ex alloys has been introduced by some producers in order to eliminate the shrinkage porosity which can occur in conventional gravity die-castings. At the same time the effect of the higher heat transfer rate during solidification is to modify phase relationships somewhat and to refine the microstructure. With the absence of shrinkage defects the full benefit of lowered impurity levels and heat treatment can be realized on a consistent basis.

Squeeze-casting facilitates the incorporation of casting inserts for the control of expansion etc., and a natural development is to consider the incorporation of ceramic fibre reinforcement in those points of the piston operating under the

TABLE 17.4

Alloy	Cu%	Mg%	Si%	Ni%	Fe%	Coefficient of thermal expansion per °C (20–100°C)
LM13	0.7–1.5	0.8–1.5	10.0–12.0	1.5max	1.0max	19.0 $\times$ 10^{-6}
LM28	1.3–1.8	0.8–1.5	17.0–20.0	0.8–1.5	0.7max	17.5 $\times$ 10^{-6}
LM29	0.8–1.3	0.8–1.3	22.0–25.0	0.8–1.3	0.7max	16.5 $\times$ 10^{-6}

highest cyclic temperatures and stresses. The squeezing process ensures good infiltration of the ceramic preform and its incorporation into the casting.

This application represents one of the first successful uses of MMCs. In 1982, the Toyota Motor Company first produced pistons for supercharged diesel engines by squeeze-casting aluminium alloy into an alumina preform, placed at the piston crown, where the requirements for wear resistance and high temperature strength are highest. The MMC replaced a Ni-resist^{TM} cast iron insert, used to prevent seizure of the piston ring with the top ring groove and bore. Compared with the Ni-resist^{TM}, the MMC-containing piston offers increased thermal conductivity and thermal fatigue life. Annually, millions of units of these aluminium-MMC pistons are produced. Pistons have also been fabricated by die-casting aluminium alloy reinforced with particulate SiC.

Ceramic usage with pistons

In a modern turbocharged diesel engine the brake thermal efficiency is in the region of 40%, with the remaining 60% being energy rejected to the cooling system and to the exhaust, although some of the latter may be recovered by the turbocharger. There may be some energy benefit in operating at higher cylinder temperatures with efficient turbo compounding, although emission control in relation to nitrogen oxides has to be taken into account. Quite apart from any potential energy saving which might be achieved in the engine, however, if the weight and energy usage involved in cooling could be reduced there would be an indirect saving from this. Partially insulated engines have been tested with low heat-loss pistons employing air gap thermal barriers below heat-resisting steel or superalloy crowns, which are then ceramic spray-coated (partially stabilized zirconia). Monolithic ceramic piston crowns (silicon nitride) or crown inserts (partially stabilized zirconia) have also been made. (see Figure 17.3) Wholly ceramic crowns have not performed particularly well at the present stage of development, with a tendency to brittle fracture from small flaws not readily detected in production.

The crankshaft

This component has of necessity a rather complex shape, but not to such an extent as to make casting an unavoidable choice for manufacturing

Figure 17.3 Air gap and ceramic insulated pistons from T&N Technology. (From Parker[8].)

method. In fact, forged steel has been the traditional material, probably because it was felt that a higher intergrity was ensured thereby, and it is only comparatively recently that cast crankshafts have come to be used widely in engines of small to moderate size.

The duty performed by the crankshaft is arduous and its shape, producing stress concentrations that are difficult to avoid, renders it vulnerable to failure by fatigue. The material and the treatment given to it depends upon the duty required. Cast crankshafts are generally made from spheroidal graphite cast iron and for motor car engines, especially if these are greater than one litre capacity, there must be careful control of hardness and microstructure. No flake graphite can be permitted and both very soft and very hard constituents, such as ferrite and cementite, must be strictly limited, the matrix being ideally totally pearlitic.

Forged steel crankshafts can be made from a variety of steels depending upon the application. The success of SG cast iron crankshafts now largely restricts forged versions to large or high-performance engines. Diesel engine crankshafts are usually forged.

The two main factors determining the choice of steel are strength and massiveness of sections. Strength requirements are determined by the rated power of the engine. The importance of massiveness stems from the fact that the properties of heat-treated steels depend upon the cooling rate from the austenitizing temperature, even where the heat treatment is a simple normalizing process. In the UK the massiveness of a section is described in terms of 'ruling section',[9] which is the diameter of a round bar which will cool at the same rate as the section concerned. For example, the ruling section for a rectangular section measuring $1\frac{1}{2} \times 3$ in is 2.124 in diameter. The properties of a given steel worsen as the ruling section increases and the magnitude of the effect can be appreciated when it is realized that a monobloc crankshaft for a marine diesel could have a journal diameter as large as 400 mm (16 in).[6] High strength in heavy sections requires high hardenability, which in turn requires the steel to be alloyed, but the

problems associated with the manufacture and treatment of alloy steels in large sections means that normalized plain carbon steels are still used in large slow-running engines.

The relevant UK standard is BS970 Parts 1, 2 and 6, containing many steels which could be used for crankshafts. A few examples which have been used for crankshafts are given in Table 17.5. A more complete summary of data relating to automotive and engineering steels has been given by Fox[10]. It should be noted that 722M24 is a nitriding steel; this would be suitable for a racing engine: the crankshaft would be machined all over and then nitrided to give maximum resistance to fatigue.

The connecting rod

This component is easier to make than the crankshaft and, like that component, can be cast or forged. For small low-duty engines an aluminium die-casting in the non-heat-treatable Al–3.5Cu–8.5Si alloy to LM24 in BS1490 would be suitable. For somewhat higher duty, aluminium alloy stampings may be used.

For road vehicles in the UK there is a reluctance to use cast connecting rods, even where the crankshaft is of SG cast iron, and normally wrought steels are used. Cast connecting rods seem to be used more widely in other countries. The factors governing choice of steel are then identical to those already discussed for the wrought crankshaft. A typical choice for a motor car of up to 2-litre capacity would be hardened and tempered carbon-manganese steel 150M36. For higher performance engines medium-carbon low-alloy steels would be used culminating in 817M40 (1.5Ni–Cr–Mo) heat-treated to a tensile strength of 1000 MPa.

Steel-connecting rods have been produced by powder-forging for some automobile engines. Not only does such a route reduce material wastage and overall labour costs, but with clean powder is capable of producing components of improved fatigue life, largely as a result of the very fine grain size which can result. The main

TABLE 17.5 Steels for engine components (Data selected from BS970: 1983)

Limiting Ruling Section mm / in	13 / 0.5			19 / 0.75			29 / 1.14			63 / 2.5			100 / 3.9			150 / 5.9			250 / 9.8			Steel type
	σ_{TS}	σ_{YS} min	Iz min	σ_{TS}	σ_{YS} min	Iz min	σ_{TS}	σ_{YS} min	Iz min	σ_{TS}	σ_{YS} min	Iz min	σ_{TS}	σ_{YS} min	Iz min	σ_{TS}	σ_{YS} min	Iz min	σ_{TS}	σ_{YS} min	Iz min	
080M36 N										550	280	20							490	245	–	0.36C
080M36 HT							625–775	400	25													
080M50 N																620	310	–	570	295		0.50C
080M50 HT	850–1000	570					775–925	495	–	700–850	430					625–775	390	–				
150M36 N																620	385	–	600	355	–	0.36C–1.5Mn
150M36 HT	850–1000	635	25				775–925	555	30	700–850	480	30				625–775	400	35				
605M36 HT				1000–1150	850	35	925–1075	755	35	850–1000	680	40	775–925	585	40	700–850	525	40	770–850	495	25	0.36C–1.5Mn–0.25Mo
709M40 HT				1075–1225	940	30	1000–1150	850	35	925–1075	755	35	850–1000	680	40	775–925	585	40	775–925	555	20	0.4C 1Cr 0.3Mo
722M24 HT													925–1075	755	35	925–1075	755	35	850–1000	650	30	0.24C–3 Cr–0.5Mo
817M40 HT							1225–1375	1095	18	1000–1150	850	35	925–1075	755	35	850–1000	680	40	850–1000	650	30	0.4C–1.2Cr–0.3Mo

Note: N = Normalized, HT = Hardened and Tempered, σ_{TS} = Tensile strength (MPa), σ_{YS} = Minimum yield stress (MPa), Iz = Izod impact value (ft. lbf.) Extracts from British Standards are reproduced by permission of the British Standards Institution, 2 Park St. London W1A 2BS from whom complete copies of the documents can be obtained.

problem that remains is ensuring that no non-metallic particle is consolidated with the powder, to subsequently exist as a major inclusion, at which fatigue failure could initiate readily.

The camshaft

The camshaft is fairly lightly stressed and the main requirement of wear resistance may be achieved either by case-hardening a wrought low-carbon steel bar or by using a nodular cast iron which is chilled on the cam faces. Cams produced by powder metallurgy in suitable ferrous systems may also be bonded into position on tubular steel shafts.

Using a slider follower, porosity in the P/M camshaft lobes may be beneficial, retaining oil lubricant. However, roller followers are generally favoured, due to reduced engine noise, lower friction and improved fuel economy, which demands a high surface quality and thus low surface porosity.

Inlet and exhaust valves

Temperature and corrosion resistance are the main choice-determining factors for cylinder valves and these determine that chromium is an important constituent in the alloys used. Valves are produced by hot extrusion or hot upset forging and are sometimes of compound construction. Of the two valves, the exhaust valve is subjected to the more arduous conditions of service since it may have to operate at temperatures in the region of 700–800°C: it is therefore usually made from a more advanced material than the inlet valve.

Early materials were ferritic and, as engines developed, came to have progressively greater additions of chromium and silicon, culminating in the 0.5C–8Cr–3Si alloy (401S45 in BS970, Part 4) known as Silchrome 1. This is still used for some inlet valves. For exhaust valves it became necessary to use austenitic steels such as 0.45C–14Ni–14Cr–2.5W–2Si (331S40) although,

after 1945, much use was made of the 0.8C–20Cr–1.3Ni–2Si alloy known as Silchrome XB (443S65). This is also still used. Modern developments have included the precipitation-hardening austenitic steels and the nickel-based Nimonic alloys. The former alloys are hardened by carbon and nitrogen and a popular example is 21–4N (0.5C–9Mn–3.75Ni–21Cr–0.4N, 349S52 in BS970, Part 4). For best properties this alloy should be solution-treated at 1180°C, quenched and aged for 12 hours at 760°C.[11]

There is much interest in the introduction of ceramic components into the valve train. The main objectives are the reduction of weight and the improvement in durability, which will increase engine performance. Tappet inserts and poppet valves have been manufactured from sialon and reaction-bonded and sintered silicon nitride, and the reduction in wear compared with standard materials is encouraging. Metal valve trains are copiously lubricated to prevent failure by scuffing at the cam–tappet interface. The oil also acts as a heat transfer medium, cooling the components. The use of ceramics with good friction and wear characteristics and good high temperature properties offers the prospect of a dry valve train. Testing has shown that even with ceramic parts, some lubrication is necessary, although it can be reduced in comparison with metal systems.[8]

Ceramic materials used include silicon carbide, sialon, silicon nitride, partially-stabilized zirconia and zirconia toughened alumina. However, it is unlikely that ceramic valves or roller followers will be used in mass production applications before the twenty-first century.

Piston rings

Most frequently these are of cast iron containing fine flake graphite. The easiest way to make the rings is to part them off from a centrifugally-cast shell. Sometimes, but not often, steel is used and in small engines, sintered iron. Very large piston rings must be individually cast and are often hard-chromium plated.

Turbochargers

Car makers have turned to turbocharging as a method of boosting the power output of conventional i.c. engines. Turbochargers can be a 'bolt-on' added extra, and the rotors are conventionally manufactured from high chromium steels and superalloys (e.g. Inconel 738). Ceramic exhaust rotors are approximately one-third the weight of a metal rotor, and they maintain their strength at significantly higher operating temperatures[12]. This allows for greater fuel economy and reduced engine exhaust emissions. High-

pressure slip-casting and low-pressure injection moulding have been employed, with silicon nitride and silicon carbide the candidate materials. Shrink fitting, sleeve joints and brazing have all been considered for fixing the rotor to the shaft, although continuous ceramic shafts are also being considered. Several Japanese manufacturers now produce silicon nitride rotors for commercial use in Japan.

The potential ceramic parts in a turbocharged diesel engine are illustrated in Figure 17.4.

The gas turbine

The simplest possible type of gas turbine has three main components, the compressor, the combustion chamber and the turbine. Figure 17.5 shows a machine consisting of a single-stage radial (centrifugal) compressor and a single-stage axial turbine mounted back-to-back on the same shaft. This single-shaft arrangement is simple but too inflexible for an automotive or industrial gas turbine. The advantages of the radial compressor stem from its robustness, reliability and relative ease of manufacture, but it is not particularly efficient and whilst it easily achieves a compression ratio of 3:1 it cannot greatly exceed 4:1. The temperature rise imparted to the compressed gas is correspondingly low being about 140°C for a 3:1 ratio rising to 180–200°C for a ratio of 4:1. The air from the compressor passes to the combustion chamber. The efficiency of the engine is improved if there is interposed at this point a heat-exchanger which adds to the air heat derived from the engine exhaust. Heat-exchangers are regarded as essential for automotive gas turbines to maximize fuel efficiency, but this has been a severe limitation to the success of gas turbines in automotive applications and the increased weight makes them less attractive for aero-engines. In the combustion chamber the fuel is burnt with the compressed air producing temperatures around 1000°C. In an industrial gas turbine a single-stage turbine would be adequate, most conveniently produced as a monobloc with integral blades. The material chosen for the turbine rotor depends upon the turbine entry tem-

Figure 17.4 Potential ceramic parts in turbocharged diesel engine. (From Kirk[12].)

Figure 17.5 Diagram of simple gas turbine engine with single-stage radial (centrifugal) compressor and single-stage axial turbine.

perature (TET) which is determined by the amount of cooling air mixed with the gases delivered from the combustor. In a simple engine of the type being described this would be about 900°C.

Gas turbines for use in aerospace are much more complex than the foregoing because of the need to maximize thrust-weight ratios. Single-shaft engines have severe limitations because when an engine is running at less than full power the compressor must run more slowly also, and, since decreased speeds produce smaller compression ratios, efficiency suffers. It is preferable, therefore, to have the power turbine on a separate shaft. The next step is to replace the centrifugal compressor with the axial type since these are more efficient, and also capable of higher compression ratios (the compression ratio in the Rolls-Royce RB211 is 25:1). Higher compression ratios produce higher air temperatures and advanced axial compressors may deliver air at temperatures as high as 500°C. This makes materials selection much more difficult. An axial compressor operates rather like a turbine in reverse – it is always multi-stage, since the compression per stage is rather small and blades must therefore be fitted individually to the disc or hub, adding to manufacturing costs.

The search for greater propulsive efficiencies has led to other complications. Turbo-prop engines employ shaft power to drive a propeller (aero or marine), whereas turbo-jet engines are propelled by the jet. In the latter case, propulsive efficiency is increased if, for a given airspeed, the emergent velocity of the jet can be decreased. Ducted-fan and by-pass engines achieve this (in different ways), by driving additional masses of low-speed air through the engine to join the exhaust gases. This requires an additional fan or low-power compressor so that advanced engines may be multi-shafted with more than one compressor unit and more than one turbine. A seemingly infinite number of materials has been developed for use in these engines and it would not be possible to describe them all in an introductory text. However, the number of classes into which these materials fall is more limited, and an attempt to exemplify each of them is made in the following paragraphs.

Materials for gas turbines

It may be helpful to realize that within an engine there is a range of harshness of environment in respect of the functions to be performed by the major components. The most demanding conditions relate to the blades in the high-power turbine: the least demanding apply to the disc in the fan or low-power compressor. As engines become more highly rated, established materials may move in relation to the zone of service: turbine materials come to be used in high-power compressors; blade alloys get used for discs (with modifications in composition and heat-treatment where necessary to suit them for their new function). It also happens that materials which are superseded in high-performance engines continue to be used in engines of lower rating. Because of this, examples of all of the main classes of materials developed for use in engines continue to be used in one application or another.

Engine materials can also be placed in a hierarchy. At the bottom lie the reinforced resins (glass-epoxy, glass-polyimide, carbon-fibre-epoxy). Then, in ascending order of performance, come magnesium alloys (used only for casings), aluminium alloys (mostly 2618 Al–2.2Cu–1.5Mg–1Fe–1Ni), titanium alloys, steels (initially low-alloy, such as 1Cr–Mo, occasionally austenitic, but mainly based on S62, the martensitic 12Cr steel, with additions of carbide-formers and solid-solution strengthening elements), cobalt-based alloys, and finally nickel–iron-based alloys (the Inconels and Incoloys) and nickel-based alloys (the Nimonics). This order of performance corresponds very roughly to position in the engine from the cold (or fan) end to the hot (or high-power turbine) end.

The compressor

Apart from the need for resistance to creep, compressor disc materials need high proof stress, some ductility at blade fixings and resistance to low-cycle fatigue (LCF). Compressor blades need strength, stiffness, resistance to fatigue and also to erosion and impact damage.[13] Short-time, low-temperature operation, as in lift engines for VTOL aircraft, allows the successful use of glass-reinforced resin (epoxy or polyimide) for first-stage compressor blades.[14] (CFRP (Hyfil) was developed for fan blades in the RB211 engine but

proved insufficiently resistant to impact damage.) Aluminium alloys, such as the heat-resisting 2618, are still used for radial compressors (and also in lift engines) but have been superseded in most aero-engines by steels, titanium alloys and even, in high-power compressors, by nickel-base alloys. Steels allow temperatures up to 500°C in discs and possibly 600°C for blades e.g. Jethete (Fe–12Cr–2.5Ni–1.75Mo + V, N) and FV535 (Fe–10Cr–6Co+Mo, Nb, V) but, except for a few special cases, they are too heavy for aero engines – Ti alloys are superior in this respect. In land-based engines weight is less important and steels may still be cost-effective.

Titanium alloys cannot be used above 550°C, 450°C is a safer upper limit – and they are therefore most valuable in the cooler parts of high-performance engines such as the fan, low-power or intermediate-power compressors. In order of increasing temperature suitable alloys are IMI318 (Ti–6A1–4V), IMI550 (Ti–4A1–4Mo–2Sn), IMI679 (Ti–11Sn–5Zr–2.25A1–1Mo) and IMI685 (Ti–6A1–5Zr–0.5Mo).

For high-power compressor discs it may be necessary to use nickel–iron or nickel-base materials, such as Incoloy 901 (Ni–34Fe–13Cr–6Mo–2.5Ti–0.25A1+C, B) or Waspaloy (Ni–20Cr–14Co–4Mo–3Ti–1.3A1+C, Zr, B) which were previously turbine disc materials.

Compressor stator vanes, which are stationary, are stressed less highly than the rotating blades – the martensitic 12Cr steel types may be retained in this application although titanium alloys are mostly preferred.

The requirements of the compressor casing are low weight and stiffness. Competing materials for this application are aluminium casting alloys, magnesium casting alloys and pressed steel; the last of these is currently finding most favour.

Combustion chambers and flame tubes

Sheet materials are used for this purpose and, apart from the obvious requirement of temperature resistance, suitable material must be readily formable and weldable. Unfortunately, efforts to improve temperature resistance tend to produce difficulties in forming and welding. Nickel-base alloys are invariably employed: the simple Nimonic 75 and Hastelloy X (Ni-22Cr–18.5Fe–9Mo–1.5Co–0.6W+C) are very suitable for relatively undemanding engine conditions, but for higher-performance engines, precipitation-hardening with aluminium and titanium is necessary, leading to alloys such as C263 (Ni–20Cr–20Co–6Mo–0.45Al–2.15Ti+C). In the USA Inconel 718 (Ni–19Cr–18.5Fe–3Mo–5.3(Nb+Ta)–0.5Al–0.9Ti+C) and 625 (Ni–22.5Cr–3Fe–9Mo–3.6(Nb+Ta)+Al, Ti, C) have been used, but Nimonic 86 (Ni–25Cr–10Mo+Ce, C) and Haynes 188 (Co–22Ni–22Cr–14W+La, C) became available later.[2]

The turbine

Provided turbine discs are cooled at the rim it is still possible to use low-alloy steels (e.g. H40 Fe–3Cr+Mo, W, V), although corrosion-resisting steels are more likely A286 (Fe–15Cr–25Ni–1.2Mo+Ti, Al, V) and FV535 (Fe–10Cr–6Co+Mo, Nb, V); these may be used in industrial gas turbines. In high-performance engines, steels were superseded by Incoloy 901 and Waspaloy (originally a blade material). An important requirement in discs is resistance to low-cycle fatigue (LCF)[2] and attempts to improve this, through reduced grain size, have led to the use of powder metal technology to produce forged discs in IN100 (Ni–10Cr–15Co–3Mo–1V–4.7Ti–5.5Al+B, Zr) by superplastic forging (Gatorizing). Current interest in furthering disc technology centres around the production of 'super-clean' alloys to minimize the risk of premature failure from non-metallic inclusions.

Where high-power turbine blades are concerned it must be recognized that turbine entry temperatures long ago exceeded the upper temperature limits of the intrinsic materials: internal blade cooling is essential to lower the blade temperature some 200°C below that of the ambient gas temperature, and this is much easier to achieve in cast blades than forged blades. For moderately low TETs forged Nimonic alloys may be used, e.g. Nimonic 115 (Ni–15Cr–15Co–4Mo–9(Ti+Al)) and Nimonic 118 (Ni–15Cr–14.5Co–4Mo–9(Ti+Al)+B, Zr), but for the highest attainable TETs attention is focused on cast directionally solidified nickel-base alloys such as MarM002 (Ni–9Cr–10Co–10W–5.5Al–2.5Ta+Hf, Zr, B, C) and Mar 200 (Ni–9–Cr–10Co–13W–7(Al+Ti)+Nb, Zr, B). More significantly, perhaps, production is proceeding rapidly with directionally solidified single crystal blades, avoiding the creep mechanisms inherent in the presence of grain boundaries and altering the approach to the alloy design. The major failure mode for high-pressure turbine blades is cracking along transverse grain boundaries in equiaxed investment castings. Dispersion-strengthened nickel-base materials are also areas of current development interest, (see page 129).

Although subjected to higher temperatures than the rotating blades, turbine nozzle guide vanes are less highly stressed and cobalt-base alloys are commonly employed for this purpose[13] e.g. MarM509 (Co–24Cr–10Ni–7W–4Ta+Ti Zr) but nickel-base alloys may also be used, often cast.

Whilst the performance of the alloy system employed for turbine blades has steadily improved, as described on p. 140, protective coatings are now mandatory for at least the first stage of turbine rotor blades against oxidation, corrosion and sometimes, high temperature erosion.

In the conventional aero-engine, much of the weight is comprised of mechanical fixings and blade spacers. One-piece bladed discs (bliscs) can reduce the parasitic weight, but weight saving can be even greater using composite spacers or composite rings manufactured from hoop-wound titanium based composites, reinforced with SiC monofilaments[15]. These materials have the potential for high specific hoop strength and stiffness combined with reasonable radial properties (about 50% lower), which eliminates the need for weighty support structures. However, these MMC components are still difficult and expensive to manufacture and protective fibre coatings must be developed to provide satisfactory resistance to interfacial degradation in service[16].

A number of alpha-2 and gamma titanium alumide (Ti_3Al and $TiAl$ respectively) jet engine components have been produced using casting and forging techniques, including low-pressure turbine blades and high-pressure compressor blades. However, further work is required to improve the oxidation resistance of the titanium alumides for use at temperatures above 850°C.

17.2 External combustion

External combustion systems employ steam as the working fluid and this is generated in a unit known as a boiler or steam generator which is separate from the power unit. Steam is conveyed from the boiler via steam pipes and headers to the shaft power machine, which may be either a reciprocating steam engine or a steam turbine. After doing work, the steam is exhausted from the machine and is then condensed back to water in another unit called a condenser (Figure 17.6). The need for all this equipment means that external combustion systems are very bulky, and this is partly why road transport systems favour the i.c. engine. Further, for large-scale power generation the cooling needs of the condenser call for cheap water, as at a coastal or estuarine location. The reciprocating steam engine is largely of historic interest and need not be discussed.

In the boiler of conventional power plant, heat is obtained from the combustion of a fossil fuel and transferred to the working fluid through the walls of tubular heat-exchangers. Modern boilers are of the water-tube type in which the working fluid circulates through tubes arranged in parallel arrays to form the walls of the combustion chamber and also as pendants within it. (In contrast, the traditional steam locomotive employs a fire-tube boiler – the hot flue-gases from the fire-box at the cab end flow through tubes towards the smoke-box end, the furnace and tubes being surrounded by water.)

The heat source in nuclear power generating plant is the nuclear reactor, and the means whereby the heat is transferred to the water to form steam varies according to the reactor system employed. In the boiling water reactor the reactor *is* the boiler, and no separate boiler as such is needed. In the steam generators of pressurized water reactors the reactor coolant water flows through tubes which are surrounded by the working fluid, but other systems prefer to circulate the working fluid through tubes which are surrounded by the reactor coolant (liquid sodium, helium or carbon dioxide).

Figure 17.6 Fossil-fuel plant for power generation.

Materials selection for steam generation is governed largely by steam temperature and working pressure, and it is the desire of the design engineer to increase both of these. In fossil fuel plant, materials limitations and considerations of cost limit steam temperatures mostly to 540°C (1004°F), although 600°C (1112°F) is technically possible and 650°C is anticipated. Gas-cooled reactors achieve similar temperatures but pressurized water reactors and boiling water reactors are limited to less than 300°C (572°F) because of the steel pressure vessel. Pressures have risen steadily over the years from around 0.3 MPa (50 psi) in early Cornish-type shell boilers to 30 MPa (4350 psi) in modern fossil plant. Gas-cooled reactors operate at similar pressures but again pressurized and boiling water reactors are limited to smaller values, in these cases around 6 MPa (1000 psi).

Water-tube boilers can be exemplified as either 'once-through' or 'recirculating' types. In principle, a once-through boiler is no more than a single long water-tube, although total tube length may approach 120 km (75 miles). Feed water enters at ambient temperature and is heated to about 250°C (482°F) in the low-temperature part of the boiler known as the economizer. It is further heated in the evaporator to 350°C (650°F), and in the form of steam is raised to its final temperature in the superheater, which may have several stages. The steam then passes to the high-pressure (HP) turbine. Its name describes the principal feature of the recirculating boiler: from the point of view of materials selection its distinguishing feature is the steam-water drum in which water droplets are separated from the steam by cyclones.

Because the steam in a boiler is raised at elevated pressure, any component through which it passes could strictly be described as a pressure vessel, including the boiler tubes and steam pipes. In practice, the term is reserved for cylindrical or spherical containers in which a fluid is stored or processed at elevated pressure: this includes boiler drums and reheater drums. A more specialized meaning still relates to nuclear reactors, some of which must be totally enclosed in a large pressure vessel, either of reinforced concrete (AGCR) or steel (PWR, BWR).

Boiler tubes

The function of the boiler tube is to conduct heat from the heat transfer medium (flue gas in a conventional boiler, reactor coolant in nuclear plant) to the working fluid of the turbine (steam). The material requirements are strength to resist the working pressure (creep strength if the temperature is high enough), corrosion resistance, and oxidation resistance (at temperatures above about 560°C). As the working fluid moves through the boiler its temperature rises and the materials used become progressively more highly alloyed. The economiser generally operates at ~250°C, i.e. below the creep range for ferrous materials, and carbon or carbon-manganese steels are adequate; these may also be used in later parts of the boiler if the temperature does not exceed 425°C (800°F). Above this temperature the creep range is entered and it becomes necessary to consider the low-alloy steels, beginning with 1Cr–0.5Mo and continuing with 2.25Cr–1Mo and the more highly alloyed steels. Creep rupture strength is the basis of selection and design. In general, as the temperature increases, chromium and molybdenum contents increase. The 2.25Cr–1Mo steel can be used up to 565°C but beyond this temperature oxidation resistance becomes more important than creep strength and austenitic stainless steels or 9Cr–1Mo steels are competitive. Corrosion resistance is as important as temperature resistance and the need for this is determined by the nature of the fluids circulating within and around the boiler tubes. Mineral impurities in the feed water can cause deposits to form within the tubes and in some boilers austenitic stainless steels have been disqualified by their poor resistance to stress corrosion in the presence of caustic soda or oxygenated chlorides.[17] For this reason the PWR boiler uses Inconel. In a conventional boiler it is possible, if desired, to employ lower-grade materials at the cost of short-term replacement: in nuclear plant higher-grade materials must be employed because of the impossibility of short-term replacement.

Pressure vessels

Although in other fields pressure vessels may be manufactured from non-ferrous materials, in the context of this chapter it is only necessary to consider steels (the reinforced concrete pressure vessel of the AGCR will not be considered here). With the vessel walls in tension from the positive internal pressure, the principal property requirements for operation at temperatures below the creep range are strength and fracture toughness. Above the creep range, creep rupture strength is required.

The problems of materials selection and design increase as the wall thickness and number of penetrations increase. Large wall thicknesses present the usual metallurgical problems of chemical heterogeneity and difficulty of developing full mechanical properties. Penetrations, such as nozzles, introduce complications of shape and variations in sections which complicate the stress analysis – this makes it more difficult to establish fracture toughness requirements. The welding of dissimilar materials may also present problems, and in any case the chance of incorporating defects increases as the number of weld fixings for nozzles, etc., increases.

There is a natural desire to employ materials of high strength for pressure vessels since higher stresses make for lightness; but the higher the stress in a vessel the greater is the available strain energy for propagation of a crack and since the stored energy in a vessel under pressure makes it likely that failure, if it occurs, will be dangerous it is necessary to be cautious, even with so ordinary a component as a boiler drum. Failure of a nuclear reactor pressure vessel must be not merely a remote possibility but inconceivable. It follows from this that the over-riding need for satisfactory fracture toughness rules out materials of very high strength. For this reason, many pressure vessels have been made of carbon-manganese steel, 0.2C–1.25Mn, where the yield strength would probably not exceed 280 MPa (40 ksi). The problems associated with any attempt to develop higher strengths are intensified when the steel is required in thick sections. Thick-walled pressure vessels are usually manufactured by welding together strakes which have been formed into hoops. Now, it is generally accepted that for a given strength a steel exhibits the highest toughness when its microstructure consists of a tempered martensite. But if one wishes to avoid the difficulties associated with liquid-quenching formed plates (distortion, etc.) then properties must be developed by air-cooling. It is hardly possible to produce a full martensitic structure in this way in very thick plates since the alloy content required to do this would be too great, but it is possible to add enough so that after an initial separation of pro-eutectoid ferrite the bulk of the transformation occurs at about 400°C to produce a lower bainite of similar properties to tempered martensite. However, such steels are very sensitive to cooling rate and the likelihood of a proportion of martensite being produced means that tempering is still necessary – normalize and temper is therefore the usual heat treatment. An example of this semi-air-hardening class of steels is Ducol W30 (0.17C–1.5Mn–0.7Cr–0.28Mo) which is capable of developing a yield of 450 MPa (65 ksi) and has given good service in the UK in fossil plants. However, this steel has been known to fail by brittle fracture during manufacture or testing prior to service; the toughness is sensitive to normalizing temperature and, in addition to this, chemical segregation in ingots larger than 35 tonnes can make heat-affected-zone cracking after welding a serious hazard. For this reason, the quenched and tempered ASTM533B steel (0.2C–1.3Mn–0.55Ni–0.5Mo) is preferred for light water reactor pressure vessels. The ring sections are water quenched and tempered after longitudinal welding to produce a yield stress of at least 350 MPa (51 ksi) and much higher toughness.[18] Nozzles, being of greater section thickness than the membrane regions of the pressure vessel, are manufactured from forgings in ASTM508 (0.25C–0.7Mn–0.7Ni–0.35Cr–0.6Mo) which has still greater toughness. Although a 50% increase in strength would be acceptable in allowing the development of larger vessels for nuclear reactions, it is doubtful if the problems discussed can be mastered sufficiently to make this possible. The requirements of safety demand that the chance of fast fracture occurring in a reactor vessel should be negligibly small. This requires that either

(1) the material is so tough, or the design stress so low, that a crack just small enough to be undetected when the reactor is commissioned cannot grow by fatigue processes to the critical size for fast fracture during the life of the reactor, or
(2) the critical crack size for fast fracture is so large that the vessel wall would be perforated before that size could be attained (leak-before-break criterion).

Procedure (1) is the one that is normally employed, and since precise data are often lacking concerning stress intensity factors and fracture toughness, it is necessary to take decisions on the conservative side of possible error. As a standard safety precaution, pressure vessels are always subjected to a hydraulic over-pressure test prior to entering service.

Reheater drums

Reheater drums in conventional power plant operate at temperatures within the creep range. The 2.25Cr–1Mo low-alloy steel is a satisfactory material for this purpose.

Steam pipes

Main steam pipes carry the steam at its maximum temperature and pressure from the boilers to the top valve of the HP turbine. Hot reheat pipes carry the reheated steam at rather lower pressure from the boiler to the LP turbine. These main and reheat pipes operate under creep conditions and are designed from creep rupture data to last the life of the plant, i.e. not less than 100, 000 hours.

The preferred materials for temperatures up to 540°C is the 0.1C–0.5Cr–0.5Mo–0.25V low-alloy steel. Temperatures higher than this demand an austenitic steel such as AISI316 (0.08C–17Cr–13Ni–2.5Mo).

The steam turbine

If steam expands through a suitably shaped nozzle, some of its heat energy is converted into kinetic energy and the velocity of the steam is increased. If the steam is then caused to pass through blades attached in a radial manner to a wheel it does mechanical work by imparting motion to the blades, thereby causing the wheel to rotate. In the turbine the nozzles are formed by the turbine stator vanes which are fixed and attached to the turbine casing; the rotating vanes are attached to the rotor. The combination of one ring of stator blades with one ring of rotating blades is called a stage – all steam turbines are multi-stage so that expansion of the steam occurs in a series of small steps (in an impulse turbine) or continuously (in a reaction turbine). The rotor sits in its own casing, called a cylinder, and a complete power-generation plant consists of two or three cylinders (high-pressure, HP; intermediate pressure, IP; and low-pressure, LP), together with an alternator (Figure 17.6). Steam from the boiler is admitted to the HP turbine and passes down the line at progressively lower temperatures and pressures, to be exhausted from the LP turbine to the condenser. In modern plant, steam emerging from the HP or IP turbine is returned to the boiler for reheating before being passed to the LP turbine. Choice of materials is governed largely by whether or not the working temperature is within the creep range, which for ferrous materials can be taken to commence at 425°C.

Turbine casings

Turbine casings, being of complex shape, are generally castings and are split along a horizontal plane, the two halves being bolted together in service. In fossil plant, HP turbine casings must always be of creep-resisting material. In time past, so long as the steam temperature at the HP turbine inlet did not exceed 480°C then, according to Crombie,[19] the 0.2C–0.6Mo steel was used, but the rupture ductility of this steel was rather poor and this, together with an increase in steam temperature to 525°C, led to substitution by the 0.15C–1.25Cr–0.5Mo steel. For still higher temperatures, up to 565°C, the 0.15C–1Cr–1Mo–0.3V steel was introduced, but this proved to be too sensitive to cooling rate and the 0.12C–0.4Cr–0.5Mo–0.25V steel was therefore preferred. Steam temperatures higher than 565°C require the use of austenitic steels such as AISI316 (0.08C–17Cr–13Ni–2.5Mo) but the high thermal expansion and low strength of these materials present problems,[1] and it has been preferred to retreat to steam temperatures closer to 540°C. LP casings do not require creep-resisting materials – they may be manufactured from spherulitic cast iron or welded steel plates.

Special problems of erosion are presented by saturated steam produced by some nuclear reactors. The 0.15C–2.25Cr–1Mo steel has superior resistance to erosion and may be used for HP cylinders even though the temperature is below the creep range.

The rotor

Because of its large size, consequent difficulty of manufacture, and level of stressing, the rotor is the most critical component in the turbine assembly. The difficulty of attaining chemical homogeneity in thick sections limits the amount of alloying which can be tolerated, and it has not been possible to use austenitic steels. The most satisfactory microstructure in rotors for HP and IP turbines is a continuous dispersion of carbides in a tempered granular upper bainite since this gives a higher creep strength than tempered martensite or lower bainite.[20] This can be obtained in a 0.25C–1.2Cr–1.2Mo–0.25V steel oil-quenched from 970°C and tempered at 700°C. LP rotors operate below the creep range but are more highly stressed due to the longer length of the blades. According to Reynolds et al.,[20] as well as high toughness LP rotors require a 0.2% PS of 700 MPa (100 ksi). The usual steel is the 0.25C–3.5Ni–1.7Cr–0.5Mo–0.1V, which, however, is susceptible to temper embrittlement resulting from the slow cooling from tempering which is needed to eliminate residual stresses. This problem has been minimized by careful compositional control monitored by measurements of fracture appearance transition temperature. Other variations and combinations of nickel, chromium, molybdenum and vanadium in basically similar steels may be used for LP and generator rotors.

Turbine blades

In the HP turbine the blades are short and must operate up to the maximum steam temperature: the LP turbine blades are long (in large turbines they may exceed 1 m (39 in) but are less in smaller plants) and operate at much lower temperatures. When the temperature of the steam entering the HP turbine is greater than 540°C the first few stages may be of austenitic stainless steel or Nimonic 80A; subsequent HP blades and also LP blades are generally made from variants of the 12Cr stainless steel. The straight 12Cr steels were used in the early days but as steam temperatures rose beyond 480°C they were improved; first, by the addition of molybdenum (0.1C–12.5Cr–0.75Mo) and then by the further addition of carbide formers such as vanadium and niobium (0.1C–1Mn–1Ni–11Cr–0.6Mo–0.3V–0.4Nb) taking the permissible steam temperature up to 540°C. In marine steam turbines the special problem of corrosion from deposited chlorides caused many failures of 12Cr-type blades and for a time Monel metal (66Ni–30Cu–2.5Fe–2Mn) was used.[21] 12Cr types have been reinstated following solution of this problem by chemical treatment of the feed water.

In turbines of moderate size the LP blading can also be made from the stronger 12Cr types, but these may be inadequate if the blades are very long. Controlled-transformation stainless steels such as FV520B (0.05C–5.5Ni–14Cr–1.6Mo–0.3Nb) or titanium alloys may then be resorted to; the high strength–weight ratio of the latter making them potentially attractive.

The condenser

This consists of an array of parallel tubes which are set into two tube plates, one at each end of the condenser. The joints between tubes and tube plates must be of high integrity since there must be no leakage of cooling water into the condensate.

Materials selection for condenser tubes is governed by the nature of the cooling water since different types of water produce different types of corrosion hazards: an inland power station using recirculating condenser coolant will have fewer problems than an estuarine plant using sulphide-contaminated waters bearing a very high content of abrasive solids.

Historically, copper-base alloys have been used for condenser manufacture, especially those with high corrosion resistance, either because of intrinsic nobility or the development of a protective oxide film. The main hazards are impingement corrosion, 'coppering' (dezincification, dealuminification, etc.) and galvanic corrosion. Where the corrosion resistance of an alloy depends upon a protective surface film, failure often occurs as a result of disruption of that film by entrained solids or liquid turbulence (see Chapter 11). One criterion of material performance is therefore the critical water velocity at which breakdown occurs. Several materials are available covering a wide range of performance, and the alloy chosen must be carefully matched to the location. The alloys with higher performance are also the more expensive so that selection is a critical matter. The best, but also the most expensive, material is titanium, followed by Cu–30Ni–0.7Fe–0.7Mn, the latter alloy being used extensively in plant for marine propulsion. The only ferrous alloy of which there is significant experience is AISI316 stainless steel (0.08C–17Cr–13Ni–2.5Mo) but under stagnant conditions (stand-by) it is vulnerable to failure by pitting.

References

1. L. M. WYATT: *Materials of Construction for Steam Power Plant.* Applied Science Publishers, 1976.
2. G. W. MEETHAM: *Metallurg. Mater. Technol.* September 1982; **14** (9), 387.
3. A. H. SULLY: 3rd John Player Lecture. *Proc. Inst. Mech. Eng.* Pt. I, February 1967; **181**, 877.
4. K. NEWTON, W. STEEDS and T. K. GARRETT: *The Motor Vehicle.* Newnes-Butterworths, 1972.
5. M. EBISAWA, T. HAVA T. HAYASHI and H. USHIO: *Production Process of MMC engine block.* SAE Paper 910835, 1991.
6. M. J. BRINER and W. FELIX: in *Materials for Marine Machinery,* (eds: S. H. FREDERICK and H. CAPPER). Inst. Mar. Eng/Media Management Ltd, 1976.
7. D. BOLTON: *Plastics Today,* Autumn 1986, (27), 8–11.
8. A. PARKER: *Metals and Materials.* 1990, **1** 14–20.
9. BS970: Part 2: 1970. British Standards Institution, 2 Park St, London W1.
10. J. H. E. FOX: *An Introduction to Steel Selection,* Pt. 1. Engineering Design Guide No. 34. Design Council/OUP, 1979.
11. E. J. DULIS: in *Metallurgical Developments in High Alloy Steels.* Iron and Steel Institute Publication No. 86, 1964.
12. J. N. KIRK: *Metals and Materials.* 1987, **11**, 647–52
13. R. J. E. GLENNY, J. E. NORTHWOOD and A. BURWOOD-SMITH: *Int. Met. Rev.* 1975; **20**, 1.
14. G. W. MEETHAM: *Metallurg. Mater. Technol.* November 1976; **8**(11), 589.

15. D. DRIVER: *Proc. Meeting of Australian Aeronautical Soc.* 1989.

16. T. W. CLYNE and P. J. WITHERS: *An Introduction to Metal Matrix Composites.* CUP, 1993.

17. L. M. WYATT: *Materials of Construction for Steam Power Plant.* Applied Science Publishers, 1976.

18. I. G. HAMILTON: in *Developments in Pressure Vessel Technology*, Vol. 3; *Materials and Fabrication.* Applied Science Publishers, 1980.

19. R. CROMBIE: *Metallurg. Mater. Technol.*, July 1978; **10**(7), 369.

20. P. E. REYNOLDS, J.M. BARRON and A.B. ALLEN: *Metallurg. Mater. Technol.*, July 1978; **10**(7), 359.

21. H. J. LEWIS: in *Materials for Marine Machinery* (eds: S. H. FREDERICK and H. CAPPER). Inst. Mar. Eng/Media Management Ltd., 1976.

18
Materials for automobile structures

The selection of materials for engines and power generation, to include internal combustion engines as used in automobiles, forms the subject of Chapter 17. Here, however, we consider the car body ('the body-in-white') and main associated structural components, and this context is used particularly to emphasize the competition to steel from aluminium and composite materials, particularly in the area of body panels.

18.1 The use of steel

Current technology for the manufacture of car bodies has been primarily developed around the long-established use of mild steel, but changes are taking place, both in the progression to the use of higher strength steels and also with the substitution of the alternatives, aluminium sheet alloys and polymer mouldings. Motivation has been in relation to fuel economy and performance through reduced weight, in improved durability and ease of maintenance, and improved safety in the context of increasingly severe safety standards. A 10% reduction in vehicle weight, for example, results in fuel savings of ~7% in the city and 4% on longer journeys[1].

The traditionally-employed rimming ingot steels are formed and then 'bake-hardened' during the painting process by the mechanism of strain ageing, where interstitial solutes diffuse to the new positions of an increased dislocation population, producing a higher yield stress, ~255 MPa (37 ksi). Purposefully nitrogenized low carbon steel could give yield strengths in the range 415–480 MPa (60–70 ksi). However, most low carbon steel sheet and strip is now produced by continuous casting as a 'killed' steel containing aluminium and possibly calcium, with the consequent immobility of, significantly, nitrogen,

and strain ageing at baking temperatures does not occur to the same extent. The desirability of bake-hardening during paint treatment after forming is likely to lead to the development and use of alternative steel compositions to produce the same effect. For example, interstial-free (IF) steels have been developed to exhibit press formability, high strength and damage resistance. Carbon contents are as low as 20 ppm, and high strength IF steels can be produced by additions of phosphorus and manganese, which will also lead to the desired bake-hardening of the steel sheet on curing the paint.

The general pressure to reduce weight and increase safety has led to the consideration of the use of higher strength steels. Rephosphorization (0.07–0.12 P) of low carbon steels results in substantial solid solution strengthening and will provide hot and cold-rolled sheet and strip with yield strengths in the range 275–345 MPa (40–50 ksi) for automotive uses. Formability is not as good as with the traditional low phosphorus material and springback after pressing is increased.

High strength low alloy steels are finding increasing application in automobile structures, both for critical regions in conventional steel construction and for the frame to which alternative panel materials are attached. For automotive use the HSLA steels normally contain of the order of 0.15C–1.3Mn. Ductility is lower than for the normal panel steels, but the yield stress will be in the range 300–600 MPa. Springback after forming is increased and there will be an increased tendency to buckling as compared to mild steel.

Dual phase low alloy steels (DPLA) based on manganese and silicon additions, which give a microstructure of islands of martensite in ferrite on suitable heat treatment, can exhibit both high strength and very good formability. The latter is associated with continuous yielding, i.e. no yield point, and a relatively low yield stress (300–350 MPa) together with a rapid rate of work hardening and high elongations (~ 30%).[2] As a result of the work hardening the yield stress in the final form is higher than in the HSLA steels at the compositions which give required formability and acceptable springback, and can exceed 500 MPa (72 ksi). The high UTS/YS ratio in this material gives a high potential energy absorption on impact with good formability. The required microstructure can be obtained by cooling from a temperature between Ac_1 and Ac_3, typically 790°C, at a rate commensurate with the hardenability and the section thickness, such that martensite forms from the austenite phase of the ferrite/austenite system existing at that temperature. Suitable compositions will give the required ferrite/martensite structure on air cooling, and systems have been developed which produce it during cooling after hot rolling. These steels are said to be significantly more expensive than mild steel however, and until costs are decreased, as by the avoidance of separate heat treatment as indicated, their use is likely to be confined to front and rear structural members where the greater benefit of the superior energy-absorbing characteristics can be obtained[3]

As the grades of steel and the thicknesses employed change, the assembly techniques have to be carefully considered and monitored. The increased hardenability of HSLA and dual phase steels and their response to spot welding with a potential loss of local ductility, has to be considered. This is clearly a field where full joint bonding with adhesives, with or without combination with welding, is likely to develop.

18.2 The introduction of plastics

In 1982 the Ford Sierra became the first car in Europe to be fitted with a plastic bumper, produced from a modified blend of polycarbonate and polybutylene terephthalate. Plastics now account for about 10% of the weight of the typical family car, and that figure is widely predicted to increase. While a large proportion of the plastics used are polyolefins (i.e. polypropylene and polyethylene) and plastic foams, finding application as chrome-plated interior trim, heating ducts, fuel tanks, bumpers and battery trays, an increasing amount of engineering thermoplastics, such as short-glass-fibre-filled ABS, acetals and nylons are being used in semi-structural applications, e.g. gears, window winders, instrument panels and fan blades. Stability up to about 120°C is important in these materials, as this allows the parts to be painted along with the metal components on the assembly line, which reduces cost and ensures colour matching, although oven temperatures can reach 190°C for curing paint on steel. A number of water-based systems are under development which would reduce this thermal constraint.

The use of plastics goes beyond semi-structural applications, and there are several reasons for the shift from steel to materials with higher specific properties. First, automotive manufacturers are under increasing pressure to make cars lighter and more fuel efficient. Increasing safety standards, however, are adding to the weight of cars, for example, the widespread introduction of airbags. This increases the need to reduce structural weight. Second, improvements in materials and manufacturing technologies have simplified assembly with lightweight materials, making them very competitive with steel, with the potential to increase productivity and lower manufacturing costs. Third, manufacturers are always looking to shorten product development cycles, bringing out new models and new body shapes more frequently.

The attraction of engineering plastics is apparent. Composites in particular can offer improved specific properties compared with steel, thereby saving weight. Plastic body panels resist corrosion, can have a high quality surface finish and have better resistance to stone chipping than steel. Engineering plastics also more easily allow the integration of components and design features, which helps reduce the product development cycle. Tooling costs are usually reduced for a part made of plastic compared with steel. For example, a set of production tools for conventional steel bumpers costs about $3–4 million, whereas the cost of tooling for the equivalent plastic part is just over $1 million.[4]

Plastics have been considered, and frequently rejected, on part cost alone. The automotive industry has had to re-think its approach to cost, and consider a more realistic 'whole-life' cost, including the cost of service, failure, consumer satisfaction and the recyclable value of the material. The VW/Audi Group have calculated the energy cost for components of equal stiffness over a 100,000 km lifetime (Table 18.1). According to these figures, components produced from glass-filled nylon or magnesium use less energy than steel.

The 'body-in-white' (so-called because of the appearance of an all-steel car body on the assembly line prior to painting) provides a lot of potential for materials substitution because of its size and mass. For a steel car, a body-in white will weigh between 100 and 160 kg and comprises several hundred pieces.[5]

For its latest research prototype, Ethos 3, the design house Pininfarina, best known for styling the Ferrari, chose a blend of polyphenylene ether with nylon for the body panels (Noryl™ GTX), mainly due to its strength, stiffness, low water absorption and resistance to fuel and lubricants. An extruded aluminium spaceframe chassis was used to hang the body panels. Chassis and body are both eminently recyclable.

TABLE 18.1 Relative energy cost over 100,000 km for components made from different materials to attain equal stiffness

Material	Relative cost
Steel	1.00
Aluminium	0.71
Magnesium	0.66
Nylon	0.81
Glass-filled nylon	0.66
HDPE	0.79

Source: VW/Audi Group.

Polymer composites

The use of polymer composites for automobile body components is by no means new, as instanced by the various Scimitar sports saloon models (Reliant Motor PLC, U.K.) which used such materials for the application over many years in association with a steel structural frame.

A particular development of significance in relation to the use of polymeric materials for automobile body panels has been the introduction of reaction injection moulding systems such as RIM polyurethane and polyurea, which give advantages over both thermoplastic and injection moulding and where suitable reinforcement can be introduced (RRIM) thereby satisfying structural requirements. In particular, mould shrinkage is much reduced, giving better panel sizing and fit with superior dimensional stability on heating. RIM polyurethane and ureas will readily fill the complex three-dimensional moulds for the large body sections involved, the excellent flow characteristics reflecting the much lower viscosity as compared to thermoplastics. As a result, injection pressures are lower and clamping forces reduced, with saving on both operating and capital equipment costs. The liquid reactants require minimal heating (50°C–60°C, 120°C–140°F) and, as the in-mould reaction is exothermic, only a small amount of heat is necessary to initiate the reaction, with a mould temperature of 65°C (150°F). Material costs using the RIM urethane and urea systems are also said to be lower.

Another development, already used by Lotus, is resin transfer moulding (RTM) where a liquid polymer, typically polyester or vinyl ester is injected into a cavity containing a prepared strengthening preform, to polymerize subsequently. As in the case of RRIM with the low viscosity polymerizing systems now developed, mould pressures are low and consequently there is a reduction in operating and capital cost. Since the RTM process uses a preform, fibre contents can be high. The system is under much development to reduce the cycle times of products with up to 65% fibre volume fractions[6]. The use of the

291

fibre/fabric preforms in RTM eliminates the danger of high speed impact shattering associated with short fibre reinforced composites.

The Pontiac Fiero (General Motors) uses a HSLA 'space frame' to which non-loading/bearing panels in similar-polymer composite materials are bolted, producing a 10% weight saving over conventional low carbon steel construction. RIM polyurethane is used for the rear facia, RRIM 20% glass flake reinforced polyurethane for the side fenders and doors, RRIM 15% glass fibre reinforced polyurethane for the front facia, SMC 28% glass fibre polyester sheet moulding compound for the horizontal areas and TPO nylon for the rear quarter panels (see Figure 18.1). The SMC polyester is used for its

Figure 18.1 The Pontiac Fiero, space frame and body panels. (Reproduced with permission of General Motors Corporation.)

greater stiffness on the horizontal surfaces whereas the RRIM polyurethane on the sides gives better dent resistance. Flake reinforcement enables a better surface to the side panels which is less of a concern for the front facia. Resilience is maximized in facia materials (Young's modulus: RIM ~550/MPa (80 ksi), RRIM (15% glass) ~1030 MPa (150 ksi))

In the case of the Fiero the structural frame onto which the polymer composite panels are attached is made from HSLA with a yield strength in the range 290–415 MPa (42–50 ksi) and 22–24% elongation.

Sheet moulding compound (SMC) materials are the highest performance composites in general use, and the most widely used SMC materials consist of about 25 wt% chopped glass fibres in a polyester matrix, so they are not considered high-performance composites. Apart from the applications cited above, the 1996 Ford Taurus/ Sable contains a two-piece SMC structure moulded from polyester reinforced with 40 wt% glass fibre. The SMC replaces a number of metal components, including radiator support, headlight assembly, hood latch mounting and hood bumper support, and it also controls the overall fit and co-ordination of the facia, bumper, hood and headlights. The Alfa Romeo Spider roadster uses SMC materials in the manufacture of the hood, with integral wings and headlight and grille apertures. The surface finish is an important consideration for body panels, and recent advances in the SMC materials allow for Class-A finishes.

The success of the compression moulding of SMC in the manufacture of automotive structures is partly due to an acceptable cycle time, of the order of 1–2 mins. High speed resin transfer moulding appears to be one of the other few thermoset-composite production techniques that can offer the high-volume, high-performance manufacturing process that the automotive industry demands.

Many of the disadvantages of thermoset composite materials (e.g. limited shelf life of precursor materials, long cycle times and difficult to recycle) could be removed by the use of thermoplastic matrix composites. These materials are already finding application in a number of automotive structures. Thermoplastic composite blanks, typically polypropylene reinforced with E-glass fibre, can be formed in a fast compression moulding process, often referred to as melt flow stamping. The fast cycle time allows components to be made at economical rates. Thermoplastic matrix composites can also exhibit higher impact strengths than thermoset-based materials.[4]

Thermoplastic matrix composites are to be found in the door sills of the Range Rover and the bumper beam of Ford's Lincoln Continental, where it has to meet stringent impact-resistant criteria. The front end of the VW Golf is now manufactured from glass-fibre-reinforced polypropylene, where it offers simplified assembly, parts reduction and weight reduction.

18.3 Aluminium and its alloys

Over the years there have been many attempts to introduce aluminium and aluminium sheet alloys for automobile bodies. Aluminium has, for example, been used for a variety of Land Rover applications since 1948, and the latest Land Rover Discovery V8i features aluminium side panels. Although the modulus of aluminium is approximately one-third that of mild steel, as we have seen, the important parameter for maximum stiffness at minimum mass for a flat plate loaded in bending is $E^{1/3}/\rho$. The value of this parameter for aluminium is approximately twice that of mild steel, so for equal stiffness, in aluminium a panel will weigh about 50% less than in mild steel. The aluminium will need to be thicker, however, as the stiffness in bending is proportional to Et^3, where t is the thickness. For aluminium, with one-third the elastic modulus of steel, the panel needs to be $\sqrt[3]{3} = 1.44$ times as thick as the steel. The lower drawing index (R-value – ratio of width to thickness strain) of aluminium, approximately 0.7 (c.f. mild steel ~1.7), also necessitates an increase in thickness. Formability, as in the capacity of a sheet to be drawn, is controlled by the crystallographic texture developed on rolling (see page 53) as well as by the work-hardening component, n, of the alloy and the r-value, both of which should be as large as possible. A particular attraction of aluminium is, of course, its corrosion resistance without paint, which has made it a suitable choice for vehicles operating under rough conditions where paintwork can be seriously damaged.

The economics for recycling aluminium are particularly attractive, as it takes about one-fifth of the original energy required to remanufacture the same amount from scrap[7]. This factor, together with the corrosion resistance and potential for weight savings, has resulted in considerable interest in the use of aluminium in primary structure applications.

For the primary structure of unitary bodied cars (i.e. car bodies assembled from about 300 press-formed panels – the standard production method), it is typical to select alloys from the Al–Mg 5xxx wrought alloy series of work-hardened non-heat-treatable materials. The original alloy chosen was 5251, but this has now been superseded by a high-formability variant of 5754, as alloys from the 5xxx series can form Luders bands or 'Stretcher strain' markings which spoil the surface appearance.

Aerospace alloys of the 2xxx (Al–Cu) and 7xxx (Al–Zn) are not generally used in car manufacture because of their high cost, lack of formability and stringent production control requirements.

Although Land Rover used 5xxx series alloys for reasons of cost, formability and corrosion resistance, 6xxx (Al–Mg–Si) alloys are being introduced. In fact, exterior body panels are now more usually formed from 6xxx series heat-treatable alloys, exhibiting good strength and formability and excellent corrosion resistance[7]. These alloys are bake-hardened during any painting process, just as high strength steels are, to impart acceptable dent resistance. In the USA, Alcoa 6009 (Al–1.1 Si–0.6Mg–0.35Mn) has been used for inner panels and the stronger 6010 (Al–1.1 Si–0.8Mg–0.5Mn–0.4Cu) for outer skins. As these two alloys do not differ greatly in composition, it is not necessary for scrap separation from the two applications, which would increase life-costs. Polmear[8] quoting Starke[9] and Davies and Goodyer[10] give the properties for some body sheet alloys shown in Table 18.2.

Aluminium is not unfamiliar to the automotive industry, as we have seen in the previous chapter. Castings (e.g. cylinder blocks, pistons), forgings (e.g. suspension components, wheels),

TABLE 18.2 Typical mechanical properties of some body sheet alloys: Polmear[8] from Starke[9] and Davies and Goodyer[10].

Alloy and temper	0.2% proof stress		Tensile strength		Elongation	n	r-value
	(MPa)	ksi	(MPa)	ksi	%		
2036–T4	195	28	340	49	24	0.23	0.75
5182–0	130	19	275	40	21	0.33	0.80
5251–0	60 (min)	9	160–200	23–29	18	–	–
5251–H14	175 (min)	25	225–275	32–40	4	–	–
5154A–0	85 (min)	12	215–275	31–39	14	–	–
6009–T4	125	18	230	33	25	0.23	0.70
6009–T6	325	47	345	50	12	–	–
6010–T4	180	26	290	42	24	0.22	0.70
6010–T6	370	54	385	56	11	–	–
6016–T4	110 (max)	16	210 (max)	30	27	–	–
CR1 mild steel-concast	140 (min)	20	280 (min)	41	38	0.23	1.70

extrusions (e.g. bright trim, side impact beams, air conditioning equipment) and rolled products (exterior panels) in aluminium have been used for many years in the manufacture of cars. However, in order to take full advantage of the weight savings offered, aluminium has to be used in the primary structure, which requires changes in the traditional design and production methods, which have evolved over many years around the use of steel.

Two construction methods have been proposed for all-aluminium primary structures, namely unitary and spaceframe construction. During 1994, Ford in the US built a fleet of 40 unitary bodied Aluminium Intensive Vehicles (AIVs) based on its Taurus/Sable model for field testing around the world (see Figure 18.2). The Honda NSX high performance sports car launched in 1990 also uses unitary construction, with the base structure and hang-on exterior panels constructed from sheet pressings, 5xxx series alloys are used for most of the inner stampings and 6xxx series alloys for the exterior body panels to ensure dent resistance. Welding is the predominant method of joining, although some adhesive bonding is used for attaching outer panels. Including exterior panels, the completed body weighs 210 kg, a saving of 140 kg over a steel monocoque construction.

However, unitary construction is not suited to small-to medium-volume production, as tooling costs are high. An alternative production method, more suited to these volumes is known as spaceframe construction. Essentially, this method uses complex hollow extruded section, joined either by direct welding or cast aluminium nodes, to form a rigid framework, analogous to a frame tent[7]. This method was chosen for the all-aluminium Audi A8 luxury saloon (see Figure 18.3). Maximum weight saving (estimated at 140 kg relative to a steel body) is achieved by clothing the spaceframe in pre-stressed aluminium body panels. The Lotus Elise and Renault Spider sports cars also make use of the aluminium spaceframe concept.

A number of methods are available for joining aluminium components, and in practice a mix will be used in any one structure. Welding is possible by all the standard techniques, e.g. MIG, TIG and resistance welding. A number of developments have been made in adhesive bonding techniques in order to allow for more process tolerance than typically found in aerospace applications. Special one-part epoxies have been

Figure 18.2 The unitary construction of the Ford aluminium intensive vehicle. (Courtesy of the Aluminium Extruders Association.)

Figure 18.3 The Audi A8 Luxury Saloon. All-aluminium spaceframe construction. (Courtesy of the Aluminium Extruders Association.)

devised to develop the required bond strengths and gap-filling properties. A combination of both adhesive bonding and spot welding can be particularly beneficial. The weld contributes peel strength and the adhesive protects the weld from fatigue failure and maximizes the torsional strength of the structure.

18.4 Corrosion damage to automobiles

There are two particular accelerators in relation to steel motor vehicle corrosion–differential aeration and road de-icing salts. The presence of crevices, particularly at weld lines between body panels and the accumulation of damp road dirt on retaining surfaces will lead to differential aeration and galvanic action with anodes created where oxygen access is restricted (crevice corrosion, see p. 148), most particularly in the presence of salt as a low resistance electrolyte. McArthur[11] places particular emphasis on the role of de-icing salts, maintaining that under conditions of restricted front air entry, as controlled by passengers, high speeds cause reduced pressure in the car and replacement air to be drawn in through the bottom of the doors, sills and rear panels at drainage and inside-section paint processing holes. These are directly over the front wheel track where dense salty mists are formed, some of which can be drawn into the hollow sections, producing aggressive microclimates internally. He also points out that the salt is hygroscopic and will thus maintain a moist environment at the point of deposition over a wide range of atmospheric condition.

Clearly, details of design are particularly significant in relation to corrosion. Hidden and inaccessible ledges, sandwich or overlap sections, unbonded interweld lines where electropaint cannot penetrate, are typical errors. Even the paint processing and drainage holes come in for criticism, allowing access of salt mist into areas with least protection, unless specific internal section waxing with inhibition has been carried out. The design of some door drainage holes even produces a small reservoir of water associated with them, since drainage is not at the lowest point of the seam. The injection of waxes, particularly if they are able to be drawn into crevices and interweld areas by capillarity, can give considerable protection since access of electrolyte to the steel surface is prevented and interweld crevices are blocked.

18.5 Surface treatment of steel for car bodies

The use of wax injection as a subsidiary means of internal protection is now incorporated by a number of manufacturers, but this does not eliminate the need for good intrinsic surface protection by coatings applied during manufacture. Progressively more attention has to be paid to this aspect as the thickness of steel employed decreases with increasing steel strength.

Most car bodies are spot-welded to form a rigid shell. This is then pretreated by phosphating (see p. 318) with the resultant phosphate film producing wetting and a good key for the subsequent paint application. It also gives some corrosion resistance itself where the paint film is subsequently damaged. Electropainting is used, where deposition of the paint from aqueous suspension to the car as cathode in a large paint bath produces a thickness of film governed by the distance from the anode (i.e. the 'electrolyte' resistance). Since the film is an insulator, its own growing resistance acts as a self-regulator to thickness. However, the longer resistance paths in the electrolyte to internal surfaces, even with the provision of large apertures and 'electrocoat throw holes', means that in reasonable painting times the internal box sections do not receive as thick a coat as externally. Furthermore, where crevices exist the electropaint may not be able to penetrate at all.

Development to improve the corrosion resistance of steel beyond that provided by the phosphate/paint systems have followed under the impetus of competitive 'freedom from corrosion' warranties for mass market care and the reduction of panel thicknesses. These have mostly taken the form of coatings incorporating zinc, already on the sheet as supplied for forming. Traditionally for top range cars where the extra expense is proportionately small, hot-dipped galvanized sheet, double or single side coated, has been employed. Electrolytically-deposited zinc at a somewhat lighter coating weight and less cost is used, as are resin-based zinc metal paints applied as primers. Other methods employ zinc spray. By the use of alloys of zinc it is hoped that the corrosion resistance of the coating itself (which protects the steel sacrificially – see p. 151) and its activity can be improved. An example is the use of zinc-aluminum silicon baths (Galvalume-Bethlehem Steel).

In the case of the Pontiac Fiero the lower sections of the uniframe are two-side galvanized for maximum corrosion protection, whereas the upper sections are galvanized on one side only, to improve weldability. Clearly, the presence of zinc-rich films on the contacting surfaces of a spot-weld can have deleterious effects on weld properties, the effect intensifying with coating thickness. The external surface coatings can also cause deterioration of the welding electrodes.

18.6 Future trends in body construction and materials

There is no doubt that existing trends to increased reinforced polymer use from bumpers to bonnets will continue, extending to most bolt-on assemblies. Certainly the dominance of steel in the automotive industry is being challenged by plastics and aluminium. In the longer term aluminium alloy or high strength composite spaceframe structures may be employed. The Ford Motor Company, for example, have developed a prototype minivan spaceframe in collaboration with several of its key suppliers, including BASF and Vetro tex[12]. The only trace of metal in the all-composite vehicle body is found in the front-end engine framework. The various parts – side body panel, underbody, cross beam and inner and outer roof panels – are joined entirely by adhesives. Ford also recently launched an all-composite cross beam on its 1995 Ranger and Explorer utility vehicles. The one-piece hollow beam replaces 20 steel and plastic parts.

The question remains whether car manufacturers and car buyers are widely willing to accept a more expensive, if lighter, product. The present consensus appears to be that except for certain niche products, steel remains the low cost material of choice. Much will depend on the future price structure for the individual materials, developments in materials processing, and the impetus for further weight saving in relation to fuel efficiency.

A particular problem which has to be resolved, moreover, is the impact that increased polymer use has on bulk car scrap handling and assessed value – i.e. its marketability. Shredded plastic, particularly thermoset composite waste, is difficult to dispose of and may involve a charge for dumping. Perhaps major chemical companies will be persuaded to design and install plant for return plastic scrap and its treatment, linking to automobile scrap dealers and car manufacturers, the latter in relation to the use of materials with the recycling aspect in mind.

In the meantime the use of higher strength steels at thinner gauge for panels etc., will continue to expand taking advantage of the very efficient monocoque steel sheet body structure and the high production rates possible with this system, and with increased use of coatings to provide corrosion protection.

18.7 Exhaust systems

One of the first items on a motor car requiring replacement is the exhaust system or, at least, parts of it. Although often attributed wholly to corrosion, many mild steel exhaust failures are, in fact, primarily due to fatigue or fatigue acting with corrosion (p. 152) and often associated with stress raisers or imperfections at welds. However, the conditions along the length of the exhaust system arising from elevated internal temperatures and external road spray are very exacting (Figure 18.4) and it is hardly surprising that failure is liable to occur relatively rapidly.

There are two approaches to improvement. For expensive, performance-oriented cars, the use of ferritic stainless steel provides a solution which should give at least four times the life of the mild steel unit, and some family car owners who do not change cars frequently may wish to replace the first exhaust with stainless steel to last the rest of their ownership of the car. An alternative for the car producer is to use aluminized rather than plain steel sheet for the manufacture of parts of the exhaust system (silencers), where the aluminium coating is achieved by the use of a dipping line similar to galvanizing. The aluminium increases both high temperature oxidation resistance and corrosion resistance. At a lower cost than stainless steel this will ensure longer life and is widely employed.

Figure 18.4 Typical car operating conditions.[13] (Courtesy British Steel.)

Figure 18.5 Sectional drawing of a Jaguar silencer box made from Hyform 409[13] (Courtesy British Steel.)

Where oxidation and corrosion resistance are improved and metal thinning is no longer a contributory factor to failure it is important to ensure that the design eliminates as much of the danger of straight fatigue failure as possible. For example, in stainless steel units the design should avoid stress raisers such as rapid changes in cross-section and residual stresses from welding, bending, assembly and fitting. As indicated in Chapter 9.3 the influence of welding on fatigue strength is always deleterious, but the performance can be maximized by ensuring that weld profiles are smooth and regular, without undercutting and free of defects. In relation to design of the overall

system and its mounting there is considerable significance in endeavouring to avoid resonant frequency development which will cause maximum vibration.

The Fe–18Cr–8Ni austenitic steels give excellent service in exhausts, particularly where stabilized with titanium against intergranular attack (see p. 204) and maintain a high polish. A cheaper and very satisfactory alternative is the low carbon ferritic grade as instanced by British Steel Hyform 409 grade (Fe<0.04C_{max} (typically 0.03)<0.9Si_{max}<0.6Mn_{max}<0.5Ni_{max} 11 to 13Cr–5xC to 0.70Ti).[13] As a result of the tight compositional control on carbon content it may generally be formed on tooling normally used for making

The figures above refer to average surface metal loss. In some conditions the corrosion attack may be localized.

Figure 18.6 Corrosion resistance of ferritic stainless steel[14].

mild steel components, and postwelding heat treatment is not necessary, as little or no brittle martensite should be produced. It is estimated that a complete system manufactured in Hyform 409 has an ex-works cost approximately 50% above that of a comparable aluminized/mild steel system[14]. A typical silencer box in ferritic stainless for Jaguar is illustrated in Figure 18.5. The ductility of the material in enabling complex forming to lock seam joints is clearly apparent. The original bright surface finish will not, however, be retained after long exposure, as with the more highly alloyed stainless steels (e.g. 18/8 chromium nickel). Performance is, however, very much better than mild steel, as indicated in Figure 18.6.

References

1. D. L. BLEVISS: *The New Oil Crisis and Fuel Economy Technologies*. Quorum, Westport, 1988.
2. R. W. K. HONEYCOMBE and H. K. D. H. BHADE-SHIA: *Steels, Microstructure and Properties*, 2nd. Edition Arnold, London 1995.
3. G. DAVIES and R. A. ESTERLOW: *Metals and Materials* 1985, **1** (1), 21.
4. R. MCKECHNIE: *Metals and Materials*. **1** 31–5, 1990.
5. F. R. FIELD and J. P. CLARK: Automotive Body Materials, in: *The Encyclopedia of Advanced Materials*, eds D. BLOOR, R. J. BROOK, M. C. FLEMINGS and S. MAHAJAN: Pergamon Press, 1994.
6. *ADVANCED COMPOSITES ENGINEERING* 1988, **3** (2), 10–14.
7. THE ALUMINIUM EXTRUDERS ASSOCIATION: The Aluminium Car Report. Birmingham, UK, January 1995.
8. I. J. POLMEAR: *Light Alloys-Metallurgy of the Light Metals*, Arnold, London 1981.
9. E. A. STARKE: *Jnl. Mater. Sci. Eng.* 1977, 29, 99.
10. G. M. DAVIES and B. G. GOODYER: *Metals and Materials*. 1991, **2** 86–91.
11. H. MCARTHUR: *Metals and Materials*, 1985, **1**, 28–32.
12. *Advanced Materials News*. 1995, **72** 12.
13. British Steel Data Sheet – Hyform 409 BSS 036/4.84. 4m.
14. British Steel 'Stainless Steel for Vehicle Exhaust Systems' BSS leaflet 008, August 1978.

19
Materials for bearings

An initial decision has to be made between the use of rolling bearings (ball or roller races) or plain bearings.

19.1 Rolling bearings

Rolling bearings have what is termed counter-formal contact, with only a small area involved. This means that since very high loads have to be carried on these small areas the materials used have to be hard, to resist deformation and fatigue. Gears, cams and tappets are other examples of counterformal contact and similar criteria for selection apply, although there may be a relative sliding component in their movement which will reduce the allowable load in comparison with pure rolling action.

Rolling bearings are most suited to conditions of large supported or transmitted loads in limited space, often operating at high speeds, with lubrication in sealed systems. Compared with plain bearings there is less viscous drag and a lower starting torque. Lubrication is simple and there is less loss of efficiency at high speeds. Since hard surfaces are required there is little tolerance for misalignment or for the presence of abrasive particles if wear is to be acceptable. From the design point of view there are specific advantages; a ball race, for example, will take some thrust loading whereas a straight roller or plain bearing will not.

For the reason given, to obtain the necessary high hardness and resistance to deformation it is best to use through-hardened steel; case-hardened steels are sometimes employed for large rollers. The steel normally employed for balls, rollers and races is BS970 534A99 (En 31) 1% C, 1% Cr (US designation AISI 52100). Whilst this carbon–chrome steel may be suitable for operating at temperatures up to 200°C, it would be necessary to use a more highly alloyed system with more stable carbides for service at higher temperatures, e.g. high-speed steel 0.8% C, 18% W, 4% Cr, although since lubrication has to be provided for the operation of races, temperatures in excess of 250°C should not occur in service. Silicon carbide and silicon nitride systems have been considered and devised for use without lubrication at temperatures above 250°C, but are more difficult and expensive to produce than steel.

Low carbon, free machining steels have been employed for cheap ball races where speeds are high but loads are light, primarily in domestic appliances for design reasons. Case-hardening steels are sometimes used for convenience in the manufacture of large size rolling systems. For the reasons discussed on p. 180 in relation to subsurface deformation, and for good fatigue life, deep cases and substantial core hardness may be required.

The quality of the steel from which heavily loaded races and balls or rollers are made is very high. Non-metallic inclusions result in fatigue failure, with cracks nucleating and spreading from these second-phase particles. Rolling bearings are designed to be lubricated. Adhesive wear and possibly seizure will follow the failure of lubrication, most rapidly at high loads. Abrasive wear may follow contamination of the lubricant, and this indicates the importance of oil filtration and the regular renewal of filters.

19.2 Plain bearings

In these systems the surfaces are in conformal contact, i.e. 'conform' to one another over a substantial area. It is thus possible to obtain relatively low contact loads and to be able to use a wide range of bearing materials to meet specific requirements of maximum bearing load and the velocity of relative motion.

By using split shell bearings assembly can be achieved on to a crank shaft without dismantling, and since spread loads allow the use of softer materials there is more tolerance in assembly, with varying degrees of conformability to the shaft depending on the material chosen. Frequently the bearing can be made as part of a monobloc component, where this is cast iron, bronze and even aluminium alloy such as duralumin. With a large contact surface, pressure lubrication is required.

Totally dissimilar materials may be employed across the contact, enabling deliberate control of the coefficient of friction, although the lubrication should avoid metal/metal contact except at start–stop. The requirements of materials for plain bearings are given in Table 19.1.

The metallographic structures of materials for plain bearings, as they have developed, usually consist of a combination of hard and soft phases. The hard phase provides the loaded contact area. In the more highly dynamically loaded systems materials are usually chosen where the hard phase constitutes the matrix or continuous phase, with dispersed regions of softer phase, as, for example, in leaded bronze, aluminium–tin or cast iron (graphite flakes in ferrite/pearlite matrix). In steadily and lightly loaded systems the softer phase may be continuous, containing dispersed harder separated particles. Examples

TABLE 19.1. Requirements for lubricated plain bearing materials

Loading capacity	Embeddability	Conformability	Compatibility	Corrosion resistance
Strength ranges from high tin bronzes, through medium tin content phosphor bronze to white metals. Compressive strength and fatigue strength to sustain dynamic loading at the operating temperature must be considered. A composite layered system can be designed for high loads.	The soft white metals will embed foreign particles and thus protect shafts from damage and reduce the risk of seizure through over-heating. In leaded bronzes the dispersed lead phase, and in aluminium–tin the dispersed tin phase, provide some acceptance for dirt, but with these harder materials the best solution is a soft overlay such as lead–indium.	In starting up, or in the event of short-term failure of the lubrication film between the conforming surfaces, there may be direct contact, and it is important that adhesion and seizure should not occur. Again, the traditional white metals are the best from this point of view, but overlays on copper-base or aluminium-base alloys can be provided to enhance the property. In the extreme, bearings may be expected to run without lubrication and the choice then lies between thermoplastics, polymers plus fillers (bronze, Ag, glass, etc.), graphite, metal-solid lubricant (bronze–graphite, bronze–MoS_2) and other complex systems with the choice dictated by load, maximum service temperature, heat conductivity requirements and cost.	The soft white metals are again superior in this respect, and, as with embeddability, lead–indium or lead–tin overlay on harder bearing metals will provide most requirements.	The acidic breakdown products of lubricating oils or entrained water may be corrosive to the bearing surface increasing wear. The tin–base white metals are excellent except where sea water may be entrained. Lead–base white metals and leaded bronzes are attacked, with improved behaviour by a lead–tin or lead–indium overlay. The aluminium–tin alloys have excellent corrosion resistance.

in this class are the soft tin- and lead-base alloys, (Babbitt or 'white' metals) in which hard inter-metallic compounds are formed by antimony and copper additions (Figure 19.1).

The role of the softer phase is to embed abrasive particles that get into the bearing, so that they do not wear the journal (cf. abrasive wear), to provide tolerance for some misalign-ment of the journal when initially set up, or subsequent journal deflection under load in service, and to minimize friction when lubrica-tion is occasionally incomplete and thus reduce the risk of seizure. Clearly, materials with dis-persed soft phase are less able to provide these three requirements to the same extent as white metals, and traditionally plain bearings in the copper-base alloys such as leaded bronze and more still, phosphor bronze, required very care-ful alignment and 'running-in'. Much of this difficulty in providing the soft phase require-ments in bearings for medium and high loads and with reduced emphasis on alignment toler-ance (i.e. comformability), particularly on long shafts, has been overcome by using thin overlays of lead–tin or lead–5–10% indium. A low tin,

highly leaded bronze cast on to a steel backing, with overlay, provides an ideal combination of strength and fatigue resistance for high loads and a satisfactory degree of conformability and embeddability for dirt.

The aluminium–tin alloys have been known as being suitable for plain bearings for a con-siderable time, but only in recent years have they realized their full potential through improve-ments in fabrication providing a fully desirable microstructure. In the cast 20% tin material the tin, which is virtually insoluble in aluminium, forms the continuous phase, rejected to the grain boundaries. This results in a load limit which relates to the properties of tin rather than to aluminium. By rolling the material to strip, breaking up the tin network and annealing, however, the tin is redistributed as separated regions in a continuous aluminium matrix. This material is then roll-bonded to a shell using an aluminium foil interlay, or used (as with bronzes and cast iron) as a complete component, e.g. a bearing bush. With this microstructure the high aluminium–tin bearing alloy is able to withstand higher dynamic loads than white

Figure 19.1 Microstructure of a traditional white metal. Cuboids of SnSb and needles of Cu_5Sn_6 in a predominantly lead–tin soft matrix. ×250.

metals, varying, of course, with the tin content, and fatigue strength can be similar to the bearing bronzes, particularly if the bearing surface is plated.

Powder metallurgy has made possible a range of bearing systems based either on the provision of linings to steel shells in such materials as leaded bronze, or on the ability to provide a controlled degree of porosity through sintering bronze, or iron bushes, which can then be vacuum-impregnated with oil. As the bearing runs and warms, oil exudes from the pores and provides the necessary lubrication to bring the contact friction under control. By powder metallurgy methods non-alloying metal/non-metal systems such as bronze/graphite can be produced. Further developments for light loads relate to the use of mixed plastic/metal systems, taking advantage of the low coefficient of friction of, say, PTFE and providing the necessary stiffness and heat conduction by impregnation with bronze powder and a steel back.

Selection of a material from such a wide range is not a simple matter. The correct balance has to be struck between the requirement of embeddability and conformability, which implies softness, and strength to support the load and provide the required resistance to fatigue (Table 19.1). Frequently the best solution for a wide range of loads is formed in a multilayer system with strength increasing towards the backing and with the layers of controlled thickness. With thin white metal linings or thin lead-base overlays, strength at the surface is increased by constraint provided by the underlying material and a satisfactory compromise between embeddability and strength can be achieved by thickness control of the working lining.

Aluminium–tin and the white metals are not as strong, and cover the lower range of loads. An important advantage of the white metals is that they give the best service under conditions of boundary lubrication, although the new PTFE-impregnated bronze and PTFE–glass fibre materials are clearly also suitable for such conditions. Textile-thermoset laminates (fabric) with fillers such as MoS_2, graphite and PTFE give useful service as plain rubbing bearings against steel.

A particular corrosive situation arises in marine stern tube and other bearings associated with sea water. It has been customary to employ special zinc-base alloys for use in conjunction with bronze stern tubes, to avoid galvanic corrosion.

An excellent although brief, review of the selection of materials for bearing systems has been given by Hutchings.[1]

Reference

1. I. M. HUTCHINGS: *Tribology: Friction and Wear of Engineering Materials*. Edward Arnold, London, 1992.

20
Materials for springs

A wide range of materials exhibit basic mechanical properties which are suitable for the storage of energy in a form of spring. Most will be in the form of helically wound wire or rod. Although leaf springs are somewhat less efficient in terms of energy stored per unit of material, they are still widely used where they are more suitable to the design or where they can also act as a structural member, as in cars. Suitable materials for a spring, of whatever form are all related by similar values of σ_y^2/E (see Figure 14.4) with glasses, spring steel, titanium, reinforced composites (e.g. CFRP), nylon, rubbers, all having values in the range 5–20 MJ/m^3. The higher the value the more energy can be stored, and failure is by buckling rather than by yield. Choice will then depend on other required properties and those of size, shape, weight and, of course, acceptable cost. Even wood, as used in an archery bow, comes into contention on the basis of specific modulus and specific strength, where a similar family relationship can be shown for $\sigma_y^2/E\rho$, with ceramics (in compression), titanium alloys, glasses, CFRP, GFRP, woods (directional), nylons and elastomers as possible candidates. Metals, with the possible exception of titanium where cost is the main disincentive to choice, are disadvantaged by being of higher density or of too low a modulus or strength, as in the case of aluminium. In the case of ceramics the properties have to be considered in compression. With cyclic loading the fatigue strength will be as important as the yield strength and the criteria have to be altered in this regard.

The following sections consider the materials that are widely used for springs in practice.

20.1 Steels

The high modulus and high strength levels available with elastic behaviour in steels have endured for a wide range of spring applications. Coils may be cold or hot wound, in the latter case with subsequent heat treatment. Titanium-based alloys have similar modulus and strength levels to steels and on a specific modulus basis have advantage, but are expensive.

Patented wire springs

The smaller steel springs are frequently produced by cold winding patented galvanized wire, and are effective at low cost. The process of patenting has been described an p. 56. It is often thought of as only applicable to high carbon steels, where it can give the strongest commercial material available. At fine gauges, e.g. 0.005 in, strengths as high as 4200 MPa have been obtained. In fact, the process can be applied with advantage to any steels of carbon content above ~0.25%C, as a continuous process. Patenting requires a low temperature of transformation to produce fine pearlite and no pro-eutectoid ferrite, a structure which is capable of severe draughting to obtain the high strength. Not all springs, of course are of a suitable wire form or section for patenting.

Austempered wire springs

In this the wire spring is cooled from the high temperature austeritic condition and isothermally transformed at a temperature a little above the M_s to lower bainite, usually in a molten salt and subsequently air-cooled to room tempera-

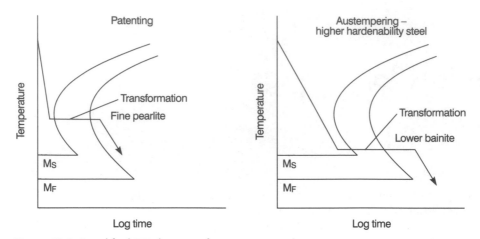

Figure 20.1 Simplified TTT diagrams for patenting and austempering.

ture. The process has the advantage over quenching the martensite and subsequent tempering in that transformation stresses, distortion and cracking are reduced, and a tougher product is obtained. It is thus also valuable for other preformed parts, but it does require a steel of higher hardenability as compared with normal quenching and tempering. Figure 20.1 gives simplified TTT diagrams illustrating the two processes.

The choice of steel for a heat-treated spring will depend on section thickness, relaxation requirements and the temperature of operation (i.e. stability of carbides). The introduction of silicon retards carbide formation, requires a higher tempering temperature and avoids the 'blue brittleness' trough in the 250–300°C range. Importantly it improves the relaxation resistance. Vanadium refines the grain and thereby improves properties (Table 20.1).

TABLE 20.1. Some heat-treated spring steels (quenched and tempered)

Type	Composition %	Properties					E%
		UTS		σ_{YS}			
		MPa	ksi	MPa	ksi		
Carbon-Manganese							
BS 970 080 A52	0.52C–0.8Mn	1175–1390	170–201	1030	150		7
DIN 50 Mn7	0.5C–1.7 Mn (increased hardenability)	1175–1390	170–201	1030	150		7
BS 970 070 A72	0.72C–0.7Mn	–	–	–			
BS 970 060 A96	0.96C–0.6Mn	1760–2310	255–336	1665	242		5
Silicon-Manganese							
BS 250 A53	0.53C–0.7/1Mn–1.7/2.1Si	1280–1465	186–213	1080	157		6
BS 735 A50	0.5C–0.6/0.7Mn–0.8/1.1Si–0.15minV	1330–1540	193–224	1175	171		6

Stainless steel springs are usually produced from the cutlery grade BS 420 S45 containing 0.3C–0.5Mn–13Cr, oil or water quenched from 975°–1040°C and tempered at 400–450°C, giving a tensile strength of 1470 MPa (213 ksi).

With increasing section of the spring form, steels of increased hardenability are required e.g. 0.5Ni–0.5Cr–0.2Mo (e.g. BS 805 A60), with the introduction of higher Cr, Si/Cr, Cr/V, Cr/Mo, Cr/Mo/V and Cr/W/V steels for components at progressively higher temperatures of usage, requiring more stable carbides.

20.2 Non-ferrous springs

There may be a requirement for electrical conductivity, or for a specific corrosion resistance in a spring, which can be provided by the use of a non-ferrous metal. As in many steels, required spring properties at room temperature may be obtainable simply in the cold-worked condition. Titanium alloys are clearly useful in the context of corrosion and temperature resistance, whilst copper-base alloys provide the highest electrical conductivity with the elastic modulus ~$\frac{2}{3}$ and $\frac{1}{2}$ of that of steel respectively. Beryllium copper and phosphor bronze are able to withstand deflections of the same order as spring steels without permanent set. Nickel silver is less satisfactory from this point of view, but is still a very useful material to employ when freedom from corrosion is important, as in precision instruments.

Copper–beryllium alloys

These contain 1.75/2.5Be–0.5Co or 0.5Ni and are quenched from 800°C to produce a supersaturated solid solution, followed by tempering at ~300°C (Figure 20.2), possibly with a degree of

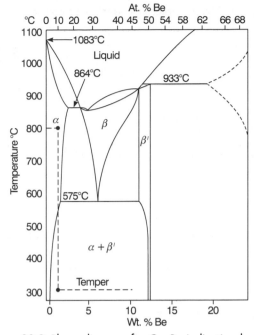

Figure 20.2 Phase diagram for Cu–Be indicating heat treatment.

cold work in between. The final condition and properties depend greatly on the time of tempering, as determining the growth of the β′ precipitate and its effectiveness in hardening the alloy. The alloy has a high fatigue strength as well as a high static strength, although not of course, as high as a spring steel. It is thus valuable for Bourdon tubes and bellows in pressure-measuring equipment and other applications of a similar sort which require good corrosion and fatigue resistance. It also has a high wear resistance and is resistant to softening up to

TABLE 20.2. Comparative non-ferrous spring properties

Property	Cu–Be–range of heat treatment/working	Hard–rolled phosphor bronze	Hard–rolled 15% Ni Nickel–silver	High C spring steel quenched/tempered
σ_{LP} MPa – limit of proportionality	700–1000	420–510	310–355	1000–1500
E GPa – elastic modulus	125–130	110	146	200–210
$\sigma_{LP} \times 10^3/E$ – average	6.6	4.3	2.4	6.1
Fatigue strength – MPa	224–280	220	170	580

TABLE 20.3. Typical properties of copper–beryllium alloys

Condition	0.1% proof stress		UTS		E%	Fatigue strength 100 × 10⁶ cycles	
	MPa	ksi	MPa	ksi		MPa	ksi
Annealed, 800°C, Quench	154	22	465	67	42	224	33
Annealed, 800°C, Q, Temper 300°C	770	112	1160	168	5		
Annealed, 800°C, Q, Cold roll	690	100	820	119	4		
Annealed, 800°C, Q, Cr, Temper 300°C	1000	145	1300	189	2	285	41

~250°C. Its hardness makes it suitable (along with aluminium bronze) for hammers and other tools where there is a non-sparking requirement, as in explosives factories and coal mines.

Phosphor bronze

Phosphorus is widely used as a deoxidant for standard bronzes at the casting stage, so that most bronzes will contain some residual phosphorus content. With more than 0.1%P the material is an intentionally alloyed phosphor-bronze, with compositions in the range 2.5/8.5Sn–0.1/0.35P, where all the phosphorus is retained in solid solution. In the as-cast condition the UTS is ~230 MPa with 18% elongation. The level of strength achieved in its fully hard drawn or rolled condition depends on composition, but at the higher levels of tin and phosphorus can reach 700–800 MPa UTS with 5% elongation, controlled by annealing (e.g. 450 MPa and 60% elongation). Because of its good electrical conductivity phosphor bronze is frequently used for brush springs in electric motors. Where it is required to maximize conductivity these alloys can be readily plated with silver. This is frequently employed for small contact springs carrying substantial current.

Nickel silvers

Also known as German silvers these are, in fact, brasses decolourized and strengthened by the addition of nickel. The α alloys contain 7/30Ni–63Cu–remZn, and above about 18% nickel appear silvery white, take a good polish and are resistant to corrosion. In an 18% Ni alloy in hard spring quality (cold drawn or cold rolled) a proof stress of ~505 MPa, a UTS of ~610 MPa, with 7% elongation and a Vickers hardness of ~190 are obtainable. Annealing for control is usually carried out at 650–800°C depending on the nickel content.

The α nickel silvers are, of course, the NS of EPNS, providing a self-coloured basis for silver electroplate, and have a wide variety of other specialized uses, apart from springs. The α/β alloys, with ~45Cu–45Zn ~10Ni, like the 60/40 brasses are available as extended sections and hot stampings for architectural use, e.g. handrails. The silvery colour has led to the use of the name 'silver bronze', although it is a nickel brass, containing no tin, or, of course, silver.

Other non–ferrous alloys which find use for springs are the Monels, e.g. Monel 400 (Ni66 Cu 31.5) and Monel K–500 (Ni65–Cu29.5–Al2.8), for applications requiring good corrosion resistance at moderately elevated temperatures. Good corrosion resistance at higher temperatures still can be provided by, for example, Inconel 600 (Ni76–Cr–15.8–Fe7.2) or Inconel 718 (Ni52.5–Cr18.6–Fe18.5).

20.3 Non-metallic springs

The use of glass filament for galvanometer suspensions and the experimental use of macro-defect-free cement springs has been mentioned already. Polymers have a high loss factor, dissipating energy during a loading and unloading cycle. They also tend to creep, making them unsuitable for applications where the spring is under a steady load. However, they are perfectly satisfactory for applications where the component is unstressed for most of the time, as in catches. Rubber springs are, of course, widely employed.

Perhaps the most significant development in the non-metallic field has been the introduction of carbon fibre reinforced plastic (CFRP) for such applications as leaf springs and archery bows.

Further detail on springs and spring materials may be obtained from the standard texts[1,2,3].

References

1. *ASM Handbook*, 10th Edition. New York, 1990
2. *Handbook of Spring Design*. Spring Manufacturers Inst., 1988.
3. A. M. WAHL: *Mechanical Springs*. 2nd Edition, McGraw Hill, 1963.

21
Investigative case studies

The dismantling of manufactured objects, with metallographic examination and/or analysis of the individual components, to determine the material used and its condition, can be of value to students in demonstrating not only examples of material selection (both good and bad), but also the manufacturing and assembly methods employed. Detailed case studies, with the artefact displayed, can also be a useful basis for lectures. The three case studies which follow formed part of a course given by JAC.

21.1 Electric chain saw (Black & Decker Ltd.*)

This wood-cutting chain saw was designed for domestic, i.e. non-industrial, use and, with blade lengths varying from 10 to 16 in (250–400 mm), can cope with logs of diameter up to twice the blade length. The cutting chain is driven through gears by a 1.3 kW electric motor operating on mains frequency and voltage. The chain saw aims at the middle of the market with, therefore, a strong emphasis on price control, and saws are produced at a rate of about 200,000 per annum. The design requirements are as follows:

Strength

Use is out-of-doors, generally by amateur operators. In the UK it must survive an impact test in which the assembled machine is dropped from a height of 3 ft on to a solid floor. It must also pass

*The authors gratefully acknowledge the assistance of Mr G. Hughes, and Mr J. Rogers in the preparation of this case study.

structural tests laid down by the European International Commission on Rules for Approval of Electrical Equipment. Sturdy construction is therefore essential.

Weight

The machine must be as light as possible to avoid operator fatigue.

Electric insulation

With the saw operating from a mains electrical supply and the likelihood of the operator standing on moist earth a very high standard of double electrical insulation is required. (Ensuring total safety through the incorporation of a step-down transformer is rarely employed for hand-held equipment because of increased weight, complexity and cost.)

Fire risk and temperature

Exposure to external heat is not a service requirement, but it is necessary to consider the effect of any sparks from the motor commutator due to motor fault conditions and of heat from bearings and motor field coils. Operation could, however, be at temperatures corresponding to several degrees of frost – this increases the need to use materials of adequate toughness.

Safety

The chain saw is fitted with a chainbrake which stops the chain immediately if kickback should occur, a bar associated with the handgrip comes into contact with the left hand and triggers the braking mechanism.

Description of machine

The general arrangement of the chain saw is shown in Figure 21.1 and some of the internal parts are shown in Figure 21.2. The cutting chain runs around the periphery of and is supported by, the chain bar which has a groove in the centre of the outside edge to provide accurate chain location. The chain slides around the bar with metal-to-metal contact, with automatic injection of oil to avoid overheating and excessive wear.

The bar is bolted on to a support frame which is provided with teeth to dig into the log being cut, to provide leverage and control for the cutting force. The chain is driven by a six-toothed sprocket on a shaft located by a bearing set in the support frame. A 53-toothed gearwheel on the shaft meshes with a 10-tooth pinion driven directly by the electric motor.

Both gearwheel and pinion are located and mounted in a gear housing, above the electric motor. All mechanical parts are mounted in, and

Figure 21.1 General arrangement of chain saw.

Figure 21.2 Internal view of chain saw.

located by, a two-part split casing held together by screws with recessed heads, driven directly into bosses in the lower part. There is a removable end-cap at the motor commutator end of the casing.

Materials of construction

Cutting chain

This is of similar construction to a bicycle chain except that a sharpened tooth projects from every other link to provide the cutting action. Maintenance of sharpness of the cutting edges means high hardness, and they are therefore manufactured from high-carbon steel strip, quenched and tempered. The provision of more sophisticated and higher-hardness systems for the teeth is not warranted in terms of service and cost.

Chain bar

The material of the chain bar must exhibit a carefully judged combination of strength and toughness, for if it were to bend the machine would be inoperable, whereas if it were to snap then considerable personal harm might come to the operator. The periphery of the chain bar must be given sufficient hardness to withstand the rubbing action of the chain and, in addition, a way must be found of forming the groove which locates the chain. The overall solution is to make a three-piece sandwich by spot-welding together three steel plates. The core is of mild steel, the outer pieces of high carbon steel (C70 from BS1499 containing 0.70% carbon). The peripheral groove is formed by making the core smaller than the covering plates, and peripheral hardness is provided by induction hardening the edge track.

Chain sprocket

Once again, because of the need for wear resistance, steel is chosen. Chain sprockets are often forged from a low-carbon steel, machined to final form, case-hardened around the meshing surfaces and then heat-treated to give the required combination of properties in core and case. It is difficult to find an alternative to this procedure when considering typical chain sprockets as used in bicycles, since these are generally thin and of rather open design. In the present instance, however, the sprocket is small in diameter and quite 'chunky' in form: this allows the option of powder metallurgy. It is, therefore, manufactured from sintered iron containing 0.6–0.9% carbon and carbonitrided to 600 DPN. It is important that the density of the product be not less than $6.8\,\text{tonnes}/\text{m}^3$ or the case-hardening will not be effective.

Reduction geartrain

The motor speed is reduced to that of the chain sprocket by means of the gearwheel and pinion, as previously described. Several materials are available for gear manufacture, but note that the pinion, rotating 5.3 times faster, must do a good deal more work than the mating gear wheel. Since operation of the chain saw will be intermittent, an essential requirement of the gear train materials is impact resistance. Several combinations of metallic gears might be considered for this application. However, there are four reasons which argue against the choice of a totally metallic gear train. One is the need for lubrication. Another is the fact that metallic gear trains are noisy and heavy. Third, the cost factor is unfavourable for some metallic gears. Finally, the need for electrical insulation suggests that at least one gear should be non-metallic. There are three main non-metallic contenders, all of which are widely used for gears: (1) acetal copolymer – a thermoplastic of good stiffness and toughness requiring minimal lubrication; (2) laminated phenolic materials – compressed layers of fabric, bonded together with a phenolic thermosetting resin; such laminated resin gears are, however, not recommended for applications involving shock loading; (3) nylon (polyamide) is widely used for gears, having good strength and toughness, no need for lubrication and good noise reduction characteristics. Its main disadvantage is its tendency to absorb moisture.

In the present instance, trials with the gearwheel manufactured from polyacetal indicated that this material had insufficient resistance to shock loading, with a tendency for teeth to break. Unfilled nylon was unsatisfactory since the teeth bent and standard glass-filled nylon had insufficient impact resistance. Although the toughness of the latter could be improved by conditioning with 2% of moisture, this unfortunately caused rusting of the steel shaft to which the gearwheel was bonded. After initial difficulties with brittleness, the problem was eventually solved by developing a grade of glass-filled nylon with sufficient toughness in the as-moulded condition.

Where a pinion mates with a gearwheel under fairly heavy-duty conditions, it is preferable that the two are not of similar materials, since this leads to wear. Because the pinion suffers more wear than the wheel, the former is usually made from a harder, more wear-resistant material. In the present case, therefore, it is sensible to manufacture the pinion from a metallic material. This must not be too hard. Mild steel would be functionally satisfactory but involves machining costs. The material chosen was sintered iron containing 0.45% phosphorus which in solution gives a minimum hardness of 120 DPN.

Gear housing

The function of the gear housing is to provide accurate dimensional location for the gear and motor shafts. This requires stiffness rather than strength and freedom from residual stresses such as may cause distortion of the as-manufactured article. Undoubtedly, metals would perform the required mechanical functions very satisfactorily but they are rejected because of their high densities and electrical conductivities. Among plastics the thermosets are too brittle, as are several of the thermoplastics, such as polystyrene, SAN and the acrylics. Better impact resistances are found with polycarbonate, ABS, polypropylene and nylon. Of these, polypropylene is the cheapest but has the lowest elastic modulus. Although the price of ABS is twice that of polypropylene it is only half that of nylon. However, ABS has a low modulus, as also does polycarbonate. It is clear that for the present purpose all of these materials need improvement of the elastic modulus, and this can be done with some form of reinforcement. In injection-moulded thermoplastics, the commonest form of reinforcement is of glass fibres, each being about 0.5 mm (0.020 in) in length. Unfortunately, reinforcement with such short fibres imparts a variable anisotropy to injection-moulded parts, depending upon the flow of the material as it enters the mould. This presents problems in the present application since it causes unacceptable distortions and loss of flatness in the part as it leaves the die. These defects are reduced if some of the reinforcement takes the form of glass beads rather than fibres. Glass beads are rather less effective than fibres in increasing the elastic modulus and heat deflection temperature so that reinforcing solely with beads is not acceptable. A good compromise is glass-reinforced nylon containing 20% fibres and 15% beads.

Bar support frame

This component has the function of transmitting, from the operator to the log, the force which produces the cutting action: it therefore requires higher levels of strength and stiffness than any other part in the chain saw apart from the chain bar. These requirements mean that no polymeric material would be satisfactory and it is therefore necessary to select a metallic alloy. Several combinations of materials and methods of manufacture can be envisaged. However, inherently expensive materials cannot be considered and the part is too large to be manufactured by powder metallurgical methods. Table 21.1 shows several possibilities.

The cheapest aluminium alloy to use would be a general-purpose diecasting alloy such as LM2 or LM24 in BS1490 (Al–10Si–2Cu–1Fe and Al–8Si–3Cu–1Fe, respectively). Of these non-heat-treatable alloys. LM24 provides higher strength

313

TABLE 21.1.

	Sheet stamping	Drop forging	Sand casting	Die casting
Iron or steel	1	3	3.4	–
Aluminium alloy	2	4	4	6
Zinc base	–	–	–	3
Magnesium base	–	4	4	5

1. Possible, but would probably require assembly by spot-welding. Would need protection from corrosion.
2. Inadequate stiffness.
3. Too heavy.
4. Too expensive.
5. Possible, but alloys with adequate properties are metallurgically complex and the fire precautions necessary with magnesium swarf are a disincentive.
6. Satisfactory.

at the expense of some ductility. Still higher mechanical properties can be obtained from the heat-treatable LM25 (Al–7Si–0.3Mg) but this alloy is not recommended for high-pressure diecasting. The frame is therefore pressure die-cast in one or other of LM2 or LM24.

Casing

The main service requirement of the casing is resistance to impact abuse, but there is also need for strength, stiffness and moderate resistance to elevated temperature. Metallic materials are not considered for construction of the casing for the reasons previously discussed, and thermoplastics are the main contenders. The needs for good strength and elastic modulus eliminate polypropylene and toughened polystyrene but tend to favour acrylic, nylon, polyacetal, thermoplastic polyester, modified PPO and SAN. However, the acrylics, GP polystyrene and SAN must be rejected because they have insufficient resistance to impact abuse. In contrast, ABS, nylon and polycarbonate are excellent in this respect.

The need for temperature resistance is governed by the heat generated in the electric motor. The simplest way of assessing this property is in terms of the 'deflection temperature under load' test (ASTM 648) in which a loaded test piece is progressively heated and the temperature recorded at which a prescribed deformation is obtained. In the present application, a suitable material needs a heat-deflection temperature not less than 100°C and on this account it is necessary to reject nylon 6, unplasticized PVC and thermoplastic polyester. The remaining contenders are polycarbonate. ABS, nylon 66, polyacetal and PPO.

It is necessary at this point to consider the possibility of reinforcement to increase strength, modulus and heat deflection temperature although it may influence toughness adversely.

As always, the final choice must be a compromise between desirable properties and price. For the first production run the casing was constructed of glass-fibre-reinforced nylon 66, this being readily available and marginally cheaper than the otherwise satisfactory polycarbonate. However, in a subsequent cost reduction exercise the choice moved to ABS, which was deemed satisfactory after some redesign to take account of its lower heat-deflection temperature as compared with g.f. nylon. This redesign entailed the provision of an air gap between the wall of the casing and the field coils of the motor together with improved motor cooling. The motor field coils were also provided with heat shields of g.f. nylon. The heat-deflection temperature of the chosen grade of ABS was 102°C (216°F), rather higher than normal, and sufficient to cope with motor faults.

The casing end cap in the vicinity of the motor commutator is made of a fire-retardant grade of polycarbonate to control any commutator sparking due to motor faults.

Bearings

A needle bearing was chosen for the most heavily loaded position immediately below the chain sprocket. An oil-impregnated sintered bronze bearing is provided at the commutator end of the armature, surrounded by a substantial mass of die-cast zinc to provide a heat sink to prevent hot running. A ball bearing is employed at the fan end since a sintered bearing would be inadequate and a needle bearing would not control end float.

Motor fan

This item needs strength and temperature resistance and is therefore made of glass-fibre-reinforced nylon.

21.2 The Sturmey Archer gear*

The various parts of the Sturmey Archer bicycle gear are shown in Figure 21.3. The manufacture of small component parts for use in geared hubs has been gradually evolved to the present high level of automation and minimal material loss over 70 years since the hub was invented.

The production over those years has been of three-speed gear systems, and these are still the major product. However, recent competition from external gear systems has been met by the manufacture of a successful seven-gear hub. This has the advantage of single cable control and gives an equivalent gearing range to a commonly employed fifteen-gear external system. It also has the advantage, of course, that the drive chain is less likely to come off.

Originally, machining from bars and slugs on single-speed automatic lathes was the only method available. With the increase in complexity of automatic lathes, particularly as regards the increase in the number of machining points and with the development of free cutting steels, automatic multi-spindle machining for the production of small components has reached a high level of efficiency.

Unfortunately machining from solid bar gives a low material utilization; in some cases as much as 70% of the bar can be lost in the form of turnings. Consequently, wherever possible there was a move to automatic hot forging machines to produce rough component shapes and thus

* The authors gratefully acknowledge the assistance of Mr D. Rayner in the preparation of this case study.

Figure 21.3 Components of three-speed bicycle gear.

minimize losses. Final machining of the forged components was carried out on high-speed semi-automated machines. This advantage of preforming before machining is quite separate from the more familiar metallurgical one of reducing the danger of fracture at points of stress concentration due to the emergence of the fibre of the metal at the surface, as with a machined component.

Even the preforming operations have been an area of radical change over the years, and for these small steel articles cold impact extrusion has entirely replaced the conventional hot forging. This method gives very accurately sized pieces at a very high output rate (25 times conventional hot forging) with a further reduction in turning losses, partly because of the better size control and freedom from flash but also due to the better surface finish resulting from the cold operation.

Conventional presswork is used to produce components from deep drawing and pressing quality steels of low carbon content. In most cases, irrespective of the primary method of production, secondary machining operations to obtain certain critical final dimensions have to be carried out on special-purpose C.N.C. lathes and milling machines.

The majority of components in the gear are then case-hardened by gas carburizing methods. Finally where a component has to be protected from corrosion, plating operations are carried out. The internal components are phosphated and oiled for protection.

At this point it is convenient to consider generally the surface hardening processes which are available and the techniques used to produce the phosphate coatings as typically employed in this gear system.

Methods of producing a hard surface on steel components

Carburizing

A cheap mild steel (about 0.15% C) is generally used. Components are machined slightly over-size, then carburized by one of the several possible methods.

In pack carburizing the components are packed in steel boxes with a carburizing compound (commonly charcoal + coke to reduce shrinkage + 10% of an 'energizer' which is commonly $BaCO_3$). They are heated to about 900°C for long enough to give the desired thickness of case (e.g. 0.03 in (0.7 mm) in 4 hours). The boxes are withdrawn from the furnace and the components are removed from them when cold.

The carbon content of the case is about 0.9%: its hardness is only about 250 H_V. Both case and core have a coarse overheated structure. Heat treatment is therefore necessary to refine the structure and harden the case. For best results a double treatment is used. In order to refine the core the object is heated to 860/900°C for about 20 minutes and quenched in oil. This is then followed by refining and hardening the case by heating to 760/780°C and quenching in water. The case now consists of fine-grained martensite, 700 H_V. The core consists of ferrite and fine transformation products and perhaps marten-site, and has the following average properties: UTS 540 MPa (78 ksi); elongation 25%; impact strength 50 ft.lb f Izod (68J). Light tempering may follow for stress relief.

To save time and expense the first treatment is sometimes omitted. The core then has a rather coarse structure, and is therefore not so ductile or tough.

When a higher tensile strength is required in the core, a low-carbon alloy steel is used, e.g. 0.15C–3.5Ni–0.75Cr. Carburizing at 900/950°C, refining by oil quenching at about 870°C, and hardening by quenching in water or oil from 770°C give the following typical properties in the core: UTS 1000 MPa (140 ksi); elongation 17%; impact strength 50 ft.lb f Izod (68J).

In gas carburizing the components are placed on shelves in an externally heated gas-tight muffle through which the carburizing gas (e.g. CO + a small proportion of hydrocarbons) flows. After say 4 hours at 900°C, the components are cooled in the gas, then heat-treated as above. The method is clean and saves time, as the components heat much more rapidly, but the equipment is relatively expensive and skilled control of gas composition is essential.

In cyaniding or liquid carburizing the components are immersed in a molten salt mixture contained in a steel pot: a common mixture is 45% NaCN + 40% Na_2CO_3 + 15% NaCl, and the temperature is usually about 880°C. Carbon and some nitrogen dissolve in the surface of the steel. Immersion time is 15 minutes for a case depth of 0.08 mm (0.003 in) to 1 hr for a depth of 0.25 mm (0.01 in). The components are removed from the salt and quenched in water. Because of the comparatively short duration of heating, the structure is not coarse and refining treatments are not necessary. The process is rapid and results in a clean surface, but the case is relatively thin. The salt is poisonous and the bath gives off unpleasant fumes which must be extracted through a hood.

Nitriding

An alloy steel has to be used, typically one which contains about 0.4C–0.25Mo–1Al–1.5Cr. Note the relatively high carbon content: the reason for this is that the steel is heat-treated (oil quenched about 900°C and tempered at 500–650°C depending on the properties required in the core) before nitriding and not afterwards. Lower carbon contents might be employed for wear resistance under light stresses.

The components are heated to 500°C, enclosed in a gas-tight box through which a stream of ammonia gas flows. The ammonia is partly dissociated into nitrogen and hydrogen, and some of the former diffuses slowly into the steel forming minute particles of aluminium nitride and chromium nitride which increase the hardness of the surface to about 1100 DPN. It takes about 90 hours to produce a case about 0.25 mm (0.01in) thick. After nitriding the components are cooled in the gas, lightly ground to remove a thin brittle layer (about 0.025 mm (0.001 in)thick), and are ready for service without further treatment. The process thus used is slow and an expensive alloy steel is necessary, but the case is remarkably hard and retains its hardness to 500°C, whereas a carburized case begins to soften at 200°C.

Plasma in carburizing and nitriding

The use of glow discharge at reduced pressure for surface heating and the dissociation of gases at a component surface for absorbtion has been steadily developed and is becoming an important technique for both carburizing and nitriding.

In the context of carbonitriding it can replace salt bath treatment (cyanide) with a low toxicity risk from the gases used. In gas nitriding it gives more precise control of the process in terms of uniformity of hardening (e.g. blind hole penetration) and control of the composition and extent of the monophase surface layer. This coupled with lower temperatures of operation (as low as 375°C) with even less distortion, reduced pollution, lower processing time and reduced treatment gas consumption means overall treatment costs may be reduced by as much as 45%.

Flame hardening (Shorterizing)

This process may be applied to straight carbon or alloy steels, the carbon content being usually about 0.5%. The components are generally hardened and tempered to the required core strength before flame hardening: they are not heat-treated afterwards.

The surface of the component is rapidly heated above the upper critical point of the steel by means of an oxyacetylene flame, and is immediately water-quenched. The time of heating must be accurately adjusted to obtain the required depth of heating, i.e. thickness of case. The process takes only a few minutes, against hours for carburizing and days for conventional nitriding, and any thickness of case up to about ¼ in (6 mm) can be produced. The surface remains smooth and relatively clean, and needs no further treatment. Disadvantages are that the equipment is expensive and that the process is only suitable for repetition work on components of simple shape.

Induction hardening

This process is similar to flame hardening except that the surface is heated by a high-frequency electric current. The components (carbon content about 0.4%) are placed inside a close-fitting water-cooled coil, through which the HF current is passed. Induced eddy currents heat the surface of the steel with great rapidity, to a depth that increases with time and is inversely proportional to the square root of the frequency (10 kc/sec to several thousand kc/sec). After being heated for the required length of time, which varies from a fraction of a second to about a minute depending on the size of the component and the case depth, the surface is quenched by means of a water jet. Owing to the rapidity of the process the surface is undamaged, so that the components can be finish-machined before treatment. The equipment is expensive and the process is only suitable for mass-produced components of simple shape.

Hard facing

This process consists in depositing a layer of some hard material on the surface of the component. Stellite (an alloy of Co, Cr and W) and various iron alloys containing 20% or more of these elements and others, are used. The alloy is melted and flowed over the red hot (in some cases, melted) surface of the component. When the maximum possible hardness is required inserts of tungsten carbide are welded to the surface of the steel. These processes are much more expensive than case hardening and are only used for special purposes.

Hadfields manganese steel

This is a steel containing about 14% Mn and 1.4% C. Components are generally cast to shape, as the material is extremely difficult to hot work and cannot be machined with ordinary tools. The castings are heat-treated by quenching in oil from about 1000°C. In this condition the material is austenitic and therefore tough and relatively soft (about $230\,H_V$). Cold working, such as hammering or violent abrasion, distorts the surface layers thus increasing the local hardness to about $500\,H_V$. This material is widely used for railway crossings, rock crusher jaws, mechanical shovel dippers, dredge bucket lips, etc.

Phosphate coatings

The production of phosphate films on iron and steel is based on dilute phosphoric acid solutions of iron, manganese and zinc primary phosphates, separately or in combination as formulated for a number of proprietary treatments. Phosphoric acid in these, usually hot, solutions produces primary ferrous phosphate by reaction with the iron surface and liberates hydrogen

$$Fe + 2\,H_3PO_4 \rightarrow Fe\,(H_2PO_4)_2 + H_2$$

As the various phosphates now together in aqueous solution dissociate readily, the effect of the reduction in local acidity by the attack on the iron is to give a precipitation of the sparingly soluble secondary and insoluble tertiary phosphates on to the metal surface. Since the attack on the iron is an essential feature of the process the crystalline film formed always contains ferrous phosphate, the proportion depending on the initial composition of the solution with regard to iron, manganese and zinc and the extent of use as the others become used up. In ensuring that precipitation occurs on the surface and does not result in sludge formation, careful control of the free phosphoric acid content is important. Too high an acid content will retard the formation of a film; too low a content will lead to precipitation removed from the surface.

The process of attack at the steel surface is, of course an electrochemical one, with local cathodic and anodic regions. Modifications to the cathodic and anodic processes of this electrochemical attack through additions are used to accelerate the precipitation of the phosphate. Circumstances favouring the cathode process of phosphate deposition and the discharge of hydrogen ions will accelerate the coating process, and conversely factors favouring the dissolution of iron will hinder the process. For example, an oxidizing agent such as nitrate or chlorate acts as an accelerator by depolarization of the cathodes, raising the current density so that rapid passivation of the local iron anodes in pores takes place, favouring the cathodic process. With the accelerated baths, however, where attack on the iron is reduced, even light rust is not so readily removed with the less vigorous reaction, and for a good phosphate film prior cleaning or pickling could be essential.

The coatings produced from the simple treatment with phosphoric acid and iron salts consist of dark grey or black hydrated crystals of $Fe(PO_4)_2$ (tertiary ferrous phosphate), $FeHPO_4$ (secondary ferrous phosphate), and some $FePO_4$ (tertiary ferric phosphate). Where the solutions are based on manganese and zinc phosphates, these will contain the secondary and tertiary iron phosphates in solution.

Following phosphating, rinsing in dilute chromic acid (0.05%) is often used to passivate bare metal or pores, and by absorption into the coating thereby improving the corrosion resistance. After phosphating and any further treatment, thorough rinsing in water is necessary in order to remove soluble salts. This is particularly important where the phosphate film is to be used as a base or 'key' for paint, otherwise blistering of the paint film may occur.

In the context of the gear, the use of phosphating is to improve wettability by oil, and the sealed coat thus obtained provides a good resistance to corrosion in the finished product. Unsealed phosphate coatings (i.e. no oil or paint) do not provide substantial corrosion resistance. One important use of phosphate coatings, in providing a good distribution of oil over a surface, is in conjunction with deep drawing, where an initial cup from a blank may be phosphated prior to more extensive drawing of the cup requiring good lubrication in the die. An added advantage is that the maintained oil film thus produced protects the components from corrosion on handling through the production line and intermediate storage between stages.

BS 3189: 1959 describes the operation of phosphating processes to give best results and also covers the testing of phosphate coatings.

Returning to the Sturmey Archer hub:

Basically, the main components – hub ends, gear ring, planet cage and axle, are produced by cold extrusion, followed by secondary machining. The pawls are produced by powder metallurgy, with advantages in wear resistance, and other components such as planetary and sun pinions are machined directly from semi-free cutting case hardening steels; e.g. BS 970 210 M15 (En 32M) with silicon and sulphur being specially controlled within the specification range for certain applications C 0.18 max, Si 0.05–0.35, Mn 1.2–1.5, S 0.1–0.18, P 0.05 max.

The manganese content is to give hardenability to the carburized case, the level determined by the ruling section of the component. The only difference between this and the more general carbon-manganese case hardening steel is the higher sulphur content allowed for free-machining considerations (based on the presence of MnS in the microstructure).

These machined components are case hardened to various case depths, depending on their function, and hardened by quenching in oil from the carburizing temperature. The next stage is shot blasting and phosphating in an automatic conveyor transfer machine.

The driver is produced by cold impact extrusion from a straight carbon case hardening steel (e.g. BS 970 080 M15 (En 32C): C 0.1–0.18, Si 0.05–0.35, Mn 0.6–1.0, S 0.05 max., P 0.05 max.)

Since this is not a free-machining steel and it has a low sulphur content the manganese content for the required case hardenability is lower. After secondary machining gas carburizing heat treatments are applied.

Previously, the axle cones were produced from spheroidized annealed (process annealed) bar in 1% carbon-chromium steel.

(The term 'spheroidized annealed' is the term describing heating to just below the lower transformation temperature (A_{C1}) so that the pearlite carbide is redistributed as coarse spheroids, giving the softest condition for machining or cold work. It can be applied to lower carbon pearlite structures, although more usually relevant to high carbon and air-hardening steels to enable machining to be carried out prior to through-hardening. The same condition, enabling cold work, is used in the manufacture of razors, needles, ball bearings, etc.)

The axle cones are now manufactured from low-carbon case hardening steel 210 M15 by machining on multi-spindle automatic lathes followed by case hardening and quenching into oil. The ball bearings are produced by external suppliers from 1% carbon-chromium steel.

The hub shell is manufactured by cold press-work from a 15.24 mm (6 in) diameter disc in an extra-deep drawing, aluminium killed mild steel to BS 1449 Part 1, 1983, C.S.1. specification: C 0.12 max, Mn 0.5, S 0.05, P 0.05.

The disc is cupped and then gradually stretched into a tube by a series of drawing operations followed by blanking out the bottom. The ends of the tube are then turned at right angles to the tube to produce flanges. It might be asked why thin-walled tube is not used, cut into sections and flanged. Much of the cost of a tube length is inherent in getting a long continuous hollow section, and it is therefore wasteful to subsequently chop it up into short lengths. The cost differential between tube and strip stock allows the forming operation to close tolerance to be effected with cost saving. The flanges are pierced by presswork for the spokes. The outside is machine burnished, and a protective and decorative finish of 0.025 mm duplex nickel and chromium is applied by passing the hub shells through a programmable plating plant.

The small components nuts, washers, etc are barrel-plated with zinc, passivated and clear lacquered.

21.3 High-power gridded tube (English Electric Valve Co. Ltd.*)

This study is included primarily for its consideration of the use of refractory metals which will maintain their shape and integrity when operating at very high temperatures for long periods – an essential feature in the operation of high-power gridded tubes. Of secondary interest are the techniques for making hermetic seals between the metal and insulating parts of the vacuum envelope.

Contrary to a widely held view, gridded tubes are still extensively used where high-power and high-frequency requirements make it impractical or impossible to use solid-state devices. The tube to be described below, and seen in Figure 21.4 in its component sub-assemblies, is a 200 kW output triode for use in radio frequency industrial heating at frequencies from 50 kHz up to 27.12 MHz (EEV type BW 1185J2). A schematic cross-section is shown in Figure 21.5.

The cathode

Cathode types in common use in electron tubes are shown in Table 21.2. The cathode most commonly used in high-power gridded tubes, and used in this case, is made of thoriated tungsten wire directly heated. The thoria serves two purposes: it stabilizes the grains against

* The authors gratefully acknowledge the assistance of Mr D. Carr in the preparation of this case study.

Figure 21.4 Components of high-power gridded valve.

Figure 21.5 Schematic cross-section of gridded valve.

grain growth at the high temperature of operation, which would otherwise greatly weaken the filament and it increases the emission from the cathode by reducing the work function as described further below.

The dispersion of thoria is obtained by power metallurgy, and the tungsten billet is then

TABLE 21.2. Cathodes commonly used in electron tubes

Cathode type	Operating temperature (°K)	Work function (eV)	Emission density (A/cm²)	Heater power (W/cm²)	Emission efficiency (mA/cm²/W)	Application
Ba, Sr, Ca oxides on activated nickel substrate indirectly heated	1000	1.0	0.1*	2	50*	Small valves, magnetrons, low-power TWTs, camera tubes, CRTs, klystrons
Ba aluminate impregnated into tungsten matrix indirectly heated	1300	2.0	0.2*	6	33*	High-power beam tubes
Thoriated tungsten directly heated	2000	2.7	1.0*	25	40*	High-power tubes
Silicated tungsten directly heated	2600	4.5	0.4	75	5	Obsolete high-power gridded tubes

* The figures given are for long-life cathodes (5–10,000 hours). Larger currents can be taken in pulses with low-duty cycles, or continuously with sacrifice of life.

reduced to wire form by swageing and drawing. This process produces a fibrous structure of long grains which gives the material its strength, and the thoria helps maintain this structure at high temperature by impeding recrystallization. If wire of a special sag resistance is required, for a helical filament for instance, a sample will be tested by heating to 2600 K for 10 minutes and subsequently sectioned to show the crystal structure. To pass this test the crystal length-to-width ratio must exceed four. Without doping, the crystal structure after heating is poor.

The thoria improves the cathode emission by acting as a source of thorium for the monatomic layer on the wire surface which reduces the work function from 4.5 to 2.7 eV. Two other conditions are required to maintain stable emission at 2000 K: the tube must have a good vacuum to avoid excessive back bombardment by gas ions which could sputter away the thorium layer faster than it can be replaced; the tungsten must be carburized in its surface layers.

The exact function of the carburization is not fully understood. It is not known if the thoria is all reduced during the carburizing operation, or if it happens gradually during the running life of the tube. However, it is known that thorium steadily diffuses to the surface during running, and replaces material lost by evaporation or ion bombardment. We also know that a decarburized filament loses its emission. The filament temperature must be carefully chosen to ensure that the carbon does not diffuse or evaporate away during a life of 1000–20,000 hours.

Most modern cathodes are made in cylindrical mesh form with one layer of helical wires laid over a layer of the opposite hand and spot-welded at each intersection. The cathode is supported inside the anode on coaxial molybdenum tubes, to which it is welded by arcing with a carbon electrode in a protective (hydrogen or cracker gas) atmosphere. The molybdenum support tubes are joined in turn to the cathode feed-through contacts at the opposite end of the tube.

The control grid

Between the cathode and the anode is the control grid, which is usually produced as a welded wire cage in molybdenum or tungsten, so as to maintain its shape during the high-temperature operation. The cage is usually mounted on the end of a molybdenum tube, which locates the grid around the tungsten wire mesh cathode. On most grids there is a 20% shadow area, and some direct electron emission results from the heating, particularly as thoria evaporates from the cathode and tends to deposit on the grid. If the grid emits, the tube loses efficiency. Apart from minimizing the shadow area, this is reduced by plating the grid with platinum, which can absorb thorium, can maintain a high work function (5.4 eV) and so holds down the electron emission. Unfortunately, at these temperatures the platinum and the molybdenum or tungsten can alloy through diffusion during use, and distortion may occur. To prevent this it is necessary to create a barrier between the platinum surface and the refractory metal substrate. One method is first to coat the latter with zirconium carbide by cataphoretic coating and vacuum sintering at 2000°C. This minutely rough surface is then platinum-plated and the whole assembly fired again in vacuum at 1500°C. The resultant grey platinum surface has high thermal emissivity and low primary electron emissivity.

Like the cathode, the grid is supported from the contact feed-through by a molybdenum tube (see Figure 21.5 for section.)

Modern high power ultra high frequency tubes need closely spaced grids that radiate incident power well, do not emit electrons, and do not distort at high operating temperatures. Pyrolytic carbon is an almost ideal material for this purpose. It is deposited from methane at 2000°C onto a suitably shaped graphite substrate mandrel. A thin-walled one piece pyrolytic carbon shell comprising grid cap, cylindrical active area and grid support skirt separates from the mandrel by differential contraction on cooling. The grid pattern is then machined in the cylindrical active area by shot-blast erosion through a metal mask; bolt holes in the support skirt can be machined in the same way.

Some of the metal parts in the tube do not get hot, and as we move away from the electrodes, so less refractory metals can be used which are easier to form and less expensive. Such parts are usually copper or nickel. In some tubes even high-purity iron is employed – in fact one type of tube, the Magnetron, has a strong magnetic field inside, developed from the iron pole pieces of an electromagnet.

Another metal in the valve is the getter which is put in to remove the last traces of reactive gases in the envelope whilst running. The metal is usually zirconium or titanium. In many of the big valves the getter is mounted as a piece of foil or sheet on the cathode support, or it may be as a coating on the molybdenum mount for the grid.

Although it has been implied that the refractory metals are difficult to fabricate this is only a relative assessment, and molybdenum is readily formed (e.g. spun) as is obvious from the tube form often adopted for the tube mount for the grid.

The anode

Gridded valves are not 100% efficient in converting the electrical energy supplied into radio frequency power; 10–40% of the power input is dissipated as heat, most of which is on the anode. In a high-power valve it is not reasonable to put the anode inside an envelope as in small valves, and instead the anode is made part of the envelope so that the heat developed can be got away and overheating does not develop. In order to transfer heat rapidly to external cooling the anode is made as a thick-walled copper cylinder with the outside surface shaped according to the intended cooling method. For water cooling the anode surface will be scrolled (deep multi-start square thread), and the finished tube will be fitted in a water jacket. For boiling water cooling, holes parallel to the tube axis and open at both ends are drilled in the considerable thickness of the anode wall; the anode is then stood in a boiler jacket for operation and fed with purified water while steam is removed for condensation in a separate heat exchanger. For air cooling the anode is provided with copper fins brazed to the anode cylinder with copper–silver eutectic solder.

Inside, the anode is usually plated with a rough chromium-plate deposited in a plating bath at high current density. The grey or black film produced, whilst having a high thermal resistance, helps the anode to absorb heat, and not reflect it back on the grid and centre cathode.

The tube envelope

The tube envelope must be hermetically sealed so that, after evacuation, a satisfactory working pressure of 10^{-5} torr can be maintained. The anode, closed at one end, and the grid and cathode feed-through contacts, are joined to, and electrically isolated by, sections of impervious alumina insulator.

As sub-assemblies are joined to make larger assemblies, each successive join must be made with a lower-temperature solder to avoid dismantling earlier work. Generally tungsten and molybdenum are joined by arc welding, but they can also be brazed with platinum. Nickel can be used for brazing molybdenum. Nickel, iron and nickel–cobalt–iron alloy (Ni–Co–Fe) can be brazed with copper. However, where copper components are involved, copper–gold, copper–silver–palladium, and copper–silver eutectic alloys can be used in descending order of temperature. Brazing is carried out in a reducing atmosphere, and no flux is required.

Hermetic alumina ceramics are almost invariably used as envelope insulators in new tube designs, but borosilicate glass is still used in some of the smaller types in current production. The techniques for joining these insulators to the metal components are described further below.

Ceramic to metal seals

After washing the ceramic, the ends to be joined are first painted with finely divided molybdenum. After drying, the ceramic is furnaced at 1460°C in a protective atmosphere to bond the molybdenum to the glassy constituents of the ceramic. The metallizing is then nickel-plated and furnaced at 800°C in a protective atmosphere to check for adhesion. The insulator can now be brazed with copper–silver eutectic solder to adjacent metal parts.

Glass to metal seals

Assuming that the metal is rigid and does not deform plastically, it is possible to match a glass to any metal as regards expansion coefficient at any particular temperature, but the requirements are much more complex. There are limitations on the sort of glass that can be used. It has to be baked at, say, 450°C and must also not soften in use. It must also not electrolyse as a supercooled liquid, with the migration of sodium, for example, with resultant failure. There are also, therefore, restrictions on the metals that can be used to match useful glasses as regards expansion coefficient at a specific temperature. Suitable glasses may be found to match tungsten and molybdenum, but these metals themselves, produced by powder metallurgy, may not be sufficiently pressure-tight.

In any case, whereas the expansion for a pure metal will be linear with temperature, for a glass it is likely to diverge considerably within the range up to the temperature where it expands to the same total extent. There will therefore be temperature ranges in which stresses may be set up at the joint, before the matching temperature is reached.

Beyond a certain temperature, of course, the glass will become soft and will move with the metal. At such a temperature, however, it loses its usefulness as an envelope material.

There are, however, alloys available which are not only produced in absolutely sound form by casting and mechanical working methods, but which due to micro-structural changes also have a shaped expansion curve of a similar form to borosilicate glass, so that the expansion coefficients match closely over the whole range up to the glass softening point (e.g. Covar, Nilo K, 29 Ni–17 Co–54 Fe). The cast and wrought form may be preferred for vacuum tightness. For absolute uniformity of composition, and therefore of physical properties, the powder metallurgy manufacturing route may be advantageous. Full density and freedom from any porosity, and lack of pressure tightness, is, however, essential. A second form of seal is with copper or nickel, where the metal is machined off to a thin lip or flange so that it deforms with the glass or ceramic. Stresses induced by the glass are sufficient to extend or compress the metal non-elastically. In this case it is not rigid and merely changing shape by thermal expansion alone.

The metals are oxidized at the jointing points prior to sealing, either on a glass lathe or in a special air furnace. In the latter case the surface may be sprayed with ground glass and thus glazed in the one furnace operation. The glass partially absorbs the tenacious metal oxide and so makes a strong hermetic seal. Much of this work is still carried out on glass lathes with oxy-gas torches and ring-fires.

After sealing into sub-assemblies the Ni–Co–Fe components to be used are usually silver-plated or copper-plated, principally for reasons of electrical and thermal conductivity. Such platings significantly increase the heat dissipation and the power loading of a tube. Copper

plate is satisfactory internally, but oxidation on the outside reduces its effectiveness. Silver plate forms sulphide as the only atmospheric deterioration product, and this has virtually no effect on the surface conductivity.

Evacuation

The finished tube assembly is usually evacuated using diffusion and rotary pumps via an annealed copper tube. During pumping each component of the tube will be heated to a temperature higher than it will meet in life, in order that most gas can be de-sorbed and pumped away. Finally the tube is 'pinched off' from pumps by squeezing the copper tube with hydraulically moved hardened steel rollers, which make a cold weld either side and then part the tube.

22
Problems

Static strength

1. Compare and contrast the stress-strain response of polymers and ceramics, taking into consideration the effects of loading rate and temperature. Discuss the deformation mechanisms operating in each case.

   Comment on the use of ceramics in structural engineering, emphasizing their advantages and disadvantages and highlighting how modern research is looking to improve their capability.

2. A structure is designed as a horizontal cantilever beam, of circular cross-section, and length 1 m. One end is built into a wall and the other supports a load of 3 kN. Applying a safety factor of 2, what bar diameter is required to prevent surface yielding, if the material is:

   (a) mild steel of yield strength 225 MPa?
   (b) maraging steel of yield strength 2250 MPa?

   [Hint: Second moment of area $I = \dfrac{\pi d^4}{64}$

   for round bar of diameter d.

   Bending moment = force × moment arm

   $\therefore \mathrm{d}M = \mathrm{d}A \times x$]

3. Define the terms *principal stress* and *principal strain*.

   Calculate the principal strains produced in a piece of silicon-iron sheet by the 2-D stress system: $\sigma_{11} = 10\,\mathrm{MPa}$; $\sigma_{22} = 4\,\mathrm{MPa}$; $\sigma_{12} = 16\,\mathrm{MPa}$.

   Determine the magnitude of the stress σ_{33} to reduce the through-thickness strain ϵ_3 to zero.

Why might your calculations be inappropriate for rolled and annealed silicon iron sheet?

Assume: $E = 200\,\mathrm{GPa}$ and $\nu = 0.3$.

[Hint: Construct Mohr's circle to calculate principal stresses σ_1 and σ_2 and use

$$\epsilon_1 = \frac{1}{E}[\sigma_1 - \nu(\sigma_2 + \sigma_3)], \text{ etc.}]$$

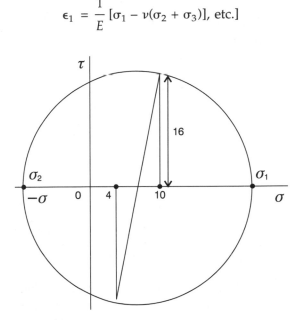

Figure 22.1

4. You have been asked to select material for the manufacture of the hull of a nuclear submarine, which is designed to have a maximum speed of 40 knots and a maximum operating depth of 150 fathoms (275 m). The hull is to be 6 m in diameter and good quality plate can be guaranteed only for thicknesses up to 40 mm. The interior is to be pressurized

to 2 atmospheres and the retarding force on the vessel may be taken as an average of 200 kN/knot. Estimate the principal stresses in the hull and select a suitable material from the following:

- normalized mild steel, yield stress 260 MPa;
- weldable carbon-manganese steel, y.s. 400 MPa;
- weldable nickel-chromium steel, y.s. 540 MPa.

[Hint: for a thin-walled tube,

$$\sigma_{hoop} = \frac{Pr}{t}, \quad \sigma_{longitudinal} = \frac{Pr}{2t} \quad \text{and}$$

$$\sigma_{radial} = 0,$$

where P = internal pressure, r = vessel radius and t = wall thickness]

5. A composite material has unidirectional glass fibres surrounded by an epoxy matrix. Details of the material are given below.

Modulus of fibres	E_f	= 90 GPa
Modulus of matrix	E_m	= 3 GPa
UTS of fibres	σ_f^*	= 1000 MPa
UTS of matrix	σ_m^*	= 200 MPa
Volume fraction of fibres	V_f	= 0.25
Volume fraction of matrix	V_m	= 0.75
Interfacial shear strength	τ_i^*	= 5.5 MPa
Fibre length to radius ratio	L/r	= 150

(a) Calculate the tensile modulus of this composite when stressed parallel to the fibres using the Halpin–Tsai equation:

$$E_c = E_m \left[\frac{1 + (L/r)\eta V_f}{1 - \eta V_f} \right]$$

where $\eta = (E_f/E_m - 1)/(E_f/E_m + (L/r))$.

(b) Using this modulus, calculate the following:

(i) The tensile stress, parallel to the fibres, at which fibre pull out occurs.

[The stress on the fibres is given by $\sigma_f = \tau L/r$ where τ is the interfacial shear stress].

[Hint: $E_{fibre} = E_{matrix} = E_{composite}$]

(ii) The tensile stress, parallel to the fibres at which fibre fracture occurs.
(iii) The tensile stress at which subsequent matrix failure occurs.

Having calculated these values, determine the mechanisms of failure and the UTS of this composite.

Toughness

6. Explain how the presence of a notch in a specimen under uniaxial loading can introduce local stress triaxiality. Sketch the resulting elastic stress distribution at the notch.

7. Outline how compressive surface stresses may be used to give an apparent increase in the fracture toughness of a brittle material.

Describe how such stresses are generated in:

(a) Chemically toughened glasses;
(b) Zirconia toughened ceramics.

For case (a) explain why glasses with deep surface defects may show little or no increase in strength.

8. Ceramic bars are tested in three point loading and the following strengths are obtained (units MPa):

1240, 1290, 1180, 1360, 1110, 1200, 1250, 1270, 1310.

(a) Estimate the Weibull modulus and explain what significance it has.
(b) Find the probability of failure at service stresses of 1 GPa and 800 MPa.
(c) If the volume of the testpieces were doubled what would be the new average strength?

9. Why can the tetragonal to monoclinic transformation in zirconia increase its resistance to crack growth? Why does this also lead to an increase in the Weibull modulus?

A small Si_3N_4 tube in a longer component is designed so that under the peak operating stress of 337 MPa there is a one in a thousand probability of failure. It only costs 10p, but the cost to the company of replacing it in the whole component under warranty is £100. Tubes of calcia-stabilized ZrO_2 are available at a cost of 15p each. Testing indicates a Weibull modulus of 15 and a 50% failure probability at 650 MPa. Should the company change to the zirconia tube?

10. A weldable aluminium alloy, of yield strength 300 MPa and plane-strain fracture toughness $44.8\,MNm^{-3/2}$ is used to make large, thick-walled pressure-vessels, having pipe inlets which are circular holes of diameter 1 m. The conventional safety factor is 1.5. If the welding process produces a defect at the circumference of the hole, normal to the hoop stress, of length 8 mm, calculate the lower bound to the critical value of hoop stress at which the vessel can be safely operated. What would be the lower bound if the initial defect size were 1 mm in length?

[Hint: $K_{1c} = 1.12\,\sigma\sqrt{\pi a}$, where K_{1C} is the plane-strain fracture toughness, σ is the applied stress and a is the critical crack length. Stress at the edge of a circular hole under applied stress $\sigma = 3\sigma$.]

Stiffness

11. Explain what is meant by tensile-shear interactions during stressing of a unidirectional fibre composite sheet and give a brief outline of the factors affecting elastic distortion of laminates.

Derive the following simple relationships for the elastic constants of a single lamina where subscripts f and m below refer to the fibre and matrix respectively:

$$E_1 = V_f E_f + V_m E_m;$$
$$\frac{1}{E_2} = \frac{V_f}{E_f} + \frac{V_m}{E_m};$$
$$\frac{1}{G_{12}} = \frac{V_f}{G_f} + \frac{V_m}{G_m};$$
$$\nu_{12} = V_f \nu_f + V_m \nu_m.$$

where E_i are Young's moduli;

V_i are volume fractions;

G_i are shear moduli;

ν_i are Poisson's ratios.

Hence calculate the percentage decrease in tensile stiffness (relative to the stiffness parallel to the fibre axis) for a lamina of 60 vol% carbon fibres in an epoxy resin, loaded in tension at an angle of 10° to the fibre axis.

$$E_f = 380\,GPa; \quad \nu_f = 0.35;$$
$$E_m = 4\,GPa; \quad \nu_m = 0.39$$

[Hint: the expression for the first element of the transformed compliance matrix is:

$$S_{11} = \left(\frac{1}{E_1}\right)\cos^4\phi + \left(\frac{1}{G_{12}} - \frac{2\nu_{12}}{E_1}\right) \times$$
$$\sin^2\phi\cos^2\phi + \left(\frac{1}{E_2}\right)\sin^4\phi,$$

$$\text{and} \quad \nu_{12} = \frac{-\epsilon_2}{\epsilon_1}.]$$

12. A cast-iron tube, 3 m long, is completely filled with concrete and used as a vertical strut. If the external diameter of the tube is 450 mm and the wall is 35 mm thick, determine the maximum compressive load the composite strut can support if the stress in the concrete is not to exceed 2 MPa. By how much will the strut shorten under this load?

For cast-iron $E = 100\,GPa$;
for concrete, $E = 10\,GPa$.

13. Using the slab model, show that the shear modulus G_{12} of a long fibre composite is given by

$$G_{12} = \left[\frac{f}{G_F} + \frac{(1-f)}{G_m} \right]^{-1}$$

where f is the volume fraction of the fibres, G_f and G_m are the shear moduli of the fibre and matrix respectively and the fibres are lying in the l-direction.

14. (a) Show that, for a horizontal cantilever beam of length l, with one end built into a wall, subject to a load P at its free end, the maximum deflection is given by

$$\frac{Pl^3}{3EI}$$

where E is Young's modulus and I is the second moment-of-area.

[Hint: $\dfrac{d^2y}{dx^2} = \dfrac{-M}{EI}$

where M = bending moment.]

(b) Show that the lightest solid cylindrical tie rod of given length to give the minimum extension under a given uniaxial tensile load will have the maximum value of E/ρ, where ρ is the rod density.

(c) Also show that, to minimize the mass and central deflection of a beam fixed at each end of fixed length l under bending moment M, where the beam width, $W = 2B$, where B is the beam height, the value of E/ρ should be maximized if B is fixed and E/ρ^2 should be maximized if B varies.

[Hint: $\dfrac{d^2y}{dx^2} = \dfrac{-M}{EI}$ and $I = \dfrac{BW^3}{12}$.]

(d) From the following list of materials, select the best for minimum weight/ minimum extension applications under uniaxial load.

Also select the best material for a minimum weight/minimum deflection beam of specified shape but free section area.

	E (GPa)	ρ (Mg/m^3)
Wood (spruce parallel to grain)	13.0	0.50
Aluminium	69.9	2.70
Titanium	121.0	4.53
Glass fibre/resin composite	20.0	2.00
Kevlar fibre/resin composite	130.0	1.45

15. Give reasons for choosing composites in preference to monolithic materials, commenting on limitations and demonstrating how these might be alleviated. Describe the essential elements of a composite system giving examples where appropriate.

A uni-directional composite is made with 60% by volume carbon fibre with a modulus of 338 GPa embedded in an epoxy matrix with a modulus of 3.8 GPa. Comment on what modulus values you might expect the composite to have parallel and perpendicular to the fibres.

Fatigue

16. In fatigue testing, distinguish between the terms 'endurance limit' and 'fatigue limit'. Use the diagram below.

Figure 22.2

17. A critical value of the elastic stress intensity factor, K_{IC}, is used to describe brittle fracture, and the stress intensity factor range, ΔK, is used to characterize fatigue crack propagation, a predominantly plastic process. Justify the use of K in both these cases, and hence explain the apparent contradiction.

An aircraft wing spar, made of a 7475 Al–Mg–Zn alloy, is inspected for cracks every 20 flights. The non-destructive inspection (NDI) technique employed can detect surface defects greater than 1 mm deep. During a flight the stress in the spar fluctuates between 20 and 150 MPa, on average 1200 times. The fatigue crack growth behaviour of the alloy can be described by

$$\frac{da}{dN} = 3.10^{-11}\ \Delta K^{3.6}$$

(for a in m, ΔK in MPa m$^{1/2}$)

Comment on the inspection interval.

Would it be better to use a higher toughness alloy for the component or to purchase new NDI equipment with a resolution of 750 μm? Give reasons for your answer.

[Data for the 7475 alloy: E = 71 GPa, 0.2% proof stress = 540 MPa, K_{IC} = 43 MPa m$^{1/2}$]

18. What is meant by high cycle fatigue and low cycle fatigue and what two equations are used to describe smooth surface component life in these two situations? Suggest typical components which might be subject to these types of fatigue.

The lifetime of an aero-engine is limited by the failure of some titanium alloy compressor blades. The rotation of the compressor induces a constant centrifugal stress of 200 MPa in the blades as well as vibrational stresses with $\Delta\sigma$ = 30 MPa at a frequency of 10 Hz. During flight, thermal effects also result in an additional tensile stress of 300 MPa in some regions of the blade. Estimate the number of six–hour flights the aircraft can make before failure of the blade occurs, given that Basquin's Law is

$$\Delta\sigma\ (N_f)^b = B$$

where the constants b and B have values of 0.1 and 700 MPa respectively and N_f is the number of cycles to failure.

On take off, a bird is ingested into the aero-engine and strikes a compressor blade causing an edge crack of length 1 mm. Estimate the remaining flight time for the compressor blades, given that the increase in crack length per cycle da/dN is related to the difference between the maximum and minimum stress intensity factors ΔK according to

$$\frac{da}{dN} = 3.5 \times 10^{-9}\ \Delta K^{2.5}$$

where the units of da/dN are m/cycle and those of ΔK are MPa $\sqrt{m}$.

[For the titanium alloy: K_{IC} = 100/MPa m$^{1/2}$; E = 100 GPa; α = 7.10^{-6} K^{-1}.]

[Hint: Use Miner's Law $\Sigma\ \dfrac{n}{N_f} = 1$

to determine the number of flights until failure under the combination of thermal, mechanical and vibrational loading.]

19. Subcritical crack growth can be described in terms of the crack tip stress intensity factor, K, in a wide range of loading conditions. Briefly describe the variation of crack growth rate with K in a metallic alloy under conditions of (a) fatigue and (b) stress corrosion cracking. How are the two approaches combined to predict corrosion fatigue behaviour?

(c) An airline uses a fleet of similar planes on short inland routes. The engine support spars, of forged 7075 aluminium alloy, experience a major loading cycle on every flight. At a particular location on the spar the stress rises rapidly to 250 MPa (tension) for the duration of the flight and drops again to zero on landing. If the resolution of the NDI technique used both before and during service is 0.75 mm, recommend suitable inspection intervals for this component.

(b) Following several years of service and after fitting with new engine support spars, the planes are sold on to another company for island hopping between coastal airports. When the new owners request your advice on inspection intervals, what do you recommend? (The average flight duration between islands is 30 minutes.) State any simplifying assumptions which you make in your calculations.

[For the alloy and conditions in question:

0.2% proof stress = 430 MPa

K_{Ic} = 35 MPa $\sqrt{m}$

K_{Iscc} (sea water) = 9.2 MPa $\sqrt{m}$

ΔK_{th} (air) = 2.4 MPa $\sqrt{m}$

$\dfrac{da}{dt}$ = $1.67.10^{-17} K^6$ in sea water

$\dfrac{da}{dN}$ = $6.78.10^{-12} \Delta K^{3.2}$ in air

In the above two expressions, a is in m, t is in s and K is in MPa $\sqrt{m}$.

You may assume that $K = 1.12\,\sigma\,\sqrt{(\pi a)}$ for a surface crack.]

20. The so-called 'Paris Law', which treats the propagation of fatigue cracks, states that the crack growth increment per cycle, da/dN, is related to the stress-intensity range, ΔK, through the expression

$$da/dN = A\,\Delta K^m$$

where A and m are constants. Explain the fact that the value of m is close to 2 for ductile materials, but is often much higher for brittle materials.

A large, thick-walled pressure vessel (made in a quenched-and-tempered forged steel possessing a 0.2% proof stress of 1 GPa) is periodically cycled from zero stress to half its proof stress. The vessel contains a keyway, with sharp corners, of depth 2 mm and cracks are observed to initiate from these stress raisers after the first cycle. Two grades of steel might be employed: an air-melted version having a toughness of 80 MPa m$^{1/2}$ and a vacuum-melted version with a toughness of 160 MPa m$^{1/2}$. Given that the rate of fatigue-crack propagation is

$$da/dN = 10^{-11}\,\Delta K^3 \text{ m/cycle}$$

calculate the fatigue lives for these vessels. One vessel was found to fail after 800 cycles. What value of fracture toughness does this imply? What metallurgical factors could contribute to this value?

Creep and temperature resistance

21. Explain the conditions and materials for which creep behaviour is important and discuss the mechanisms involved, e.g. visco-elastic and plastic deformation, thermally activated processes, diffusion or dislocation climb. What is secondary creep and how can creep curves be interpreted for polymers and for metals?

22. Explain the meaning of viscoelasticity in polymeric materials, commenting on the practical significance of this concept in engineering design.

Discuss the theoretical basis of creep and stress relaxation responses in plastics through consideration of spring and dashpot models, including basic mathematical derivations used to describe linear viscoelastic behaviour.

Illustrate common ways for representing creep data suitable for design purposes under constant and intermittent loading conditions.

23. (a) With reference to Figure 22.3, describe how creep curves can be transformed into (i) isochronous, (ii) isometric and (iii) creep modulus – time curves.

(b) Your remit is to consider the design of a 1.0 metre diameter spherical, polypropylene water storage tank for underground use. The design requirements are that the tank should resist an external pressure of 20 kPa for a period of 3 years.

Calculate a suitable value for the wall thickness using the data provided in Figure 22.3.

Additional information:

The value of the external pressure, (P_c), which would cause a thin wall sphere of radius, (R), to collapse is given by $P_c = 0.365\ E\ (h/R)^2$ where h = wall thickness and E is the modulus.

Figure 22.3 Creep curves for polypropylene at 20°C.

Corrosion and wear

24. Describe what is meant by oxidation and reduction in the context of corrosion. Write down and explain the significance of two reactions of each type.

Explain why the corrosion of zinc in dilute hydrochloric acid is increased by:

(a) raising the dissolved oxygen content of the acid;

(b) adding ferric ions to the acid.

25. Write on the following:

(i) The characteristic features of cleavage fracture and the factors that affect the likelihood of this type of fracture occurring in low-carbon (structural) steel plate.

(ii) Hydrogen induced cracking (HIC) of steel pipe-lines carrying sour gas/oil.

(iii) Characteristics and examples of stress-corrosion cracking.

26. Discuss the major methods which are available to protect metallic alloys from corrosion.

Identify for each of the following applications what method of protection you would select, and justify your choice:

(a) Cold drawn α-brass tubes which are to be used in a chemical laboratory where ammonium hydroxide solution is regularly decanted.

(b) Cast iron water mains with power cables buried nearby.

(c) Legs of a North Sea oil platform.

27. Over recent years, motor vehicle manufacturers have become more aware of corrosion related problems. Briefly discuss the main problem areas.

Discuss how the use of materials, construction methods and finishing techniques have developed to provide enhanced corrosion protection.

Discuss the relevance of corrosion warranties to both the vehicle manufacturer and the customer.

28. Identify the corrosion mechanism and suggest suitable countermeasures in each of the following cases.

(a) When restoring his 'Classic' sports car the owner decided to fix the plastic fasteners for his soft top to the steel body of the car using brass nuts and bolts because these 'would not go rusty'. Within weeks rings of rust had broken through the paint immediately surrounding the fasteners.

(b) A copper lightning conductor was mounted on a brick chimney above an unpainted aluminium roof in such a way that rainwater falling onto the conductor subsequently ran over the roof. No electrical connection existed between the conductor and the roof. The aluminium roof was designed to last for decades but in fact it corroded severely and perforated within months.

(c) An unpainted aluminium aerial support situated behind a ship's funnel displayed severe pitting corrosion. The funnel acted as an exhaust for the sooty combustion products of the fuel oil used to heat the ship's boiler.

(d) Mild steel sheets were stored in a warehouse where they were stacked horizontally. The atmosphere in the warehouse was damp and periodically droplets of water would form on the sheets by condensation. The exposed surfaces of the sheets were soon covered with circular scabs or tubercles of rust. When the rust was removed with a wire brush the underlying metal was seen to be deeply pitted.

(e) The doors of a certain saloon car were made from thin sheet steel lined with sound absorbent foam. It was a common occurrence to find rows of blisters in the external paintwork of the door after only two or three years service.

29. Describe the tribological interactions that can take place in the piston/cylinder bore area of the internal combustion engine.

Arising out of this basic understanding, discuss the main features of cast iron which has led to its extremely wide use in this area.

What materials/processes are available for improving the wear resistance of cast iron.

Selection and processing

30. During the process of selecting a material for a new mechanical application, it is important to use a sound methodology. Briefly outline each stage of a suitable methodology.

Describe in greater detail the first stage in this methodology.

31. Describe and compare critically the possible methods of selecting materials. Include in your answer a discussion of databases containing properties and costs, material selection diagrams, figures of merit and expert systems.

Specify, giving reasons, the material and fabrication route you would recommend to make each of the following.

(a) A screw to be used in constructing the hull of a wooden boat.
(b) A spanner.
(c) A curtain track.
(d) A cylinder block for a car engine.

32. (a) When trying to minimize component weight what are optimal strength and optimal stiffness and how are they quantified?
(b) Why is it often difficult to apply these criteria in materials selection problems for:
 - metals
 - polymers
 - ceramics
 - composite materials?
(c) If you do not have a choice of materials and are constrained to use only one particular material how would you go about achieving optimal strength and stiffness?

33. The properties of various materials which can be used to fabricate traditional tubular bicycle frames are listed in Table 22.1 below.

Comment, without detailed calculations, on the likely modes of mechanical failure of such frames. Hence suggest relative significance factors (with values between 1 and 10) for each of the properties in Table 22.1. Use these factors to derive an overall rating for each material, and comment on the implications of your results for both the manufacture and use of bicycles.

34. (a) Describe briefly selection techniques which can be used to identify suitable polymers for a specified application.
(b) List and comment briefly on factors which have to be considered in selecting polymers for use in the main body of a domestic kettle.
(c) Propose and justify your selection for polymers which can be used in the main body of a domestic kettle.

TABLE 22.1.

Material	Yield stress σ_y MPa	Fracture toughness K_c MPa m$^{1/2}$	Young's modulus E GPa	Density ρ Mg m^{-3}	Cost £ Kg^{-1}
Al alloy	350	45	70	2.7	4
Ti alloy	900	70	120	4.4	35
Steel	1100	110	190	7.8	0.7
C-fibre composite	600	40	220	1.5	120

35. Figure 22.4 shows the constant cross-section of a blade for the impeller of a fan which is designed to operate at 10 000 rpm and is used for ten hours per day during which the temperature may rise to 60°C.

Figure 22.4 Impeller blade

(a) (i) Show from first principles that the maximum tensile stress in the blade due to centrifugal forces occurs at the root and is given by $\sigma = \rho\omega^2 L(a + L/2)$ where ρ is the density and ω is the angular velocity.

[Hint: $\int d\sigma = \rho\omega^2 \int_a^R l\,dl$]

(ii) Calculate the maximum tensile stress in the blade for an impeller with a hub outer diameter of 50 mm and a blade length of 100 mm when made from a material of density = 1.2 Mg/m³.

(b) (i) If the impeller is closely fitted into an annular shroud suggest a method for determining the minimum clearance between the impeller and shroud due to lengthening of the blade by centrifugal forces. You may assume that complete recovery of the material occurs during the period of the day when the fan is not in use.

(ii) Briefly state what other parameters need to be considered when determining the clearance between the impeller and the shroud.

(iii) Suggest a suitable material for the impeller giving reasons for your choice.

36. Discuss the advantages and the drawbacks of choosing a thermosetting material in preference to a thermoplastic when selecting a polymeric material for an engineering application.

A manufacturing company wishes to make the base tray for a fume cupboard from a polymeric material. It is planned that the company will be manufacturing about 1000 of these trays using either injection moulding or compression moulding. The trays will have to support the weight of full chemical jars and withstand articles being dropped onto them.

From the list of materials in Table 22.2, decide which should be used. Give all your reasons for selecting or eliminating materials.

37. Suggest a methodology which would assist in the selection of the appropriate polymeric material for an engineering application, through consideration of property and design requirements, together with manufacturing and economic criteria.

Consider ways in which the stiffness and strength of polymeric products can be influenced by material composition, component shape, design, and manufacturing procedure (particularly in relation to effects on polymer microstructure).

38. Compared to metals and ceramics, the use of polymeric materials in engineering applications is often restricted by their limited toughness, stiffness and strength.

Consider approaches commonly adopted to extend the range of useful mechanical

TABLE 22.2.

	PE	PS	Nylon	ABS	GRP*
Tensile strength (MPa)	15	35	45	45	100
Modulus (GPa)	0.8	2.8	1.6	3.2	12
Impact strength (Jm^{-1})	100	10	150	150	200
Creep resistance	P	G	M	G	G
Cost (£/kg)	1.0	0.8	2.8	1.4	2.1
Injection mouldability	G	G	G	G	P
Compression mouldability	M	G	M	M	G
Resistance to acids	G	G	P	G	G
Resistance to solvents	G	P	G	M	G

*Glass fibre reinforced polyester. G: Good, M: Moderate, P: Poor

properties available from plastics, through the following modifications:

(i) introduction of fibrous reinforcing additives,
(ii) blending with rubbery modifiers,
(iii) preparation of integral skin plastics foams

Your answer should include specific examples relevant to each of these approaches, an indication of mechanical property benefits which can result from the modifying process, and a brief mechanistic explanation accounting for the observed changes in behaviour.

39. What are the advantages and disadvantages of adhesive bonding compared with other joining methods?

Describe the mechanical, electrostatic and diffusion theories of adhesion and define their limitations.

40. Rapid prototyping (colloquially known as desk-top manufacturing) is an emerging technique. Describe the three generic types of rapid prototyping (photopolymer, powder and sheet). What are the advantages and disadvantages of this technology?

41. Describe, giving reasons, the processing routes and materials you would use for the following items:

(a) the cap and bottle for a liquid detergent;
(b) the casing for a washing machine;
(c) a child's plastic duck;
(d) a police riot shield;
(e) a bearing to support a rotating shaft.

42. Describe, giving reasons, the processing routes and materials you would use for the following items:

(a) the skin of a light aircraft;
(b) the leading surface of a space re-entry vehicle;
(c) an electric light bulb filament;
(d) a car crankshaft;
(e) the body of a spark plug.

43. Identify the following materials. Discuss the significance of the composition in relation to the properties and microstructure commonly observed. Indicate typical commercial uses for each material. (% = wt%)

(a) 30% Zn, bal. Cu;
(b) 4.3% Cu, 0.6% Mg, 0.8% Si, 0.75% Mn, bal. Al;
(c) 6% Al, 4% V, bal. Ti;
(d) 14% H, 86% C;
(e) 0.5% C, 0.9% Mn, 0.04% S, bal. Fe.

44. Identify the following materials. Discuss the significance of the composition in relation to the properties and microstructure commonly

observed. Indicate typical commercial uses for each material. (% = wt%).

(a) 1% Mg, 0.6% Si, 0.25% Cu, 0.2% Cr, bal. Al;
(b) 0.42% C, 1.6% Si, 1.8% Ni, 0.8% Cr, 0.4% Mo, 0.1% V, bal. Fe;
(c) 12% Cr, 0.5% Ni (max), 0.6% Mn (max), 0.04% C(max), %Ti = 5 × %C, bal. Fe;
(d) 18% Cr, 8% Ni, 0.8% Mn, 0.5% Ti, 0.08% C, bal. Fe;
(e) 11% Si, 0.01% Na, bal. Al.

Answers to selected problems

2. (a) 6.5 cm
(b) 3.0 cm

3. σ_1 = 23.3 MPa; σ_2 = −9.3 MPa.
ϵ_1 = $1.3.10^{-4}$; ϵ_2 = $-0.81.10^{-4}$;
ϵ_3 = $-0.21.10^{-4}$.
σ_{33} = 4.2 MPa.

4. σ_1 = 0; σ_2 = −104.4 MPa;
σ_3 = −187.5 MPa.

Tresca yield criterion: yielding if y < 187.5 MPa. Von-Mises yield criterion: yielding if y < 162.7 MPa. Material with higher yield stress permits the use of thinner sheet with an adequate safety margin. (Weldable nickel-chromium steels are used in practice.)

5. (a) 22 GPa
(b) (i) 200 MPa
(ii) 244 MPa
(iii) 200 MPa

UTS of composite is 244 MPa.

6.

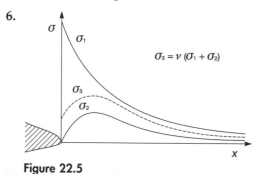

Figure 22.5

8. (a) m ≈ 18
(b) $P_{1\,GPa}$ = 0.02 = 2%
$P_{800\,MPa}$ = 0 = 0%
(c) 1200 MPa

9. Average cost per Si_3N_4 tube = 20p; Average cost per calcia-stabilized ZrO_2 tube = 15.4p; (taking number of £100 failures into account).

10. σ_{hoop} = 84.1 MPa.

For 1 mm, σ_{hoop} = 237.9 MPa, i.e. > ⅔ yield stress, i.e. > design stress ∴ design stress will determine operating stresses.

11. 65% decrease

12. Maximum load = 1.15 MN (compressive)
ϵ = −0.0002 ∴ shorten by 0.6 mm.

15. $E_{||}$ = 204.3 GPa (Rule of mixtures)
$E_\perp$ = 9.3 GPa (Reuss model)
= 14.6 GPa (Halpin–Tsai)

17. a_f = 20.85 mm; N_f = 20 276 cycles
Improving NDI → N_f = 25 905 cycles
Increasing K_{Ic} → N_f = 21 754 cycles
(assuming a_f = ∞) ∴ new NDI equipment recommended.

18. High cycle fatigue: Basquin's Law $\Delta\sigma N_f^a$ = B. Low cycle fatigue: Coffin–Manson $N_f^{1/2}\Delta\epsilon_p$ = constant (both for 'smooth-surfaced' component life).

Number of flights before failure = 4702. a_{crit} = 9.03 mm; under vibrational loading, N_f ≈ 1.10^5 (i.e. halfway between London and New York!)

19. (a) N_f = 2.96×10^4 flights

Inspect at $\dfrac{1}{10}$ to $\dfrac{1}{4}$ of fatigue life.

(b) SCC crack growth dominates.

N_f = 1.93×10^3 cycles

∴ inspection frequency must be increased by at least a factor of ten.

20. N_f = 2037 cycles;
N_f = 3305 cycles;
K_{Ic} = 53.8 MPa m$^{1/2}$.

23. 7 mm

31. (a) Alpha brass (70 wt% Cu; 30 wt% Zn). Rod cut to length, head upset. Threads likely to be cut.
(b) Steel containing ~1 wt% Cr and ~2 wt% Mo. Closed die forged from blanks cut from rod.
(c) Aluminium (6xxx series containing Mg and Si). Extruded. Also PVC, extruded.
(d) Grey cast iron or aluminium silicon alloy.

33. Consider possible failure modes of struts.

Consider weighting factors. For each property, obtain a relative rating (i.e. 0 < relative rating ≤ 1).

For a traditional bicycle where cost is important and weight unimportant, steel is a good choice. For a performance machine where cost is not an issue, C-fibre composite is very attractive with high stiffness to weight ratio.

34. (c) Early plastic kettle bodies were manufactured from acetal resins, but the material of choice now is polypropylene – 20% talc filled.

35. (a) (ii) 13.2 MPa
(b) (i) Creep strain after 10 hours at 13.2 MPa tensile stress at 60°C (plus safety factor).
(ii) Thermal expansion.
Bending moments.
Fatigue.
(iii) Creep resistance; toughness; cost. Filled polymers (e.g. HDPE, ABS, nylons). Cast magnesium or aluminium alloys. Possibly polymer matrix composites.

[Postscript: In the real world, a model for an automotive cooling fan is used with 25 equations and 39 variables! See: R. BROOKS, D. G. ELLIMAN and T. B. CHIA, *2nd Intl. Symp. on Automotive Technology and Automation.* Aachen, Germany, Sept 1993.]

Acknowledgements

The problems have been adapted from University examination papers, and the authors thank the following for providing questions, or their bases.

University of Cambridge (numbers 2, 3, 4, 6, 9, 10, 11, 13, 14, 16, 17, 18, 19, 20, 31 and 33);

Brunel University (numbers 8, 12, 21, 22, 23, 24, 27, 29, 30, 34, 37, 38, 39);

University of Sheffield (numbers 7, 25, 26, 35);

University College, Swansea (numbers 1, 5, 15, 28, 32, 36, 40).

Useful texts

There are, of course, a great many texts that give additional information, helpful in answering the questions fully and in extending the knowledge base for effective material selection. Below are just a few.

Static strength, toughness, stiffness, fatigue and creep

G. E. DIETER: *Mechanical Metallurgy.* McGraw-Hill Book Co, 1986.

A. H. COTTRELL: *Mechanical Properties of Matter.* John Wiley, 1964.

S. P. TIMOSHENKO and J. N. GOODIER: *Theory of Elasticity.* McGraw-Hill Book Co, 1982.

J. F. KNOTT: *Fundamentals of Fracture Mechanics.* Butterworths, 1978.

N. P. CHEREMISINOFF ed.: *Handbook of Ceramics and Composites.* Marcel Dekker, 1990.

D. HULL: *An Introduction to Composite Materials.* Cambridge University Press, 1981.

A. KELLY ed: *Concise Encyclopedia of Composite Materials.* Pergamon Press, 1989.

Corrosion and wear

U. R. EVANS: *Metallic Corrosion, Passivity and Protection.* Edward Arnold.

L. L. SHREIR, R. A. JARMAN and G. T. BURSTEIN eds: 'Corrosion', 3rd Edition, Butterworth-Heinemann, Oxford, 1994.

I. M. HUTCHINGS: *Tribology: Friction and Wear of Engineering Materials.* Edward Arnold, 1992.

Selection and processing

N. A. WATERMAN and M. F. ASHBY eds: *Materials Selector.* Chapman and Hall, 1996.

M. F. ASHBY: *Materials Selection in Mechanical Design.* Pergamon Press, 1992.

G. E. DIETER: *Engineering Design, a Materials and Processing Approach.* McGraw-Hill, 1983.

G. LEWIS: *Selection of Engineering Materials.* Prentice Hall, 1990.

Specific materials

R. W. K. HONEYCOMBE and H. K. D. H. BHADESHIA: *Steels, Microstructure and Properties*, 2nd Edition. Edward Arnold, 1995.

D. T. LLEWELLYN: *Steels: Metallurgy and Applications.* Butterworth-Heinemann, 1992.

I. J. POLMEAR: *Light Alloys: The Metallurgy of the Light Metals.* Edward Arnold, 1995.

N. J. MILLS: *Plastics: Microstructure and Engineering Applications*, 2nd Edition. Edward Arnold.

H. DOMININGHAUS: *Plastics for Engineers.* Hauser Publishers, 1993.

General

R. H. HIGGINS: *Engineering Metallurgy*, 6th Edition. Edward Arnold, 1993.

V. J. COLANGELO and F. A. HEISER: *Analysis of Metallurgical Failures*, 2nd Edition. Wiley, 1987.

D. R. ASKELAND: *The Science of Engineering Materials*, 2nd Edition. Chapman and Hall, 1990.

Conversion Factors for SI Units

Quantity	To convert from	To	Multiply by
Angle	Degree	Rad	1.745×10^{-2}
Area	in^2	m^2	6.452×10^{-4}
Density	lb/in^3	Mg/m^3	27.68
Force	lbf	N	4.448
Fracture toughness	ksi $\sqrt{in}$	MPa $\sqrt{m}$	1.099
Stress	lbf/in^2	Pa	6.889×10^3
	ksi	MPa	6.889
Temperature	°F	°C	$T_C = (T_F - 32)/1.8$

Index